THE INTERNATIONAL SERIES OF MONOGRAPHS ON CHEMISTRY

GENERAL EDITORS

R. BRESLOW
J. B. GOODENOUGH
J. HALPERN, FRS
J. S. ROWLINSON, FRS

THE INTERNATIONAL SERIES OF MONOGRAPHS ON CHEMISTRY

1. J. D. Lambert: *Vibrational and rotational relaxation in gases*
2. N. G. Parsonage and L. A. K. Staveley: *Disorder in crystals*
3. G. C. Maitland, M. Rigby, E. B. Smith, and W. A. Wakeham: *Intermolecular forces: their origin and determination*
4. W. G. Richards, H. P. Trivedi, and D. L. Cooper: *Spin-orbit coupling in molecules*
5. C. F. Cullis and M. M. Hirschler: *The combustion of organic polymers*
6. R. T. Bailey, A. M. North, and R. A. Pethrick: *Molecular motion in high polymers*
7. Atta-ur-Rahman and A. Basha *Biosynthesis of indole alkaloids*
8. J. S. Rowlinson and B. Widom: *Molecular theory of capillarity*
9. C. G. Gray and K. E. Gubbins: *Theory of molecular fluids. Volume 1: Fundamentals*
10. C. G. Gray and K. E. Gubbins: *Theory of molecular fluids. Volume 2: Applications*
11. S. Wilson: *Electron correlation in molecules*
12. E. Haslam: *Metabolites and metabolism*
13. G. R. Fleming: *Chemical applications of ultrafast spectroscopy*
14. R. R. Ernst, G. Bodenhausen, and A. Wokaun: *Principles of nuclear magnetic resonance in one and two dimensions*
15. M. Goldman: *Quantum description of high-resolution NMR in liquids*
16. R. G. Parr and W. Yang: *Density-functional theory of atoms and molecules*
17. J. C. Vickerman, A. Brown, and N. M. Reed: *Secondary ion mass spectrometry: Principles and applications*

DENSITY-FUNCTIONAL THEORY OF ATOMS AND MOLECULES

ROBERT G. PARR and WEITAO YANG
University of North Carolina

OXFORD UNIVERSITY PRESS · NEW YORK
CLARENDON PRESS · OXFORD

Oxford University Press

Oxford New York Toronto
Delhi Bombay Calcutta Madras Karachi
Kuala Lumpur Singapore Hong Kong Tokyo
Nairobi Dar es Salaam Cape Town
Melbourne Auckland Madrid

and associated companies in
Berlin Ibadan

Copyright © 1989 by Oxford University Press

First published in 1989 by Oxford University Press, Inc.,
198 Madison Avenue, New York, New York 10016-4314

First issued as an Oxford University Press paperback in 1994

Oxford is a registered trademark of Oxford University Press

All rights reserved. No part of this publication may be reproduced,
stored in a retrieval system, or transmitted, in any form or by any means,
electronic, mechanical, photocopying, recording, or otherwise,
without the prior permission of Oxford University Press, Inc.

Library of Congress Cataloging-in-Publication Data
Parr, Robert G., 1921–
Density-functional theory of atoms and molecules/
Robert G. Parr and Weitao Yang.
p. cm.—(International series of monographs on chemistry: 16)
Bibliography: p. Includes index.
ISBN 0-19-504279-4   ISBN 0-19-509276-7 (pbk.)
1. Electronic structure. 2. Density functionals.
3. Quantum theory. 4. Quantum chemistry.
I. Yang, Weitao. II. Title.
III. Series.
QC176.8.E4P37 1989 530.4'1—dc19 88-25157

1001284928

4 6 8 9 7 5 3
Printed in the United States of America

PREFACE

This book is an exposition of a unique approach to the quantum theory of the electronic structure of matter, the density-functional theory, designed to introduce this fascinating subject to any scientist familiar with elementary quantum mechanics.

It has been a triumph of contemporary quantum chemistry to solve accurately Schrödinger's equation for many-electron systems, and useful predictions of facts about molecules are now routinely made from quantum-mechanical calculations based on orbital theories and their systematic extensions. Thus, we have restricted and unrestricted Hartree–Fock models, configuration-interaction and many-body perturbation methods for computing correlation effects, and so on. For small molecules, the accuracy achieved is phenomenal. Excited states as well as ground states can be handled, as can potential-energy surfaces for chemical reactions. Standard program packages are available.

Our subject here is not the systematic calculations of traditional quantum chemistry, however, but something quite different. We shall be primarily concerned with ground states, and for ground states there exists a remarkable special theory, the density-functional theory. This constitutes a method in which without loss of rigor one works with the electron density $\rho(\mathbf{r})$ as the basic variable, instead of the wave function $\psi(\mathbf{r}_1, s_1, \mathbf{r}_2, s_2, \ldots, \mathbf{r}_n, s_n)$. The density ρ is just the three-dimensional single-particle density evinced in diffraction experiments and so readily visualized, and the quantum theory for ground states can be put in terms of it. The simplification is immense. The restriction to ground states is what makes density-functional theory possible, the minimum-energy principle for ground states playing a vital role. This is reminiscent of thermodynamics, which is largely a theory of equilibrium states.

The various terms that enter density-functional theory directly, or pop up in it naturally, are quantities of great intuitive appeal, mostly long well-known to chemists in one guise or another. These include the electronegativity of Pauling and Mulliken, the hardness and softness of Pearson, and the reactivity indices of Fukui. These concepts are prominent in our presentation.

Density-functional theory offers a practical computational scheme, the Kohn–Sham equations. These equations are similar to Hartree–Fock equations, yet in principle they include both exchange and correlation effects. This method is employed for very many contemporary calculations done for solids, and it has been increasingly applied to atoms and molecules. Large challenges remain (as discussed in this book), having to do with the need to improve the approximate form of the energy functional. As these challenges are met, the importance for computational chemistry will increase. For larger molecules, density-functional methods may well prove superior to conventional methods.

Density-functional theory has its roots in the papers of Thomas and Fermi in the 1920s, but it became a complete and accurate theory (as opposed to a model) only with the publications in the early 1960s of Kohn, Hohenberg, and Sham. In this book, we include many references to these and other authors, but our plan is to give a coherent account of the theory as it stands today without special regard for the historical development of the subject.

The table of contents indicates the specific topics covered. We emphasize systems with a finite number of electrons, that is, atoms and molecules. Time-dependent phenomena are discussed, as are excited states and systems at finite ambient temperature. We attempt to be fairly rigorous without emphasizing rigor, and we try to be fairly complete as regards basic principles without being all-encompassing. Our bibliography should be particularly helpful to new workers in the field. Appendix G is a guide to other expositions of the subject.

The first two chapters contain background material only; the exposition of the subject of density-functional theory begins with the third chapter. Many of the more mathematical arguments throughout the book can be glossed over lightly by readers not interested in details of the theory. But we would urge every reader not previously exposed to density-functional theory to dwell at length over the entire §§3.1–3.4 and §§7.1–7.4. The Kohn–Sham concept of noninteracting reference system, first introduced in §7.1, is hard for some to grasp, but it is a beautiful idea absolutely essential for appreciating what contemporary density-functional theory is all about.

We are greatly indebted to Professor Mel Levy of Tulane University for many useful comments on the manuscript of this book, and to Ms. Evon Ward for her expert typing of it. Members of the UNC quantum chemistry group, past and present, have been helpful in many ways. The senior author gratefully acknowledges research support from the National Science Foundation and the National Institutes of Health, over a number of years.

Chapel Hill, N.C.   R. G. P.
June 1988   W. Y.

CONTENTS

1. Elementary wave mechanics   3
   1.1 The Schrödinger equation   3
   1.2 Variational principle for the ground state   5
   1.3 The Hartree–Fock approximation   7
   1.4 Correlation energy   13
   1.5 Electron density   14
   1.6 Hellmann–Feynman theorems and virial theorem   16

2. Density matrices   20
   2.1 Description of quantum states and the Dirac notation   20
   2.2 Density operators   24
   2.3 Reduced density matrices for fermion systems   27
   2.4 Spinless density matrices   32
   2.5 Hartree–Fock theory in density-matrix form   35
   2.6 The N-representability of reduced density matrices   40
   2.7 Statistical mechanics   44

3. Density-functional theory   47
   3.1 The original idea: The Thomas–Fermi model   47
   3.2 The Hohenberg–Kohn theorems   51
   3.3 The v- and N-representability of an electron density   53
   3.4 The Levy constrained-search formulation   56
   3.5 Finite-temperature canonical-ensemble theory   60
   3.6 Finite-temperature grand-canonical-ensemble theory   64
   3.7 Finite-temperature ensemble theory of classical systems   66

4. The chemical potential   70
   4.1 Chemical potential in the grand canonical ensemble at zero temperature   70
   4.2 Physical meaning of the chemical potential   74
   4.3 Detailed consideration of the grand canonical ensemble near zero temperature   75
   4.4 The chemical potential for a pure state and in the canonical ensemble   81
   4.5 Discussion   84

5. Chemical potential derivatives — 87
   5.1 Change from one ground state to another — 87
   5.2 Electronegativity and electronegativity equalization — 90
   5.3 Hardness and softness — 95
   5.4 Reactivity index: the Fukui function — 99
   5.5 Local softness, local hardness, and softness and hardness kernels — 101

6. Thomas–Fermi and related models — 105
   6.1 The traditional TF and TFD models — 105
   6.2 Implementation — 110
   6.3 Three theorems in Thomas–Fermi theory — 114
   6.4 Assessment and modification — 116
   6.5 An alternative derivation and a Gaussian model — 118
   6.6 The purely local model — 123
   6.7 Conventional gradient correction — 127
   6.8 The Thomas–Fermi–Dirac–Weizsacker model — 132
   6.9 Various related considerations — 136

7. The Kohn–Sham method: Basic principles — 142
   7.1 Introduction of orbitals and the Kohn–Sham equations — 142
   7.2 Derivation of the Kohn–Sham equations — 145
   7.3 More on the kinetic-energy functional — 149
   7.4 Local-density and $X\alpha$ approximations — 152
   7.5 The integral formulation — 157
   7.6 Extension to nonintegral occupation numbers and the transition-state concept — 163

8. The Kohn–Sham method: Elaboration — 169
   8.1 Spin-density-functional theory — 169
   8.2 Spin-density functionals and the local spin-density approximations — 174
   8.3 Self-interaction correction — 180
   8.4 The Hartree–Fock–Kohn–Sham method — 183
   8.5 The exchange-correlation-energy functional via the exchange-correlation hole — 186
   8.6 The exchange-correlation-energy functional via wave-vector analysis — 194
   8.7 Other studies of the exchange-correlation-energy functional — 197

9. Extensions — 201
   9.1 Finite-temperature Kohn–Sham theory — 201
   9.2 Excited states — 204

9.3	Time-dependent systems	208
9.4	Dynamic linear response	210
9.5	Density-matrix-functional theory	213
9.6	Nonelectronic and multicomponent systems	215

10. **Aspects of atoms and molecules** — 218
    - 10.1 Remarks on the problem of chemical binding — 218
    - 10.2 Interatomic forces — 219
    - 10.3 Atoms in molecules — 221
    - 10.4 More on the HSAB principle — 224
    - 10.5 Modeling the chemical bond: The bond-charge model — 229
    - 10.6 Semiempirical density-functional theory — 234

11. **Miscellany** — 237
    - 11.1 Scaling relations — 237
    - 11.2 A maximum-entropy approach to density-functional theory — 239
    - 11.3 Other topics — 243
    - 11.4 Final remarks — 244

Appendix A	Functionals	246
Appendix B	Convex functions and functionals	255
Appendix C	Second quantization for fermions	259
Appendix D	The Wigner distribution function and the $\hbar$ semiclassical expansion	265
Appendix E	The uniform electron gas	271
Appendix F	Tables of values of electronegativities and hardnesses	276
Appendix G	The review literature of density-functional theory	281

Bibliography — 285

Author index — 319

Subject index — 325

Density-Functional Theory of Atoms and Molecules

1
ELEMENTARY WAVE MECHANICS

1.1 The Schrödinger equation

The principles of density-functional theory are conveniently expounded by making reference to conventional wave-function theory. Therefore, this first chapter reviews elementary quantum theory (Levine 1983, Merzbacher 1970, Parr 1963, McWeeny and Sutcliffe 1969, Szabo and Ostlund 1982). The next chapter summarizes the more advanced techniques that we shall need, mainly having to do with density matrices.

Any problem in the electronic structure of matter is covered by Schrödinger's equation including the time. In most cases, however, one is concerned with atoms and molecules without time-dependent interactions, so we may focus on the time-independent Schrödinger equation. For an isolated N-electron atomic or molecular system in the Born–Oppenheimer nonrelativistic approximation, this is given by

$$\hat{H}\Psi = E\Psi \qquad (1.1.1)$$

where E is the electronic energy, $\Psi = \Psi(\mathbf{x}_1, \mathbf{x}_2, \ldots, \mathbf{x}_n)$ is the wave function, and $\hat{H}$ is the Hamiltonian operator,

$$\hat{H} = \sum_{i=1}^{N} (-\tfrac{1}{2}\nabla_i^2) + \sum_{i=1}^{N} v(\mathbf{r}_i) + \sum_{i<j}^{N} \frac{1}{r_{ij}} \qquad (1.1.2)$$

in which

$$v(\mathbf{r}_i) = -\sum_{\alpha} \frac{Z_\alpha}{r_{i\alpha}} \qquad (1.1.3)$$

is the "external" potential acting on electron i, the potential due to nuclei of charges Z_α. The coordinates $\mathbf{x}_i$ of electron i comprise space coordinates $\mathbf{r}_i$ and spin coordinates s_i. Atomic units are employed here and throughout this book (unless otherwise specified): the length unit is the Bohr radius $a_0(=0.5292 \text{ Å})$, the charge unit is the charge of the electron, e, and the mass unit is the mass of the electron, m_e. When additional fields are present, of course, (1.1.3) contains extra terms.

We may write (1.1.2) more compactly as

$$\hat{H} = \hat{T} + \hat{V}_{ne} + \hat{V}_{ee} \qquad (1.1.4)$$

3

where

$$\hat{T} = \sum_{i=1}^{N} (-\tfrac{1}{2}\nabla_i^2) \qquad (1.1.5)$$

is the kinetic energy operator,

$$\hat{V}_{ne} = \sum_{i=1}^{N} v(r_i) \qquad (1.1.6)$$

is the electron–nucleus attraction energy operator, and

$$\hat{V}_{ee} = \sum_{i<j}^{N} \frac{1}{r_{ij}} \qquad (1.1.7)$$

is the electron–electron repulsion energy operator. The total energy W is the electronic energy E plus the nucleus–nucleus repulsion energy

$$V_{nn} = \sum_{\alpha<\beta} \frac{Z_\alpha Z_\beta}{R_{\alpha\beta}} \qquad (1.1.8)$$

That is,

$$W = E + V_{nn} \qquad (1.1.9)$$

It is immaterial whether one solves (1.1.1) for E and adds V_{nn} afterwards, or includes V_{nn} in the definition of $\hat{H}$ and works with the Schrödinger equation in the form $\hat{H}\Psi = W\Psi$.

Equation (1.1.1) must be solved subject to appropriate boundary conditions. Ψ must be well-behaved everywhere, in particular decaying to zero at infinity for an atom or molecule or obeying appropriate periodic boundary conditions for a regular infinite solid. $|\Psi|^2$ is a probability distribution function in the sense that

$$|\Psi(\mathbf{r}^N, s^N)|^2\, d\mathbf{r}^N = \text{probability of finding the system}$$
$$\text{with position coordinates between } \mathbf{r}^N \text{ and}$$
$$\mathbf{r}^N + d\mathbf{r}^N \text{ and spin coordinates}$$
$$\text{equal to } s^N \qquad (1.1.10)$$

Here $d\mathbf{r}^N = d\mathbf{r}_1, d\mathbf{r}_2, \ldots, d\mathbf{r}_N$; $\mathbf{r}^N$ stands for the set $\mathbf{r}_1, \mathbf{r}_2, \ldots, \mathbf{r}_N$, and s^N stands for the set $s_1, s_2, \ldots, s_N$. The spatial coordinates are continuous, while the spin coordinates are discrete. Because electrons are fermions, Ψ also must be antisymmetric with respect to interchange of the coordinates (both space and spin) of any two electrons.

There are many acceptable independent solutions of (1.1.1) for a given system: the eigenfunctions Ψ_k, with corresponding energy eigenvalues E_k. The set Ψ_k is complete, and the Ψ_k may always be taken to be orthogonal and normalized [in accordance with (1.1.10)],

$$\int \Psi_k^* \Psi_l\, d\mathbf{x}^N = \langle \Psi_k | \Psi_l \rangle = \delta_{kl} \qquad (1.1.11)$$

We denote the ground-state wave function and energy by Ψ_0 and E_0. Here $\int d\mathbf{x}^N$ means integration over $3N$ spatial coordinates and summation over N spin coordinates.

Expectation values of observables are given by formulas of the type

$$\langle \hat{A} \rangle = \frac{\int \Psi^* \hat{A} \Psi \, d\mathbf{x}}{\int \Psi^* \Psi \, d\mathbf{x}} = \frac{\langle \Psi | \hat{A} | \Psi \rangle}{\langle \Psi | \Psi \rangle} \tag{1.1.12}$$

where $\hat{A}$ is the Hermitian linear operator for the observable A. Many measurements average to $\langle \hat{A} \rangle$; particular measurements give particular eigenvalues of $\hat{A}$. For example, if Ψ is normalized, expectation values of kinetic and potential energies are given by the formulas

$$T[\Psi] = \langle \hat{T} \rangle = \int \Psi^* \hat{T} \Psi \, d\mathbf{x} \tag{1.1.13}$$

and

$$V[\Psi] = \langle \hat{V} \rangle = \int \Psi^* \hat{V} \Psi \, d\mathbf{x} \tag{1.1.14}$$

The square brackets here denote that Ψ determines T and V; we say that T and V are *functionals* of Ψ (see Appendix A).

1.2 Variational principle for the ground state

When a system is in the state Ψ, which may or may not satisfy (1.1.1), the average of many measurements of the energy is given by the formula

$$E[\Psi] = \frac{\langle \Psi | \hat{H} | \Psi \rangle}{\langle \Psi | \Psi \rangle} \tag{1.2.1}$$

where

$$\langle \Psi | \hat{H} | \Psi \rangle = \int \Psi^* \hat{H} \Psi \, d\mathbf{x} \tag{1.2.2}$$

Since, furthermore, each particular measurement of the energy gives one of the eigenvalues of $\hat{H}$, we immediately have

$$E[\Psi] \geq E_0 \tag{1.2.3}$$

The energy computed from a guessed Ψ is an upper bound to the true ground-state energy E_0. Full minimization of the functional $E[\Psi]$ with respect to all allowed N-electron wave functions will give the true ground state Ψ_0 and energy $E[\Psi_0] = E_0$; that is,

$$E_0 = \min_{\Psi} E[\Psi] \tag{1.2.4}$$

Formal proof of the minimum-energy principle of (1.2.3) goes as follows. Expand Ψ in terms of the normalized eigenstates of $\hat{H}$, Ψ_k:

$$\Psi = \sum_k C_k \Psi_k \tag{1.2.5}$$

Then the energy becomes

$$E[\Psi] = \frac{\sum_k |C_k|^2 E_k}{\sum_k |C_k|^2} \tag{1.2.6}$$

where E_k is the energy for the kth eigenstate of $\hat{H}$. Note that the orthogonality of the Ψ_k has been used. Because $E_0 \leq E_1 \leq E_2 \leq \cdots$, $E[\Psi]$ is always greater than or equal to E_0, and it reaches its minimum E_0 if and only if $\Psi = C_0 \Psi_0$.

Every eigenstate Ψ is an extremum of the functional $E[\Psi]$. In other words, one may replace the Schrödinger equation (1.1.1) with the variational principle

$$\delta E[\Psi] = 0 \tag{1.2.7}$$

When (1.2.7) is satisfied, so is (1.1.1), and vice versa.

It is convenient to restate (1.2.7) in a way that will guarantee that the final Ψ will automatically be normalized. This can be done by the method of Lagrange undetermined multipliers (§17.6 of Arfken 1980, or Appendix A). Extremization of $\langle \Psi | \hat{H} | \Psi \rangle$ subject to the constraint $\langle \Psi | \Psi \rangle = 1$ is equivalent to making stationary the quantity $[\langle \Psi | \hat{H} | \Psi \rangle - E \langle \Psi | \Psi \rangle]$ without constraint, with E the Lagrange multiplier. This gives

$$\delta [\langle \Psi | \hat{H} | \Psi \rangle - E \langle \Psi | \Psi \rangle] = 0. \tag{1.2.8}$$

One must solve this equation for Ψ as a function of E, then adjust E until normalization is achieved. It is elementary to show the essential equivalence of (1.2.8) and (1.1.1). Solutions of (1.2.8) with forms of Ψ restricted to approximate forms $\tilde{\Psi}$ of a given type (that is, a subset of all allowable Ψ) will give well-defined best approximations $\tilde{\Psi}_0$ and $\bar{E}_0$ to the correct Ψ_0 and E_0. By (1.2.3), $\bar{E}_0 \geq E_0$, and so convergence of the energy, from above, is assured as one uses more and more flexible $\tilde{\Psi}$. Most contemporary calculations on electronic structure are done with this variational procedure, in some linear algebraic implementation.

Excited-state eigenfunctions and eigenvalues also satisfy (1.2.8), but the corresponding methods for determining approximate $\tilde{\Psi}_k$ and $\bar{E}_k$ encounter orthogonality difficulties. For example, given $\tilde{\Psi}_1$, $\bar{E}_1$ is not necessarily above E_1, unless $\tilde{\Psi}_1$ is orthogonal to the exact Ψ_0.

To summarize: For a system of N electrons and given nuclear potential $v(\mathbf{r})$, (1.2.8) defines a procedure for going from N and $v(\mathbf{r})$ to the ground-state wave function Ψ and hence through (1.1.12) to the ground-state energy $E[N, v]$ and other properties of interest. Note that in this statement there is no mention of the kinetic-energy or electron-repulsion parts of $\hat{H}$, because these are universal in that they are determined by N. We say that E is a *functional* of N and $v(\mathbf{r})$.

1.3 The Hartree–Fock approximation

Suppose now that Ψ is approximated as an antisymmetrized product of N orthonormal spin orbitals $\psi_i(\mathbf{x})$, each a product of a spatial orbital $\phi_k(\mathbf{r})$ and a spin function $\sigma(s) = \alpha(s)$ or $\beta(s)$, the *Slater determinant*

$$\Psi_{HF} = \frac{1}{\sqrt{N!}} \begin{vmatrix} \psi_1(\mathbf{x}_1) & \psi_2(\mathbf{x}_1) & \cdots & \psi_N(\mathbf{x}_1) \\ \psi_1(\mathbf{x}_2) & \psi_2(\mathbf{x}_2) & \cdots & \psi_N(\mathbf{x}_2) \\ \vdots & \vdots & & \vdots \\ \psi_1(\mathbf{x}_N) & \psi_2(\mathbf{x}_N) & \cdots & \psi_N(\mathbf{x}_N) \end{vmatrix}$$

$$= \frac{1}{\sqrt{N!}} \det [\psi_1 \psi_2 \cdots \psi_N] \tag{1.3.1}$$

The *Hartree–Fock approximation* (Roothaan 1951) is the method whereby the orthonormal orbitals ψ_i are found that minimize (1.2.1) for this determinantal form of Ψ.

The normalization integral $\langle \Psi_{HF} | \Psi_{HF} \rangle$ is equal to 1, and the energy expectation value is found to be given by the formula (for example, see Parr 1963)

$$E_{HF} = \langle \Psi_{HF} | \hat{H} | \Psi_{HF} \rangle = \sum_{i=1}^{N} H_i + \tfrac{1}{2} \sum_{i,j=1}^{N} (J_{ij} - K_{ij}) \tag{1.3.2}$$

where

$$H_i = \int \psi_i^*(\mathbf{x}) [-\tfrac{1}{2} \nabla^2 + v(\mathbf{x})] \psi_i(\mathbf{x}) \, d\mathbf{x} \tag{1.3.3}$$

$$J_{ij} = \iint \psi_i(\mathbf{x}_1) \psi_i^*(\mathbf{x}_1) \frac{1}{r_{12}} \psi_j^*(\mathbf{x}_2) \psi_j(\mathbf{x}_2) \, d\mathbf{x}_1 \, d\mathbf{x}_2 \tag{1.3.4}$$

$$K_{ij} = \iint \psi_i^*(\mathbf{x}_1) \psi_j(\mathbf{x}_1) \frac{1}{r_{12}} \psi_i(\mathbf{x}_2) \psi_j^*(\mathbf{x}_2) \, d\mathbf{x}_1 \, d\mathbf{x}_2 \tag{1.3.5}$$

These integrals are all real, and $J_{ij} \geq K_{ij} \geq 0$. The J_{ij} are called *Coulomb integrals*, the K_{ij} are called *exchange integrals*. We have the important equality

$$J_{ii} = K_{ii} \tag{1.3.6}$$

This is the reason the double summation in (1.3.2) can include the $i = j$ terms.

Minimization of (1.3.2) subject to the orthonormalization conditions

$$\int \psi_i^*(\mathbf{x})\psi_j(\mathbf{x})\,d\mathbf{x} = \delta_{ij} \tag{1.3.7}$$

now gives the Hartree–Fock differential equations

$$\hat{F}\psi_i(\mathbf{x}) = \sum_{j=1}^{N} \varepsilon_{ij}\psi_j(\mathbf{x}) \tag{1.3.8}$$

where

$$\hat{F} = -\tfrac{1}{2}\nabla^2 + v + \hat{g} \tag{1.3.9}$$

in which the Coulomb-exchange operator $\hat{g}(x_1)$ is given by

$$\hat{g} = \hat{j} - \hat{k} \tag{1.3.10}$$

Here

$$\hat{j}(\mathbf{x}_1)f(\mathbf{x}_1) \equiv \sum_{k=1}^{N} \int \psi_k^*(\mathbf{x}_2)\psi_k(\mathbf{x}_2)\frac{1}{r_{12}}f(\mathbf{x}_1)\,d\mathbf{x}_2 \tag{1.3.11}$$

and

$$\hat{k}(\mathbf{x}_1)f(\mathbf{x}_1) \equiv \sum_{k=1}^{N} \int \psi_k^*(\mathbf{x}_2)f(\mathbf{x}_2)\frac{1}{r_{12}}\psi_k(\mathbf{x}_1)\,d\mathbf{x}_2 \tag{1.3.12}$$

with $f(\mathbf{x}_1)$ an arbitrary function. The matrix ε consists of Lagrange multipliers (in general complex) associated with the constraints of (1.3.7). Also,

$$\varepsilon_{ji}^* = \varepsilon_{ij} \tag{1.3.13}$$

so that ε is Hermitian (Roothaan 1951).

Multiplying (1.3.8) by ψ_i^* and integrating, one obtains the formula for "orbital energies,"

$$\varepsilon_i \equiv \varepsilon_{ii} = \langle \psi_i | \hat{F} | \psi_i \rangle = H_i + \sum_{j=1}^{N} (J_{ij} - K_{ij}) \tag{1.3.14}$$

Summing over i and comparing with (1.3.2), we find

$$E_{\mathrm{HF}} = \sum_{i=1}^{N} \varepsilon_i - V_{ee} \tag{1.3.15}$$

where the symbol V_{ee} stands for the total electron–electron repulsion energy

$$V_{ee} = \int \Psi_{\mathrm{HF}}^*(\mathbf{x}^N)\Big(\sum_{i<j}\frac{1}{r_{ij}}\Big)\Psi_{\mathrm{HF}}(\mathbf{x}^N)\,d\mathbf{x}^N$$

$$= \tfrac{1}{2}\sum_{i,j=1}^{N} (J_{ij} - K_{ij}) \tag{1.3.16}$$

For the total molecular energy including nuclear–nuclear repulsion, one has from (1.1.9),

$$W_{HF} = \sum_{i=1}^{N} \varepsilon_i - V_{ee} + V_{nn} \quad (1.3.17)$$

$$= \sum_{i=1}^{N} H_i + V_{ee} + V_{nn} \quad (1.3.18)$$

Note that neither E_{HF} nor W_{HF} is equal to the sum of orbital energies.

Solution of (1.3.8) must proceed iteratively, since the orbitals ψ_i that solve the problem appear in the operator $\hat{F}$. Consequently, the Hartree–Fock method is a nonlinear "self-consistent-field" method.

For a system having an even number of electrons, in what is called the *restricted Hartree–Fock method* (RHF), the N orbitals ψ_i are taken to comprise $N/2$ orbitals of form $\phi_k(\mathbf{r})\alpha(s)$ and $N/2$ orbitals of form $\phi_k(\mathbf{r})\beta(s)$. The energy formula (1.3.2) becomes

$$E_{HF} = 2\sum_{k=1}^{N/2} H_k + \sum_{k,l=1}^{N/2}(2J_{kl} - K_{kl}) \quad (1.3.19)$$

where

$$H_k = \int \phi_k^*(\mathbf{r})[-\tfrac{1}{2}\nabla^2 + v(\mathbf{r})]\phi_k(\mathbf{r})\,d\mathbf{r} \quad (1.3.20)$$

$$J_{kl} = \iint |\phi_k(\mathbf{r}_1)|^2 \frac{1}{r_{12}} |\phi_l(\mathbf{r}_2)|^2\,d\mathbf{r}_1\,d\mathbf{r}_2 \quad (1.3.21)$$

$$K_{kl} = \iint \phi_k^*(\mathbf{r}_1)\phi_l(\mathbf{r}_1)\frac{1}{r_{12}}\phi_k(\mathbf{r}_2)\phi_l^*(\mathbf{r}_2)\,d\mathbf{r}_1\,d\mathbf{r}_2 \quad (1.3.22)$$

while the Hartree–Fock equations (1.3.8) now read

$$\hat{F}\phi_k(\mathbf{r}) = \sum_{l=1}^{N/2}\varepsilon_{kl}\phi_l(\mathbf{r}) \quad (1.3.23)$$

with the operator $\hat{F}$ given by (1.3.9) and (1.3.10), with (1.3.11) and (1.3.12) replaced by

$$\hat{j}(\mathbf{r}_1)f(\mathbf{r}_1) \equiv 2\sum_{m=1}^{N/2}\int |\phi_m(\mathbf{r}_2)|^2 \frac{1}{r_{12}}\,d\mathbf{r}_2 f(\mathbf{r}_1) \quad (1.3.24)$$

$$\hat{k}(\mathbf{r}_1)f(\mathbf{r}_1) \equiv \sum_{m=1}^{N/2}\int \phi_m^*(\mathbf{r}_2)f(\mathbf{r}_2)\frac{1}{r_{12}}\,d\mathbf{r}_2 \phi_m(\mathbf{r}_1) \quad (1.3.25)$$

The determinantal wave function (1.3.1) for this "closed-shell" case is explicitly

$$\Psi_{HF} = \frac{1}{\sqrt{N!}} \begin{vmatrix} \phi_1(\mathbf{r}_1)\alpha(s_1) & \phi_1(\mathbf{r}_1)\beta(s_1) & \cdots & \phi_{N/2}(\mathbf{r}_1)\beta(s_1) \\ \phi_1(\mathbf{r}_2)\alpha(s_2) & \phi_1(\mathbf{r}_2)\beta(s_2) & \cdots & \phi_{N/2}(\mathbf{r}_2)\beta(s_2) \\ \vdots & \vdots & \vdots & \vdots & & \vdots & \vdots \\ \phi_1(\mathbf{r}_N)\alpha(s_N) & \phi_1(\mathbf{r}_N)\beta(s_N) & & & & \phi_{N/2}(\mathbf{r}_N)\beta(s_N) \end{vmatrix}$$

(1.3.26)

An important property of this wave function [and also of the more general (1.3.1)] is that a unitary transformation of the occupied orbitals ϕ_k (or ψ_i) to another set of orbitals η_m leaves the wave function unchanged except possibly by an inconsequential phase factor. The operators $\hat{j}$, $\hat{k}$, and $\hat{F}$ of (1.3.23) through (1.3.25) [or of (1.3.9) through (1.3.12)] are also invariant to such a transformation (Roothaan 1951, Szabo and Ostlund 1982, page 120). That is to say, if we let

$$\eta_m = \sum_k \mathbf{U}_{mk} \psi_k \qquad (1.3.27)$$

where $\mathbf{U}$ is a unitary matrix,

$$\mathbf{U}^+\mathbf{U} = 1 \qquad (1.3.28)$$

then (1.3.23) becomes

$$\hat{F}\eta_m = \sum_{n=1}^{N/2} \varepsilon_{mn}^{\eta} \eta_n \qquad (1.3.29)$$

where

$$\boldsymbol{\varepsilon}^{\eta} = \mathbf{U}\boldsymbol{\varepsilon}\mathbf{U}^+ \qquad (1.3.30)$$

This exhibits the considerable freedom that exists in the choice of the matrix $\boldsymbol{\varepsilon}$.

Since the matrix $\boldsymbol{\varepsilon}$ is Hermitian, one may choose the matrix $\mathbf{U}$ to diagonalize it. The corresponding orbitals λ_m, called the *canonical Hartree–Fock orbitals*, satisfy the *canonical Hartree–Fock equations*,

$$\hat{F}\lambda_m(\mathbf{r}) = \varepsilon_m^{\lambda} \lambda_m(\mathbf{r}) \qquad (1.3.31)$$

Equation (1.3.31) is considerably more convenient for calculation than (1.3.23). Furthermore, the orbitals that are solutions of (1.3.31) are uniquely appropriate for describing removal of electrons from the system in question. There is a theorem due to Koopmans (1934) that if one assumes no reorganization (change of orbitals) on ionization, the best (lowest-energy) single-determinantal description for the ion is the determinant built from the canonical Hartree–Fock orbitals of (1.3.31). One then finds, approximately,

$$\varepsilon_m^{\lambda} = -I_m \qquad (1.3.32)$$

where I_m is the ionization energy associated with removal of an electron from the orbital λ_m. This equation is in error, because it ignores both reorganization and errors in the Hartree–Fock description (called correlation energy: see the next section); fortunately these tend to cancel.

The orbital energies for the canonical Hartree–Fock orbitals also control the long-range behavior of the orbitals. Naively, one would expect, from the one-electron nature of (1.3.31), $\lambda_m \sim \exp[-(-2\varepsilon_m^\lambda)^{1/2} r]$ for large r. This is correct for atoms with s electrons only, but not in general. Instead, in general the maximum (least-negative) of all of the occupied ε_m^λ determines the long-range behavior of *all* of the orbitals:

$$\lambda_m \sim \exp[-(-2\varepsilon_{\max})^{1/2} r] \qquad \text{for large } \mathbf{r} \qquad (1.3.33)$$

The long-range properties of the exchange part of $\hat{F}$ are responsible for this remarkable behavior (Handy, Marron, and Silverstone 1969). The operator $\hat{F}$ is *not* a Sturm–Liouville operator.

For the closed-shell case, entirely equivalent to the canonical Hartree–Fock description are the circulant Hartree–Fock description and the localized Hartree–Fock description. *Circulant Hartree–Fock orbitals* (Parr and Chen 1981, Nyden and Parr 1983) are orbitals the absolute squares of which are as close to each other as possible in a certain sense; for them, the matrix $\boldsymbol{\varepsilon}$ of (1.3.29) is a circulant matrix (diagonal elements all equal, every row a cyclic permutation of every other). *Localized Hartree–Fock orbitals* (Edmiston and Ruedenberg 1963) are orbitals with maximum self-repulsion or minimum interorbital exchange interaction. The electron repulsion part of (1.3.19) is, from (1.3.6),

$$V_{ee} = J - K \qquad (1.3.34)$$

where

$$J = \sum_{k,l=1}^{N/2} 2J_{kl} = \sum_k J_{kk} + \left[\sum_k J_{kk} + \sum_{k \neq l} J_{kl} \right] \qquad (1.3.35)$$

and

$$K = \sum_{k,l=1}^{N/2} K_{kl} = \sum_k J_{kk} + \left[\sum_{k \neq l} K_{kl} \right] \qquad (1.3.36)$$

J and K are each invariant to unitary transformation, but the terms in square brackets in these equations are not; the unitary transformation to localized orbitals can therefore be effected by maximizing

$$J(\text{self}) \equiv \sum_k J_{kk} = K(\text{self}) \qquad (1.3.37)$$

or, equivalently, by minimizing

$$\sum_{k \neq l} K_{kl}$$

Circulant orbitals are important because they are orbitals that have equivalent mathematical and physical footing, each very close to the square root of the electron density per particle. Localized orbitals are important because their existence reconciles molecular-orbital theory with the more traditional descriptions of molecules as held together by localized chemical bonds.

If from the beginning one neglects all interorbital exchange terms in the Hartree–Fock method, which corresponds to using a product of orbitals as the wave function in place of the antisymmetrized product of (1.3.1) or (1.3.26), one gets the *orthogonalized Hartree method*. The closed-shell equation (1.3.23) is replaced by

$$\hat{F}_k^H \phi_k^H = \sum_l \varepsilon_{kl}^H \phi_l^H \qquad (1.3.38)$$

where

$$\hat{F}_k^H = -\tfrac{1}{2}\nabla^2 + v + \hat{j}_k \qquad (1.3.39)$$

in which

$$\hat{j}_k(\mathbf{x}_1) = \int \left[|\psi_k(\mathbf{x}_2)|^2 + 2\sum_{m \neq k} |\psi_m(\mathbf{x}_2)|^2 \right] \frac{1}{r_{12}} d\mathbf{x}_2 \qquad (1.3.40)$$

This method gives orbitals even more localized than the localized Hartree–Fock orbitals; these are useful for some purposes (Levy, Nee, and Parr 1975).

When the number of electrons is not even, the standard Hartree–Fock scheme is what is called the *unrestricted open-shell Hartree–Fock method* (UHF) (Szabo and Ostlund 1982, pages 205–229). Spatial parts of spin orbitals with α spin are allowed to be different from spatial parts of spin orbitals with β spin, even within a single "pair" of electrons. Noting that orthogonality between all α-spin spin orbitals and all β-spin spin orbitals is still preserved, we see that the only problem in implementation is the complication associated with handling all N orbitals in the Hartree–Fock equations. The mathematical apparatus is (1.3.8) to (1.3.12). The UHF method can also be used for an even number of electrons. Often, indeed usually, the UHF method then gives no energy lowering over the restricted HF method. But there are important cases in which energy lowering is found. For example, the UHF description of bond breaking in H_2 gives the proper dissociation products, while the RHF description of H_2 gives unrealistic ones.

Many physical properties of most molecules in their ground states are well accounted for by use of Hartree–Fock wave functions (Schaefer 1972).

In actual implementation of Hartree–Fock theory (and also in calculations of wave functions to an accuracy higher than those of Hartree–Fock), one usually (though not always) employs some set of fixed, one-electron basis functions, in terms of which orbitals are expanded and many-electron wave functions are expressed. This transforms the mathematical problem into one (or more) matrix eigenvalue problems of high dimension, in which the matrix elements are calculated from arrays of integrals evaluated for the basis functions. If we call the basis functions $\chi_p(\mathbf{r})$, one can see from (1.1.2) what the necessary integrals will be: *overlap integrals*,

$$S_{pq} \equiv \int \chi_p^*(\mathbf{r})\chi_q(\mathbf{r})\, d\mathbf{r} \qquad (1.3.41)$$

kinetic energy integrals,

$$T_{pq} \equiv \int \chi_p^*(\mathbf{r})(-\tfrac{1}{2}\nabla^2)\chi_q(\mathbf{r})\, d\mathbf{r} \qquad (1.3.42)$$

electron–nucleus attraction integrals,

$$(A \mid pq) \equiv \int \chi_p^*(\mathbf{r}_1)\frac{1}{r_{1A}}\chi_q(\mathbf{r}_1)\, d\mathbf{r}_1 \qquad (1.3.43)$$

and *electron–electron repulsion integrals*,

$$(pq \mid rs) \equiv \iint \chi_p^*(\mathbf{r}_1)\chi_q(\mathbf{r}_1)\frac{1}{r_{12}}\chi_r(\mathbf{r}_2)\chi_s^*(\mathbf{r}_2)\, d\mathbf{r}_1\, d\mathbf{r}_2 \qquad (1.3.44)$$

Sometimes these are all computed exactly, in which case one says that one has an *ab initio method*. (For reviews, see Schaefer 1977 and Lawley 1987. The term "ab initio" was used first, though intended to have a different meaning, in Parr, Craig, and Ross 1950.) Sometimes these are determined by some recourse to experimental data, in which case one has a *semiempirical method* (Parr 1963, Segal 1977). Such details are of course vital, but here they will not be of much concern to us in the present exposition.

1.4 Correlation energy

When one is interested in higher accuracy, there are straightforward extensions of the single-determinantal description to simple "multiconfiguration" descriptions involving few determinants (for example, Section 4.5 of Szabo and Ostlund 1982).

The exact wave function for a system of many interacting electrons is never a single determinant or a simple combination of a few determinants, however. The calculation of the error in energy, called *correlation energy*, here defined to be negative,

$$E_{\text{corr}}^{\text{HF}} = E - E_{\text{HF}} \qquad (1.4.1)$$

is a major problem in many-body theory on which there has been a vast amount of work and much progress has been made. The methods employed include the linear mixing of many determinants (millions!), called configuration interaction (Chapter 4 of Szabo and Ostlund 1982), and many-body perturbation techniques (Chapter 6 of Szabo and Ostlund 1982). For comprehensive reviews, see Sinanoğlu and Brueckner (1970), Hurley (1976), and Wilson (1984).

Correlation energy tends to remain constant for atomic and molecular changes that conserve the numbers and types of chemical bonds, but it can change drastically and become determinative when bonds change. Its magnitude can vary from 20 or 30 to thousands of kilocalories per mole, from a few hundredths of an atomic unit on up. Exchange energies are an order of magnitude or more bigger, even if the self-exchange term is omitted.

1.5 Electron density

In an electronic system, the number of electrons per unit volume in a given state is the *electron density* for that state. This quantity will be of great importance in this book; we designate it by $\rho(\mathbf{r})$. Its formula in terms of Ψ is

$$\rho(\mathbf{r}_1) = N \int \cdots \int |\Psi(\mathbf{x}_1, \mathbf{x}_2, \ldots, \mathbf{x}_N)|^2 \, ds_1 \, d\mathbf{x}_2 \cdots d\mathbf{x}_N \quad (1.5.1)$$

This is a nonnegative simple function of three variables, x, y, and z, integrating to the total number of electrons,

$$\int \rho(\mathbf{r}) \, d\mathbf{r} = N \quad (1.5.2)$$

There has been much attention paid to the electron density over the years (Smith and Absar 1977). Maps of electron densities are available in many places (for example, Bader 1970). For an atom in its ground state, the density decreases monotonically away from the nucleus (Weinstein, Politzer, and Srebrenik 1975), in approximately piecewise exponential fashion (Wang and Parr 1977). For molecules, at first sight, densities look like superposed atomic densities; on closer inspection (experimental or theoretical), modest (but still quite small in absolute terms) buildups of density are seen in bonding regions.

At any atomic nucleus in an atom, molecule, or solid, the electron density has a finite value; for an atom we designate this $\rho(0)$. In the neighborhood of a nucleus there always is a cusp in the density owing to the necessity for Hamiltonian terms $-\frac{1}{2}\nabla^2 - (Z_\alpha/r_\alpha)$ not to cause blowups in $\hat{H}\Psi$ there. The specific cusp condition is (for example, see Davidson

1976, page 44)

$$\frac{\partial}{\partial r_\alpha} \bar{\rho}(r_\alpha)\big|_{r_\alpha=0} = -2Z_\alpha \bar{\rho}(0) \quad (1.5.3)$$

where $\bar{\rho}(r_\alpha)$ is the spherical average of $\rho(r_\alpha)$.

Another important result is the long-range law for electron density,

$$\rho \sim \exp[-2(2I_{min})^{1/2} r] \quad (1.5.4)$$

where I_{min} is the exact first ionization potential (Morrell, Parr, and Levy 1975; this paper also contains a generalization of Koopmans' theorem). The corresponding Hartree–Fock result will be, from (1.3.33),

$$\rho_{HF} \sim \exp[-2(-2\varepsilon_{max})^{1/2} r] \quad (1.5.5)$$

where ε_{max} approximates I_{min} by (1.3.32).

Finally, we record here certain results about electron density from the standard first-order perturbation theory for a nondegenerate state. Suppose the state Ψ_k^0 is perturbed to the state $\Psi_k = \Psi_k^0 + \Psi_k^1$ by the one-electron perturbation $\Delta V = \sum_i \Delta v(\mathbf{r}_i)$. The energy change to first order is

$$E_k^{(1)} = \int \Psi_k^{0*} \Delta V \Psi_k^0 \, dx^N = \int \rho_k(\mathbf{r}_1) \Delta v(\mathbf{r}_1) \, d\mathbf{r}_1, \quad (1.5.6)$$

while the perturbed wavefunction is, to first order,

$$\Psi_k = \Psi_k^0 + \sum_{j \neq k} \frac{\langle \Psi_j^0 | \Delta V | \Psi_k^0 \rangle}{E_k^0 - E_j^0} \Psi_j \quad (1.5.7)$$

The electron-density change is then, to first order in Δv,

$$\Delta \rho_k(\mathbf{r}_1) = N \iint (\Psi_k^* \Psi_k - \Psi_k^{0*} \Psi_k^0) \, ds_1 \, dx_2 \, dx_3 \cdots dx_N$$

$$= 2N \operatorname{Re} \left\{ \sum_{j \neq k} \frac{\langle \Psi_j^0 | \Delta V | \Psi_k^0 \rangle}{E_k^0 - E_j^0} \int \cdots \int \Psi_k^{0*} \Psi_j^0 \, ds_1 \, dx_2 \, dx_3 \cdots dx_N \right\}$$

$$= \int \frac{\delta \rho_k(\mathbf{r}_1)}{\delta v(\mathbf{r}_2)} \Delta v(\mathbf{r}_2) \, d\mathbf{r}_2 \quad (1.5.8)$$

where the "functional derivative" $\delta \rho / \delta v$ is defined by

$$\frac{\delta \rho_k(\mathbf{r}_1)}{\delta v(\mathbf{r}_2)} = \frac{\delta \rho_k(\mathbf{r}_2)}{\delta v(\mathbf{r}_1)} \equiv 2N^2 \times$$

$$\sum_{j \neq k} \frac{\left[\int \cdots \int \Psi_j^{0*} \Psi_k^0 \, ds_2 \, dx_1 \, dx_3 \cdots dx_N \right] \left[\int \cdots \int \Psi_k^{0*} \Psi_j^0 \, ds_1 \, dx_2 \, dx_3 \cdots dx_N \right]}{(E_k^0 - E_j^0)}$$

$$(1.5.9)$$

This quantity is called the *linear response function*. The symmetry represented in (1.5.9) is important. If a perturbation at point 1 produces a density change at point 2, then the same perturbation at point 2 will produce at point 1 precisely the same density change. Note that

$$\int \frac{\delta \rho_k(\mathbf{r}_1)}{\delta v(\mathbf{r}_2)} d\mathbf{r}_1 = 0 \tag{1.5.10}$$

All of these formulas assume that the number of electrons is fixed. For a general discussion of functional derivatives, see Appendix A.

1.6 Hellmann–Feynman theorems and virial theorem

Let λ be a parameter in the Hamiltonian and $\Psi(\lambda)$ be an eigenfunction of $\hat{H}$. Then

$$\frac{dE}{d\lambda} = \frac{\langle \Psi | \partial \hat{H}/\partial \lambda | \Psi \rangle}{\langle \Psi | \Psi \rangle} \tag{1.6.1}$$

$$E(\lambda_2) - E(\lambda_1) = \frac{\langle \Psi_2 | \hat{H}(\lambda_2) - \hat{H}(\lambda_1) | \Psi_1 \rangle}{\langle \Psi_2 | \Psi_1 \rangle} \tag{1.6.2}$$

and

$$E(\lambda_2) - E(\lambda_1) = \int_{\lambda_1}^{\lambda_2} \frac{\langle \Psi | \partial \hat{H}/\partial \lambda | \Psi \rangle}{\langle \Psi | \Psi \rangle} d\lambda \tag{1.6.3}$$

These identities are the *differential Hellmann–Feynman theorem* (formula), the *integral Hellmann–Feynman theorem* (formula), and the *integrated Hellmann–Feynman theorem* (formula) (Epstein, Hurley, Wyatt, and Parr 1967). The derivative $\partial \hat{H}/\partial \lambda$ is written as a partial derivative to emphasize that the integral $\langle \Psi | \partial \hat{H}/\partial \lambda | \Psi \rangle$ can depend on the coordinate system chosen to describe a particular situation.

The equation (1.6.1) is a direct result of the first-order perturbation formula for energy, (1.5.6) above. Integrating (1.6.1) from λ_1 to λ_2 gives (1.6.3). Theorem (1.6.2) can be put in a general form,

$$E_A - E_B = \frac{\langle \Psi_B | \hat{H}_A - \hat{H}_B | \Psi_A \rangle}{\langle \Psi_B | \Psi_A \rangle} \tag{1.6.4}$$

where $\hat{H}_A$ and $\hat{H}_B$ are different Hamiltonians acting on the same N-electron wave-function space, but they need not be related to each other by a parameter λ. Since Ψ_B and Ψ_A are eigenfunctions of $\hat{H}_B$ and $\hat{H}_A$, then

$$\hat{H}_A \Psi_A = E_A \Psi_A \quad \text{and} \quad \hat{H}_B \Psi_B = E_B \Psi_B \tag{1.6.5}$$

Multiply these equations respectively by Ψ_B^* and Ψ_A^* and integrate. Take the complex conjugate of the second result. Subtraction then gives

$$(E_A - E_B) \langle \Psi_B | \Psi_A \rangle = \langle \Psi_B | \hat{H}_A - \hat{H}_B | \Psi_A \rangle$$

Provided $\langle \Psi_B | \Psi_A \rangle \neq 0$, this is equivalent to (1.6.4) [and also (1.6.1) follows as the special case when the change is small].

Use Cartesian coordinates and in (1.6.1) let λ be the coordinate X_α of the position of nucleus α. Suppose that no fields are present except those due to the nuclei; i.e., that there are no extra terms in (1.1.3). Then the only terms in $\hat{H}$ that depend on X_α are v and V_{nn}, and (1.6.1), yields, using (1.1.9),

$$\frac{\partial W}{\partial X_\alpha} = -\sum_{\beta \neq \alpha} \frac{Z_\alpha Z_\beta}{R_{\alpha\beta}^3} (X_\alpha - X_\beta) - Z_\alpha \int \rho(\mathbf{r}_1) \frac{(x_1 - X_\alpha)}{r_{1\alpha}^3} d\mathbf{r}_1 \quad (1.6.6)$$

This is a purely classical expression. What it shows is that the force on nucleus α due to the other nuclei and the electrons, in some particular Born–Oppenheimer nuclear configuration, is just what would be computed from classical electrostatics from the locations of the other nuclei and the electronic charge density (see Deb 1981). This is the famous *electrostatic theorem* of Feynman (1939).

An application of (1.6.1) that we are interested in is the formula obtained if one replaces Z_α by λZ_α everywhere it appears in $\hat{H}$ and then computes $W(1) - W(0)$ for a ground state. Note that the ground state of N electrons in the absence of any nuclei has zero energy: $W(0) = 0$. Hence, for a ground state (Wilson 1962, Politzer and Parr 1974)

$$W = \sum_{\alpha < \beta} \frac{Z_\alpha Z_\beta}{R_{\alpha\beta}} - \sum_\alpha Z_\alpha \int_0^1 d\lambda \int \frac{\rho(\mathbf{r}_1, \lambda)}{r_{1\alpha}} d\mathbf{r}_1 \quad (1.6.7)$$

Here $\rho(\mathbf{r}, \lambda)$ is the density associated with the eigenfunction $\Psi(\mathbf{x}, \lambda)$ for the N-electron problem with scaled nuclear charges.

Note that the Hellmann–Feyman theorems (1.6.1) through (1.6.3) hold for any eigenstate, while (1.6.7) is only true for a ground state. Equation (1.6.7), as well as the electrostatic theorem (1.6.6), can be thought of as foreshadowing what we will be demonstrating at length in this book: For a ground state, the electron density suffices for the determination of *all* the properties. Another essential point is that these various theorems may or may not hold for approximate eigenfunctions. For example, (1.6.1) holds for exact Hartree–Fock wave functions (Stanton 1962).

Another important theorem is the *virial theorem*, which relates the kinetic energy and potential energy components of the energy in certain circumstances. This theorem results from homogeneity properties that may or may not be present for a particular problem in the kinetic and potential energy components of $\hat{H}$. The kinetic energy component,

$$\hat{T} = \sum_i (-\tfrac{1}{2}\nabla_i^2) \tag{1.6.8}$$

is homogeneous of degree -2 in particle coordinates. The total potential energy component,

$$\hat{V} = -\sum_{i,\alpha} \frac{Z_\alpha}{r_{i\alpha}} + \sum_{i<j} \frac{1}{r_{ij}} + \sum_{\alpha<\beta} \frac{Z_\alpha Z_\beta}{R_{\alpha\beta}} \tag{1.6.9}$$

is homogeneous of degree -1 in all particle coordinates. Assuming no additional forces are acting, we then find for any eigenstate of an atom,

$$E = -\langle T \rangle = \tfrac{1}{2}\langle V \rangle \tag{1.6.10}$$

and for any eigenstate of a molecule or solid with a particular sufficient set of internuclear distances $R_j = |\mathbf{R}_{\alpha\beta}|$,

$$\langle T \rangle = -W - \sum_j R_j \left(\frac{\partial W}{\partial R_j}\right)_{R_k} \tag{1.6.11}$$

and

$$\langle V \rangle = 2W + \sum_j R_j \left(\frac{\partial W}{\partial R_j}\right)_{R_k} \tag{1.6.12}$$

Proofs are elementary (Löwdin 1959). Given a normalized eigenstate Ψ, it makes stationary the $E[\Psi]$ of (1.2.1). Take a normalized scaled version of this Ψ,

$$\Psi_\zeta \equiv \zeta^{3N/2}\Psi(\zeta\mathbf{r}_1, \zeta\mathbf{r}_2, \ldots, \zeta R_1, \zeta R_2, \ldots) \tag{1.6.13}$$

and calculate $E[\Psi_\zeta]$. This is stationary for $\zeta = 1$, which gives (1.6.10) through (1.6.12). The scaling properties of the individuals components of $E[\Psi_\zeta]$ are important. Using (1.1.13) and (1.1.14), these are found to be

$$T[\Psi_\zeta] = \zeta^2 T[\Psi, \zeta R] \tag{1.6.14}$$

and

$$V[\Psi_\zeta] = \zeta V[\Psi, \zeta R] \tag{1.6.15}$$

respectively. The dependences on ζR are parametric. For a comprehensive review of the virial theorem, see Marc and McMillan (1985).

Note that in (1.6.13) both electronic and nuclear coordinates are scaled. Another type of scaling, of electronic coordinates only, is important for the purposes of this book. Let

$$\Psi_\lambda \equiv \lambda^{3N/2}\Psi(\lambda\mathbf{r}_1, \lambda\mathbf{r}_2, \ldots; R_1, R_2, \ldots) \tag{1.6.16}$$

Then we find

$$T[\Psi_\lambda] = \lambda^2 T[\Psi_1] \tag{1.6.17}$$

and
$$V_{ee}[\Psi_\lambda] = \lambda V_{ee}[\Psi_1] \qquad (1.6.18)$$

though this time V_{ne} for molecules does not scale simply. The scaling of (1.6.16) produces a simple dilation of the electronic cloud without changing its normalization; the scaling of (1.6.13) changes nuclear positions as well. For atoms, the two scalings are equivalent.

2
DENSITY MATRICES

2.1 Description of quantum states and the Dirac notation

In this chapter, the concepts and form of elementary quantum mechanics are generalized. This allows use of variables other than coordinates for the description of a state, permits ready discussion of physical states that cannot be described by wave functions, and prepares the way for formally considering the number of particles to be variable rather than constant. Taking advantage, as appropriate, of the identity of electrons and the fact that we are exclusively concerned with systems and equations that involve two-particle interactions at worst, several tools are developed for formal analysis: Dirac notation, density operators, and density matrices. We follow Dirac (1947) and Messiah (1961); see also Szabo and Ostlund (1982, especially pp. 9–12), and Weissbluth (1978).

We begin with the quantum state of a single-particle system. Such a state was described in Chapter 1 by a wave function $\Psi(\mathbf{r})$ in coordinate space (neglecting the spin for the moment). It can also be equivalently "represented" by a momentum-space wave function that is the Fourier transform of $\Psi(\mathbf{r})$. This, together with the quantum superposition principle, leads one to construct a more general and abstract form of quantum mechanics. Thus, one associates with each state a *ket vector* $|\Psi\rangle$ in the linear vector space $\mathcal{H}$, called the *Hilbert space* (Messiah 1961, pp. 164–166). The linearity of the Hilbert space implements the superposition principle: a linear combination of two vectors $C_1|\Psi_1\rangle + C_2|\Psi_2\rangle$ is also a ket vector in the same Hilbert space, associated with a realizable physical state.

Just as a vector in three-dimensional coordinate space can be defined by its three components in a particular coordinate system, the ket $|\Psi\rangle$ can be completely specified by its components in any particular *representation*. The difference is that the Hilbert space here has an infinite number of dimensions.

In one-to-one correspondence with the space of all kets $|\Psi\rangle$, there is a dual space consisting of *bra vectors* $\langle\Psi|$. For an arbitrary bra $\langle\Phi|$ and ket $|\Psi\rangle$, the *inner product* $\langle\Phi|\Psi\rangle$ is defined by

$$\langle\Phi|\Psi\rangle = \sum_i \Phi_i^* \Psi_i \qquad (2.1.1)$$

This is for the case that both $\langle\Phi|$ and $|\Psi\rangle$ are represented in a discrete basis with components Φ_i^* and Ψ_i. If the representation is continuous, one has an integral rather than a sum, for example,

$$\langle\Phi|\Psi\rangle = \int \Phi^*(\mathbf{r})\Psi(\mathbf{r})\,d\mathbf{r} \tag{2.1.2}$$

where the integral is equivalent to the sum of all component products with different values of $\mathbf{r}$. Thus, the inner product of a ket and a bra is a complex number and satisfies

$$\langle\Phi|\Psi\rangle = \langle\Psi|\Phi\rangle^* \tag{2.1.3}$$

If

$$\langle\Psi|\Psi\rangle = 1 \tag{2.1.4}$$

we call $|\Psi\rangle$ and $\langle\Psi|$ normalized. The bra $\langle\Psi|$ is said to be the *conjugate* of the ket $|\Psi\rangle$.

Consider now a complete basis set $\{|f_i\rangle\}$ (for example, the eigenstates of some Hamiltonian), satisfying the orthonormality conditions

$$\langle f_i|f_j\rangle = \delta_{ij} \tag{2.1.5}$$

Then any ket $|\Psi\rangle$ can be expressed in terms of the ket basis set $|f_i\rangle$ by

$$|\Psi\rangle = \sum_i \Psi_i |f_i\rangle \tag{2.1.6}$$

Taking the inner product of $|\Psi\rangle$ with a bra $\langle f_j|$, we find the jth component of $|\Psi\rangle$ in the representation of the $|f_i\rangle$,

$$\Psi_j = \langle f_j|\Psi\rangle \tag{2.1.7}$$

where (2.1.5) has been used. If the basis set is continuous, the orthonormality condition becomes

$$\langle \mathbf{r}|\mathbf{r}'\rangle = \delta(\mathbf{r} - \mathbf{r}') \tag{2.1.8}$$

where $\delta(\mathbf{r} - \mathbf{r}')$ is the Dirac delta function, and for an arbitrary ket $|\Psi\rangle$,

$$|\Psi\rangle = \int \Psi(\mathbf{r})|\mathbf{r}\rangle\,d\mathbf{r} \tag{2.1.9}$$

and

$$\Psi(\mathbf{r}) = \langle\mathbf{r}|\Psi\rangle \tag{2.1.10}$$

Here $\Psi(\mathbf{r})$ is precisely the ordinary wave function in coordinate space. If a basis set $|\mathbf{p}\rangle$ were used, one would instead get the momentum-space function. Bras may be expanded similarly.

An *operator* $\hat{A}$ transforms a ket into another ket in the Hilbert space,

$$\hat{A}|\Psi\rangle = |\hat{A}\Psi\rangle = |\Psi'\rangle \tag{2.1.11}$$

The *adjoint* of $\hat{A}$, denoted by $\hat{A}^\dagger$, transforms the corresponding bra,

$$\langle \Psi | \hat{A}^\dagger = \langle \hat{A}\Psi | = \langle \Psi' | \qquad (2.1.12)$$

An operator is *self-adjoint*, or *Hermitian*, if it equals its adjoint; operators corresponding to observables always have this property. For normalized ket and bra, (2.1.11) can be written

$$\hat{A} |\Psi\rangle = (|\Psi'\rangle \langle \Psi|) |\Psi\rangle \qquad (2.1.13)$$

and (2.1.12) as

$$\langle \Psi | \hat{A}^\dagger = \langle \Psi | (|\Psi\rangle \langle \Psi'|) \qquad (2.1.14)$$

When a bra $\langle\ |$ and a ket $|\ \rangle$ are juxtaposed, one has an inner product if $\langle\ |$ is before $|\ \rangle$, i.e. $\langle\ ||\ \rangle = \langle\ |\ \rangle$; and an operator if $|\ \rangle$ is before $\langle\ |$.

A very important type of operator is the *projection operator* onto a normalized ket $|X\rangle$:

$$\hat{P}_x = |X\rangle\langle X| \qquad (2.1.15)$$

The projection property is manifest when $\hat{P}_i$ acts on the ket $|\Psi\rangle$ of (2.1.6):

$$\hat{P}_i |\Psi\rangle = |f_i\rangle \langle f_i | \Psi\rangle$$
$$= \Psi_i |f_i\rangle \qquad (2.1.16)$$

Note that only the part of $|\Psi\rangle$ associated with $|f_i\rangle$ is left. Projection operators have the property

$$\hat{P}_x \cdot \hat{P}_x = \hat{P}_x \qquad (2.1.17)$$

For this reason, they are said to be *idempotent*.

By inserting (2.1.7) into (2.1.6), we get

$$|\Psi\rangle = \sum_i \langle f_i | \Psi\rangle |f_i\rangle = \sum_i |f_i\rangle \langle f_i | \Psi\rangle$$
$$= \left\{\sum_i |f_i\rangle \langle f_i|\right\} |\Psi\rangle \qquad (2.1.18)$$

from which follows

$$\sum_i |f_i\rangle \langle f_i| = \sum_i \hat{P}_i = \hat{I} \qquad (2.1.19)$$

where $\hat{I}$ is the identity operator. This is the *closure relation*. The corresponding expression for a continuous basis set is

$$\int d\mathbf{r}\, |\mathbf{r}\rangle \langle \mathbf{r}| = \int d\mathbf{r}\, \hat{P}_\mathbf{r} = \hat{I} \qquad (2.1.20)$$

The closure relation greatly facilitates transformation between different representations, which makes the Dirac notation so useful. As an example, we compute the inner product

$$\langle \Phi | \Psi \rangle = \langle \Phi | \hat{I} | \Psi \rangle$$
$$= \sum_i \langle \Phi | f_i \rangle \langle f_i | \Psi \rangle$$
$$= \sum_i \Phi_i^* \Psi_i \qquad (2.1.21)$$

which is identically (2.1.1). Or, consider the effect of the operator $\hat{A}$ in (2.1.11),

$$\langle f_i | \hat{A} | \Psi \rangle = \sum_j \langle f_i | \hat{A} | f_j \rangle \langle f_j | \Psi \rangle = \langle f_i | \Psi' \rangle \qquad (2.1.22)$$

where the complex numbers $\langle f_i | \hat{A} | f_j \rangle$ constitute the matrix representation of $\hat{A}$ in the basis set $|f_i\rangle$. [Such a matrix in full in fact defines the operator.] If we use a continuous basis set, (2.1.22) becomes

$$\langle \mathbf{r}' | \hat{A} | \Psi \rangle = \int d\mathbf{r} A(\mathbf{r}', \mathbf{r}) \Psi(\mathbf{r}) = \Psi'(\mathbf{r}') \qquad (2.1.23)$$

where $A(\mathbf{r}', \mathbf{r}) = \langle \mathbf{r}' | \hat{A} | \mathbf{r} \rangle$. Equation (2.1.23) indicates that an operator can be *nonlocal*. An operator $\hat{A}$ is *local* if

$$A(\mathbf{r}', \mathbf{r}) = A(\mathbf{r}) \delta(\mathbf{r}' - \mathbf{r}) \qquad (2.1.24)$$

Often the potential part of a one-body Hamiltonian $\hat{H}$ is local, in which case the Schrödinger equation (1.1.1) is just a differential equation. The Hartree–Fock exchange operator in (1.3.12) is nonlocal.

As another example of the use of (2.1.15), we may prove the formula for the decomposition of a Hermitian operator into its eigenfunctions. Let the kets $|\alpha_i\rangle$ be the complete set of eigenkets of the linear operator $\hat{A}$, with eigenvalues a_i. Then

$$\hat{A} |\alpha_i\rangle = a_i |\alpha_i\rangle, \qquad \hat{A} |\alpha_i\rangle \langle \alpha_i| = a_i |\alpha_i\rangle \langle \alpha_i|$$
$$\hat{A} = \hat{A} \sum_i |\alpha_i\rangle \langle \alpha_i| = \sum_i a_i |\alpha_i\rangle \langle \alpha_i| \qquad (2.1.25)$$

Here again the sum becomes an integral in the continuous case.

If particle spin is included in the above, then the closure relation is

$$\int d\mathbf{x} |\mathbf{x}\rangle \langle \mathbf{x}| = \sum_s \int d\mathbf{r} |\mathbf{r}, s\rangle \langle \mathbf{r}, s| = \hat{I} \qquad (2.1.26)$$

With this interpretation of integrals, all of the above equations may be regarded as including spin, with $\mathbf{r}$ replaced by $\mathbf{x}$.

We now turn to a quantum system of many identical particles, for which the foregoing concepts and formulas go through when suitably generalized. However, a new feature appears—the antisymmetry (or symmetry) of fermion (or boson) wave functions with respect to exchange of indices (coordinates) of any two particles. The antisymmetric and symmetric states span subspaces of the N-particle Hilbert space, $\mathcal{H}_N$, the subspaces denoted by $\mathcal{H}_N^A$ and $\mathcal{H}_N^S$. We focus on $\mathcal{H}_N^A$, since electrons are fermions. In $\mathcal{H}_N$, a normalized basis ket for N particles in suitably defined states $|\alpha_1\rangle, |\alpha_2\rangle, \ldots, |\alpha_N\rangle$, respectively, is

$$|\alpha_1\alpha_2\cdots\alpha_N) = |\alpha_1\rangle|\alpha_2\rangle\cdots|\alpha_N\rangle \quad (2.1.27)$$

while for fermions, a typical normalized antisymmetric basis ket would be

$$|\alpha_1\alpha_2\cdots\alpha_N\rangle = \frac{1}{\sqrt{N!}}\sum_P (-1)^P P |\alpha_1\alpha_1\cdots\alpha_N) \quad (2.1.28)$$

where the P's are operators permutating particle coordinates and $(-1)^P$ is the parity of the permutation P. The closure relation in $\mathcal{H}_N$ is

$$\sum_{\alpha_1,\alpha_2,\ldots,\alpha_N} |\alpha_1\alpha_2\cdots\alpha_N)(\alpha_1\alpha_2\cdots\alpha_N| = \hat{I} \quad (2.1.29)$$

while that in $\mathcal{H}_N^A$ is

$$\sum_{\alpha_1,\alpha_2,\ldots,\alpha_N} \frac{1}{N!}|\alpha_1\alpha_2\cdots\alpha_N\rangle\langle\alpha_1\alpha_2\cdots\alpha_N| = \hat{I} \quad (2.1.30)$$

The summations in both formulas become integrals if the indices are continuous.

Generalizing (2.1.10), the N-electron coordinate wave function is related to the abstract ket vector in $\mathcal{H}_N^A$ by

$$\Psi_N(\mathbf{x}_1\mathbf{x}_2\cdots\mathbf{x}_N) = (\mathbf{x}_1\mathbf{x}_2\cdots\mathbf{x}_N|\Psi_N\rangle \quad (2.1.31)$$

In the case that $|\Psi_N\rangle$ takes the form (2.1.28), describing N independent electrons moving in N one-electron states, one can show from (2.1.31) that Ψ_N is a Slater determinant of the form of (1.3.1).

2.2 Density operators

We now consider an even more general description of a quantum state. By (1.1.10), the quantity

$$\Psi_N(\mathbf{x}_1\mathbf{x}_2\cdots\mathbf{x}_N)\Psi_N^*(\mathbf{x}_1\mathbf{x}_2\cdots\mathbf{x}_N) \quad (2.2.1)$$

is the probability distribution associated with a solution of the Schrödinger equation (1.1.1), with the Hamiltonian operator $\hat{H}_N$. The main result of the present chapter will be to establish the utility of quantities of the type

$$\gamma_N(\mathbf{x}_1'\mathbf{x}_2'\cdots\mathbf{x}_N', \mathbf{x}_1\mathbf{x}_2\cdots\mathbf{x}_N) \equiv \Psi_N(\mathbf{x}_1'\mathbf{x}_2'\cdots\mathbf{x}_N')\Psi_N^*(\mathbf{x}_1\mathbf{x}_2\cdots\mathbf{x}_N) \quad (2.2.2)$$

which is more general than (2.2.1) in that the variables in the first factor are primed. The two sets of independent quantities $x_1' x_2' \cdots$ and $x_1 x_2 \cdots$ can be thought of as two sets of indices that give (2.2.2) a numerical value, in contrast with the single set $x_1 x_2 \cdots$ that suffices for (2.2.1). We therefore may think of (2.2.2) as an element of a matrix, which we shall call a *density matrix*. If we set $x_i = x_i'$ for all i, we get a diagonal element of this matrix, the original (2.2.1). Equivalently, (2.2.2) can be viewed as the coordinate representation of the *density operator*,

$$|\Psi_N\rangle\langle\Psi_N| = \hat{\gamma}_N \qquad (2.2.3)$$

since

$$(x_1' x_2' \cdots x_N'| \hat{\gamma}_N |x_1 x_2 \cdots x_N) = (x_1' x_2' \cdots |\Psi_N\rangle\langle\Psi_N| x_1 x_2 \cdots)$$
$$= \Psi_N(x_1' x_2' \cdots x_N')\Psi_N^*(x_1 x_2 \cdots x_N) \qquad (2.2.4)$$

Note that $\hat{\gamma}_N$ is a projection operator. We then have for normalized Ψ_N,

$$\operatorname{tr}(\hat{\gamma}_N) = \int \Psi_N(\mathbf{x}^N)\Psi_N^*(\mathbf{x}^N)\, d\mathbf{x}^N = 1 \qquad (2.2.5)$$

where the *trace* of the operator $\hat{A}$ is defined as the sum of diagonal elements of the matrix representing $\hat{A}$, or the integral if the representation is continuous as in (2.2.5). One can also verify from (1.1.12) that

$$\langle\hat{A}\rangle = \operatorname{tr}(\hat{\gamma}_N \hat{A}) = \operatorname{tr}(\hat{A}\hat{\gamma}_N) \qquad (2.2.6)$$

of which (1.1.12) is the coordinate representation.

In view of (2.2.6), the density operator $\hat{\gamma}_N$ of (2.2.3) carries the same information as the N-electron wave function $|\Psi_N\rangle$. $\hat{\gamma}_N$ is an operator in the same space as the vector $|\Psi_N\rangle$. Note that while $|\Psi\rangle$ is defined only up to an arbitrary phase factor, $\hat{\gamma}_N$ for a state is unique. $\hat{\gamma}_N$ also is Hermitian.

An operator description of a quantum state becomes necessary when the state cannot be represented by a linear superposition of eigenstates of a particular Hamiltonian $\hat{H}_N$ ("by a vector in the Hilbert space $\mathcal{H}_N$"). This occurs when the system of interest is part of a larger closed system, as for example an individual electron in a many-electron system, or a macroscopic system in thermal equilibrium with other macroscopic systems. For such a system one does not have a complete Hamiltonian containing only its own degrees of freedom, thereby precluding the wave-function description. A state is said to be *pure* if it is described by a wave function, *mixed* if it cannot be described by a wave function.

A system in a mixed state can be characterized by a probability distribution over all the accessible pure states. To accomplish this description, we generalize the density operator of (2.2.3) to the *ensemble density operator*

$$\hat{\Gamma} = \sum_i p_i |\Psi_i\rangle\langle\Psi_i| \qquad (2.2.7)$$

where p_i is the probability of the system being found in the state $|\Psi_i\rangle$, and the sum is over the complete set of all accessible pure states. With the $|\Psi_i\rangle$ orthonormal, the rules of probability require that p_i be real and that

$$p_i \geq 0, \qquad \sum_i p_i = 1 \qquad (2.2.8)$$

Note that if the interactions can induce change in particle number, the accessible states can involve different particle numbers.

For a system in a pure state, one p_i is 1 and the rest are zero; $\hat{\Gamma}$ of (2.2.7) then reduces to $\hat{\gamma}_N$ of (2.2.3). By construction, $\hat{\Gamma}$ is normalized: In an arbitrary complete basis $|f_k\rangle$,

$$\begin{aligned}
\mathrm{Tr}(\hat{\Gamma}) &= \sum_i \sum_k p_i \langle f_k | \Psi_i \rangle \langle \Psi_i | f_k \rangle \\
&= \sum_i p_i \langle \Psi_i | \sum_k |f_k\rangle\langle f_k | \Psi_i \rangle \\
&= \sum_i p_i \langle \Psi_i | \Psi_i \rangle = \sum_i p_i = 1 \qquad (2.2.9)
\end{aligned}$$

[Here and later Tr means the trace in Fock space (see Appendix C), containing states with different numbers of particles, in contrast to the trace denoted by tr in (2.2.5), in N-particle Hilbert space.] $\hat{\Gamma}$ is Hermitian:

$$\begin{aligned}
\langle f_k | \hat{\Gamma} | f_l \rangle &= \sum_i p_i \langle f_k | \Psi_i \rangle \langle \Psi_i | f_l \rangle \\
&= \sum_i p_i \{\langle f_l | \Psi_i \rangle \langle \Psi_i | f_k \rangle\}^* \\
&= \langle f_l | \hat{\Gamma} | f_k \rangle^* \qquad (2.2.10)
\end{aligned}$$

It also is positive semidefinite:

$$\langle f_k | \hat{\Gamma} | f_k \rangle = \sum_i p_i |\langle f_k | \Psi_i \rangle|^2 \geq 0 \qquad (2.2.11)$$

The p_i are the eigenvalues of $\hat{\Gamma}$.

For a system to be in a pure state, it is necessary and sufficient for the density operator to be idempotent:

$$\hat{\gamma}_N \cdot \hat{\gamma}_N = |\Psi\rangle\langle\Psi|\Psi\rangle\langle\Psi| = |\Psi\rangle\langle\Psi| = \hat{\gamma}_N \qquad (2.2.12)$$

The ensemble density operator in general lacks this property:

$$\hat{\Gamma} \cdot \hat{\Gamma} = \sum_i p_i^2 |\Psi_i\rangle\langle\Psi_i| \neq \hat{\Gamma} \qquad (2.2.13)$$

For a mixed state, the expectation value for the observable $\hat{A}$ is given by a natural generalization of (2.2.6),

$$\langle A \rangle = \text{Tr}\,(\hat{\Gamma}\hat{A}) = \sum_i p_i \langle \Psi_i | \hat{A} | \Psi_i \rangle \quad (2.2.14)$$

Note that this is very different from what (1.1.12) gives when $|\Psi\rangle$ is a linear combination $\sum_i C_i |\Psi_i\rangle$, in which case cross terms $\langle \Psi_i | \hat{A} | \Psi_j \rangle$ enter.

The foregoing definitions and properties also hold for time-dependent pure-state density operators $\hat{\gamma}_N$ and ensemble density operators $\hat{\Gamma}$. From the time-dependent Schrödinger equation,

$$i\hbar \frac{\partial}{\partial t} |\Psi_N\rangle = \hat{H} |\Psi_N\rangle \quad (2.2.15)$$

we find

$$\frac{\partial}{\partial t} \hat{\gamma}_N = \left(\frac{\partial}{\partial t} |\Psi_N\rangle\right) \langle \Psi_N | + |\Psi_N\rangle \frac{\partial}{\partial t} \langle \Psi_N |$$

$$= \frac{\hat{H}}{i\hbar} |\psi_N\rangle\langle\psi_N| - |\psi_N\rangle\langle\psi_N| \frac{\hat{H}}{i\hbar}$$

so that

$$i\hbar \frac{\partial}{\partial t} \hat{\gamma}_N = [\hat{H}, \hat{\gamma}_N] \quad (2.2.16)$$

where the brackets denote the commutator. More generally, the linearity of (2.2.7) leads to

$$i\hbar \frac{\partial}{\partial t} \hat{\Gamma} = [\hat{H}, \hat{\Gamma}] \quad (2.2.17)$$

This is clearly true if $\hat{\Gamma}$ of (2.2.7) only involves states with the same number of particles (canonical ensemble case). If, on the other hand, states with different numbers of particles are allowed, to interpret (2.2.17) one has to use the Hamiltonian in second-quantized form (see Appendix C), which is independent of the number of particles. The Hamiltonian in (2.2.17) is only for the subsystem of interest, neglecting all its interactions with the rest of the larger closed system.

For a stationary state, $\hat{\Gamma}$ is independent of time. Therefore, from (2.2.17)

$$[\hat{H}, \hat{\Gamma}] = 0 \quad \text{for a stationary state} \quad (2.2.18)$$

Accordingly, $\hat{H}$ and $\hat{\Gamma}$ can share the same eigenvectors.

2.3 Reduced density matrices for fermion systems

The basic Hamiltonian operator of (1.1.2) is the sum of two symmetric "one-electron" operators and a symmetric "two-electron" operator. It also does not depend on spin. Similarly, operators corresponding to other physical observables are of one-electron or two-electron type and often are spin free. Wave functions Ψ_N are antisymmetric. These facts mean that the expectation value formulas (1.1.12) or (2.2.6), and (2.2.14) can be systematically simplified by integrating the $\Psi_N\Psi_N^*$ product of (2.2.1), or its generalization (2.2.7), over N-2 of its variables. This gives rise to the concepts of *reduced density matrix* and *spinless density matrix*, which we now describe (Löwdin 1955a,b, McWeeny 1960, Davidson 1976).

One calls (2.2.1) the Nth *order density matrix* for a pure state of an N-electron system. One then defines the *reduced density matrix of order p* by the formula

$$\gamma_p(\mathbf{x}_1'\mathbf{x}_2'\cdots\mathbf{x}_p', \mathbf{x}_1\mathbf{x}_2\cdots\mathbf{x}_p)$$
$$= \binom{N}{p}\int\cdots\int \gamma_N(\mathbf{x}_1'\mathbf{x}_2'\cdots\mathbf{x}_p'\mathbf{x}_{p+1}\cdots\mathbf{x}_N, \mathbf{x}_1\mathbf{x}_2\cdots\mathbf{x}_p\cdots\mathbf{x}_N)\,d\mathbf{x}_{p+1}\cdots d\mathbf{x}_N$$
(2.3.1)

where $\binom{N}{p}$ is a binomial coefficient. In particular,

$$\gamma_2(\mathbf{x}_1'\mathbf{x}_2', \mathbf{x}_1\mathbf{x}_2)$$
$$= \frac{N(N-1)}{2}\int\cdots\int \Psi(\mathbf{x}_1'\mathbf{x}_2'\mathbf{x}_3\cdots\mathbf{x}_N)\Psi^*(\mathbf{x}_1\mathbf{x}_2\mathbf{x}_3\cdots\mathbf{x}_N)\,d\mathbf{x}_3\cdots d\mathbf{x}_N$$
(2.3.2)

and

$$\gamma_1(\mathbf{x}_1', \mathbf{x}_1) = N\int\cdots\int \Psi(\mathbf{x}_1'\mathbf{x}_2\cdots\mathbf{x}_N)\Psi^*(\mathbf{x}_1\mathbf{x}_2\cdots\mathbf{x}_N)\,d\mathbf{x}_2\cdots d\mathbf{x}_N \quad (2.3.3)$$

Note that the second-order density matrix γ_2 normalizes to the number of electron pairs,

$$\text{tr }\gamma_2(\mathbf{x}_1'\mathbf{x}_2', \mathbf{x}_1\mathbf{x}_2) = \iint \gamma_2(\mathbf{x}_1\mathbf{x}_2, \mathbf{x}_1\mathbf{x}_2)\,d\mathbf{x}_1\,d\mathbf{x}_2 = \frac{N(N-1)}{2} \quad (2.3.4)$$

while the first-order density matrix γ_1 normalizes to the number of electrons,

$$\text{tr }\gamma_1(\mathbf{x}_1', \mathbf{x}_1) = \int \gamma_1(\mathbf{x}_1, \mathbf{x}_1)\,d\mathbf{x}_1 = N \quad (2.3.5)$$

Note also that γ_1 can be obtained from γ_2 by quadrature,

$$\gamma_1(\mathbf{x}_1', \mathbf{x}_1) = \frac{2}{N-1}\int \gamma_2(\mathbf{x}_1'\mathbf{x}_2, \mathbf{x}_1\mathbf{x}_2)\,d\mathbf{x}_2 \quad (2.3.6)$$

Here the full four-variable $\gamma_2(\mathbf{x}_1'\mathbf{x}_2', \mathbf{x}_1\mathbf{x}_2)$ is not necessary, only the three-variable $\gamma_2(\mathbf{x}_1'\mathbf{x}_2, \mathbf{x}_1\mathbf{x}_2)$.

The reduced density matrices γ_1 and γ_2 as just defined are coordinate-space representations of operators $\hat{\gamma}_1$ and $\hat{\gamma}_2$, acting, respectively, on the one- and two-particle Hilbert spaces. Like $\hat{\gamma}_N$, these operators are positive semidefinite,

$$\gamma_1(\mathbf{x}_1, \mathbf{x}_1) \geq 0 \tag{2.3.7}$$

$$\gamma_2(\mathbf{x}_1\mathbf{x}_2, \mathbf{x}_1\mathbf{x}_2) \geq 0 \tag{2.3.8}$$

and they are Hermitian,

$$\gamma_1(\mathbf{x}_1', \mathbf{x}_1) = \gamma_1^*(\mathbf{x}_1, \mathbf{x}_1') \tag{2.3.9}$$

$$\gamma_2(\mathbf{x}_1'\mathbf{x}_2', \mathbf{x}_1\mathbf{x}_2) = \gamma_2^*(\mathbf{x}_1\mathbf{x}_2, \mathbf{x}_1'\mathbf{x}_2') \tag{2.3.10}$$

Antisymmetry of γ_N also requires that any reduced density matrix change its sign on exchange of two primed or two unprimed particle indices; thus

$$\gamma_2(\mathbf{x}_1'\mathbf{x}_2', \mathbf{x}_1\mathbf{x}_2) = -\gamma_2(\mathbf{x}_2'\mathbf{x}_1', \mathbf{x}_1\mathbf{x}_2) = -\gamma_2(\mathbf{x}_1'\mathbf{x}_2', \mathbf{x}_2\mathbf{x}_1) = \gamma_2(\mathbf{x}_2'\mathbf{x}_1', \mathbf{x}_2\mathbf{x}_1) \tag{2.3.11}$$

The Hermitian reduced density operators $\hat{\gamma}_1$ and $\hat{\gamma}_2$ admit eigenfunctions and associated eigenvalues,

$$\int \gamma_1(\mathbf{x}_1', \mathbf{x}_1) \psi_i(\mathbf{x}_1) \, d\mathbf{x}_1 = n_i \psi_i(\mathbf{x}_1') \tag{2.3.12}$$

and

$$\int \gamma_2(\mathbf{x}_1'\mathbf{x}_2', \mathbf{x}_1\mathbf{x}_2) \theta_i(\mathbf{x}_1\mathbf{x}_2) \, d\mathbf{x}_1 \, d\mathbf{x}_2 = g_i \theta_i(\mathbf{x}_1'\mathbf{x}_2') \tag{2.3.13}$$

For $\hat{\gamma}_1$, the eigenfunctions $\psi_i(\mathbf{x})$ are called *natural spin orbitals*, and the eigenvalues n_i the *occupation numbers*; these are very important concepts. From the rule for expressing an operator in terms of its eigenvectors, (2.1.25), we have

$$\hat{\gamma}_1 = \sum_i n_i |\psi_i\rangle \langle \psi_i| \tag{2.3.14}$$

or

$$\gamma_1(\mathbf{x}_1', \mathbf{x}_1) = \sum_i n_i \psi_i(\mathbf{x}_1') \psi_i^*(\mathbf{x}_1) \tag{2.3.15}$$

Similarly,

$$\hat{\gamma}_2 = \sum_i g_i |\theta_i\rangle \langle \theta_i| \tag{2.3.16}$$

where the g_i again are occupation numbers; the $|\theta_i\rangle$ are two-particle functions called *natural geminals*, which in accord with (2.3.11) are defined to be antisymmetric. From (2.3.7) and (2.3.8) also follow

$$n_i \geq 0, \qquad g_i \geq 0 \tag{2.3.17}$$

Differential equations for the natural orbitals ψ_i have been discussed briefly by Löwdin (1955a). Their long-range behavior is given by Morrell,

Parr, and Levy (1975):
$$\psi_i \sim \exp\left[-(2I_{\min})^{1/2} r\right] \tag{2.3.18}$$

where $I_{\min}$ is the smallest ionization potential of the system.

Comparing (2.3.14) and (2.3.16) with (2.2.7) and recalling the probabilistic interpretation of (2.2.7), one sees that n_i is proportional to the probability of the one-electron state $|\psi_i\rangle$ being occupied; similarly g_i is proportional to the probability of the two-electron state $|\theta_i\rangle$ being occupied.

For a mixed state, a corresponding set of definitions of reduced density matrices and operators is appropriate, and the same properties all hold. For the case in which all participating states have the same particle number, N, we denote the $\hat{\Gamma}$ of (2.2.7) as the Nth-order density operator $\hat{\Gamma}_N$. The pth-order mixed state density matrix is then

$$\Gamma_p(\mathbf{x}_1'\mathbf{x}_2'\cdots\mathbf{x}_p', \mathbf{x}_1\mathbf{x}_2\cdots\mathbf{x}_p)$$
$$= \binom{N}{p} \int \cdots \int \Gamma_N(\mathbf{x}_1'\mathbf{x}_2'\cdots\mathbf{x}_p'\mathbf{x}_{p+1}\cdots\mathbf{x}_N, \mathbf{x}_1\mathbf{x}_2\cdots\mathbf{x}_N)\, d\mathbf{x}_{p+1}\cdots d\mathbf{x}_N \tag{2.3.19}$$

corresponding to an operator $\hat{\Gamma}_p$. Similarly one has $\hat{\Gamma}_2$ and $\hat{\Gamma}_1$, the second of which will be of special importance for us. It corresponds to the matrix

$$\Gamma_1(\mathbf{x}_1', \mathbf{x}_1) = N \int \cdots \int \sum_i p_i \Psi_i(\mathbf{x}_1'\mathbf{x}_2\cdots\mathbf{x}_N)\Psi_i^*(\mathbf{x}_1\mathbf{x}_2\cdots\mathbf{x}_N)\, d\mathbf{x}_2 \cdots d\mathbf{x}_N \tag{2.3.20}$$

where the Ψ_i are the various N-electron states entering the mixed state in question. Many of the formulas below hold for mixed states as well as pure states, but we will not specify this in every case.

Now consider the expectation value, for an antisymmetric N-body wave function Ψ, of a one-electron operator

$$\hat{O}_1 = \sum_{i=1}^{N} O_1(\mathbf{x}_i, \mathbf{x}_i') \tag{2.3.21}$$

We have
$$\langle \hat{O}_1 \rangle = \mathrm{tr}\,(\hat{O}_1 \gamma_N)$$
$$= \int O_1(\mathbf{x}_1 \mathbf{x}_1')\gamma_1(\mathbf{x}_1', \mathbf{x}_1)\, d\mathbf{x}_1\, d\mathbf{x}_1' \tag{2.3.22}$$

If the one-electron operator is local in the sense of (2.1.24), as are most operators in molecular physics, we conventionally only write down the diagonal part; thus

$$\hat{O}_1 = \sum_{i=1}^{N} O_1(\mathbf{x}_i) \tag{2.3.23}$$

and the corresponding expectation-value formula is

$$\langle \hat{O}_1 \rangle = \int [O_1(\mathbf{x}_1)\gamma_1(\mathbf{x}_1', \mathbf{x}_1)]_{\mathbf{x}_1'=\mathbf{x}_1} d\mathbf{x}_1 \quad (2.3.24)$$

All two-electron operators that concern us are local, and so we may denote the operators by their diagonal part, neglecting the two delta functions. That is, we write

$$\hat{O}_2 = \sum_{i<j}^{N} O_2(\mathbf{x}_i, \mathbf{x}_j) \quad (2.3.25)$$

and obtain for the corresponding expectation value

$$\langle \hat{O}_2 \rangle = \operatorname{tr}(\hat{O}_2 \gamma_N)$$

$$= \iint [O_2(\mathbf{x}_1, \mathbf{x}_2)\gamma_2(\mathbf{x}_1'\mathbf{x}_2', \mathbf{x}_1\mathbf{x}_2)]_{\mathbf{x}_1'=\mathbf{x}_1, \mathbf{x}_2'=\mathbf{x}_2} d\mathbf{x}_1 \, d\mathbf{x}_2 \quad (2.3.26)$$

For the expectation value of the Hamiltonian (1.1.2), combining all the parts, we obtain

$$E = \operatorname{tr}(\hat{H}\hat{\gamma}_N) = E[\gamma_1, \gamma_2] = E[\gamma_2]$$

$$= \int [(-\tfrac{1}{2}\nabla_1^2 + v(\mathbf{r}_1))\gamma_1(\mathbf{x}_1', \mathbf{x}_1)]_{\mathbf{x}_1'=\mathbf{x}_1} d\mathbf{x}_1 + \iint \frac{1}{r_{12}} \gamma_2(\mathbf{x}_1\mathbf{x}_2, \mathbf{x}_1\mathbf{x}_2) \, d\mathbf{x}_1 \, d\mathbf{x}_2$$

(2.3.27)

It is because of (2.3.6) that in fact only the second-order density matrix is needed. In the next section, we will further simplify this equation by integrating over the spin variables.

One might hope to minimize (2.3.27) with respect to γ_2, thus avoiding the problem of the 4N-dimensional Ψ. This hope has spawned a great deal of work (see for example Coleman 1963, 1981, Percus 1978, Erdahl and Smith 1987). There is a major obstacle to implementing this idea, however, realized from pretty much the beginning. Trial γ_2 must correspond to some antisymmetric Ψ, that is, for any guessed γ_2 there must be a Ψ from which it comes via (2.3.2). This is the *N-representability problem* for the second-order density matrix.

It is a very difficult task to obtain the necessary and sufficient conditions for a reduced matrix γ_2 to be derivable from an antisymmetric wave function (Coleman 1963, 1981). A more tractable problem is to solve the *ensemble N-representability* problem for Γ_2; that is, to find the necessary and sufficient conditions for a Γ_2 to be derivable from a mixed-state (ensemble) Γ_2 by (2.3.19). It is in fact completely legitimate to enlarge the class of trial density operators for an N-electron problem from a pure-state set to the set of positive unit-trace density operators made up from N-electron states, because

$$E_0 = \text{tr}\,(\hat{H}\hat{\Gamma}_N^0) \leq \text{tr}\,(\hat{H}\hat{\Gamma}_N) \qquad (2.3.28)$$

That is, minimization of $\text{tr}\,(\hat{H}\hat{\Gamma}_N)$ leads to the N-electron ground-state energy and the ground state $\hat{\gamma}_N$ if it is not degenerate, or an arbitrary linear combination $\hat{\Gamma}_N$ (convex sum) of all degenerate ground states if it is degenerate. Thus, the search in (2.3.27) may be made over ensemble N-representable Γ_2.

It is advantageous for this problem that the set of positive unit operators $\hat{\Gamma}_N$ is *convex*, and so also the allowable $\hat{\Gamma}_2$. [A set C is convex if for any two elements Y_1 and Y_2 of C, $P_1Y_1 + P_2Y_2$ also belongs to C if $0 \leq P_1$, $0 \leq P_2$, and $P_1 + P_2 = 1$.] The situation for $\hat{\Gamma}_2$ has not yet been practically resolved, though there has been progress (Coleman 1981). But for $\hat{\Gamma}_1$ a complete solution has been found, as will be described in §2.6. Given a $\hat{\Gamma}_1$,

$$\hat{\Gamma}_1 = \sum_i n_i |\psi_i\rangle\langle\psi_i| \qquad (2.3.29)$$

the necessary and sufficient conditions for it to be N-representable are that

$$0 \leq n_i \leq 1 \qquad (2.3.30)$$

for all of the eigenvalues of $\hat{\Gamma}_1$ (Löwdin 1955a, Coleman 1963). This conforms nicely with the simple rule that an orbital cannot be occupied by more than one electron—the naive Pauli principle.

For states that are eigenstates of $\hat{H}$, the Schrödinger equation itself gives equations relating reduced density matrices of different orders (Nakatsuji 1976, Cohen and Frishberg 1976).

2.4 Spinless density matrices

Many operators of interest do not involve spin coordinates, for instance the Hamiltonian operators for atoms or molecules. This makes desirable further reduction of the density matrices of (2.3.2) and (2.3.3), by summation over the spin coordinates s_1 and s_2 (McWeeney 1960).

We define the first-order and second-order *spinless density matrices* by

$$\rho_1(\mathbf{r}_1', \mathbf{r}_1) = \int \gamma_1(\mathbf{r}_1's_1, \mathbf{r}_1s_1)\, ds_1$$

$$= N \int \cdots \int \Psi(\mathbf{r}_1's_1\mathbf{x}_2\cdots\mathbf{x}_N)\Psi^*(\mathbf{r}_1s_1\mathbf{x}_2\cdots\mathbf{x}_N)\, ds_1\, d\mathbf{x}_2\cdots d\mathbf{x}_N$$

$$(2.4.1)$$

and
$$\rho_2(\mathbf{r}'_1\mathbf{r}'_2, \mathbf{r}_1\mathbf{r}_2) = \iint \gamma_2(\mathbf{r}'_1 s_1 \mathbf{r}'_2 s_2, \mathbf{r}_1 s_1 \mathbf{r}_2 s_2) \, ds_1 \, ds_2$$
$$= \frac{N(N-1)}{2} \int \cdots \int \Psi(\mathbf{r}'_1 s_1 \mathbf{r}'_2 s_2 \mathbf{x}_3 \cdots \mathbf{x}_N)$$
$$\times \Psi^*(\mathbf{r}_1 s_1 \mathbf{r}_2 s_2 \mathbf{x}_3 \cdots \mathbf{x}_N) \, ds_1 \, ds_2 \, d\mathbf{x}_3 \cdots d\mathbf{x}_N \quad (2.4.2)$$

We also introduce a shorthand notation for the diagonal element of ρ_2,
$$\rho_2(\mathbf{r}_1, \mathbf{r}_2) = \rho_2(\mathbf{r}_1\mathbf{r}_2, \mathbf{r}_1\mathbf{r}_2)$$
$$= \frac{N(N-1)}{2} \int \cdots \int |\Psi|^2 \, ds_1 \, ds_2 \, d\mathbf{x}_3 \cdots d\mathbf{x}_N \quad (2.4.3)$$

and note that the diagonal element of $\rho_1(\mathbf{r}'_1, \mathbf{r}_1)$ is just the electron density of (1.5.1),
$$\rho(\mathbf{r}_1) = \rho_1(\mathbf{r}_1, \mathbf{r}_1)$$
$$= N \int \cdots \int |\Psi|^2 \, ds_1 \, d\mathbf{x}_2 \cdots d\mathbf{x}_N \quad (2.4.4)$$

Furthermore,
$$\rho_1(\mathbf{r}'_1, \mathbf{r}_1) = \frac{2}{N-1} \int \rho_2(\mathbf{r}'_1\mathbf{r}_2, \mathbf{r}_1\mathbf{r}_2) \, d\mathbf{r}_2 \quad (2.4.5)$$

which follows directly from (2.3.6). In particular,
$$\rho(\mathbf{r}_1) = \frac{2}{N-1} \int \rho_2(\mathbf{r}_1, \mathbf{r}_2) \, d\mathbf{r}_2 \quad (2.4.6)$$

The expectation value formulas of (2.3.24) and (2.3.26) now read, for spin-free operators $\mathcal{O}_1(\mathbf{r}_1)$ and $\mathcal{O}_2(\mathbf{r}_1\mathbf{r}_2)$,
$$\langle \hat{\mathcal{O}}_1 \rangle = \int [\mathcal{O}_1(\mathbf{r}_1) \rho_1(\mathbf{r}'_1, \mathbf{r}_1)]_{\mathbf{r}'_1 = \mathbf{r}_1} \, d\mathbf{r}_1 \quad (2.4.7)$$

and
$$\langle \hat{\mathcal{O}}_2 \rangle = \iint [\mathcal{O}_2(\mathbf{r}_1\mathbf{r}_2) \rho_2(\mathbf{r}'_1\mathbf{r}'_2, \mathbf{r}_1\mathbf{r}_2)]_{\mathbf{r}'_1 = \mathbf{r}_1, \mathbf{r}'_2 = \mathbf{r}_2} \, d\mathbf{r}_1 \, d\mathbf{r}_2 \quad (2.4.8)$$

The energy formula of (2.3.27) becomes
$$E = E[\rho_1(\mathbf{r}'_1, \mathbf{r}_1), \rho_2(\mathbf{r}_1, \mathbf{r}_2)] = E[\rho_2(\mathbf{r}'_1\mathbf{r}_2, \mathbf{r}_1\mathbf{r}_2)]$$
$$= \int [-\tfrac{1}{2}\nabla_\mathbf{r}^2 \rho_1(\mathbf{r}', \mathbf{r})]_{\mathbf{r}'=\mathbf{r}} \, d\mathbf{r} + \int v(\mathbf{r})\rho(\mathbf{r}) \, d\mathbf{r}$$
$$+ \iint \frac{1}{r_{12}} \rho_2(\mathbf{r}_1, \mathbf{r}_2) \, d\mathbf{r}_1 \, d\mathbf{r}_2 \quad (2.4.9)$$

The three terms in this formula represent respectively the electronic kinetic energy, the nuclear–electron potential energy, and the electron–electron potential energy. Though of course we still have the difficulties mentioned at the end of the previous section, it is salutatory that (2.4.9) involves only one function of three coordinates, $\rho(\mathbf{r})$, and two functions of six coordinates, $\rho_1(\mathbf{r}', \mathbf{r})$ and $\rho_2(\mathbf{r}_1, \mathbf{r}_2)$.

It is helpful to consider in a little more detail the third term in (2.4.9), the electron–electron repulsion energy. If this were purely classical, it would just be the self-repulsion energy of a distribution $\rho(\mathbf{r})$, the quantity

$$J[\rho] = \frac{1}{2} \iint \frac{1}{r_{12}} \rho(\mathbf{r}_1)\rho(\mathbf{r}_2)\, d\mathbf{r}_1\, d\mathbf{r}_2 \qquad (2.4.10)$$

where the factor 1/2 enters to prevent double counting. This suggests that we write

$$\rho_2(\mathbf{r}_1, \mathbf{r}_2) = \tfrac{1}{2}\rho(\mathbf{r}_1)\rho(\mathbf{r}_2)[1 + h(\mathbf{r}_1, \mathbf{r}_2)] \qquad (2.4.11)$$

where $h(\mathbf{r}_1, \mathbf{r}_2)$, defined by this formula, is the *pair correlation function*— a symmetric function that incorporates all nonclassical effects. The function $h(\mathbf{r}_1, \mathbf{r}_2)$ satisfies an important integral condition or "sum rule". Inserting (2.4.11) in the right-hand side of (2.4.6), one finds

$$\frac{N-1}{2}\rho(\mathbf{r}_1) = \tfrac{1}{2}\rho(\mathbf{r}_1)[N + \int \rho(\mathbf{r}_2) h(\mathbf{r}_1, \mathbf{r}_2)\, d\mathbf{r}_2]$$

Hence, we have

$$\int \rho(\mathbf{r}_2) h(\mathbf{r}_1, \mathbf{r}_2)\, d\mathbf{r}_2 = -1 \qquad (2.4.12)$$

which must hold for all $\mathbf{r}_1$. We shall later make use of this condition, which encodes a great deal of information. Another way to write it, going back to Slater (for example, Slater 1951), is obtained if we define the *exchange-correlation hole* (sometimes called the exchange-correlation charge) of an electron at $\mathbf{r}_1$ by

$$\rho_{xc}(\mathbf{r}_1, \mathbf{r}_2) = \rho(\mathbf{r}_2) h(\mathbf{r}_1, \mathbf{r}_2) \qquad (2.4.13)$$

From (2.4.12), this is a unit charge with sign opposite to that of the electron,

$$\int \rho_{xc}(\mathbf{r}_1, \mathbf{r}_2)\, d\mathbf{r}_2 = -1 \qquad (2.4.14)$$

In terms of ρ_{xc}, the electron-repulsion term in (2.4.9) becomes

$$V_{ee} = \iint \frac{1}{r_{12}} \rho_2(\mathbf{r}_1, \mathbf{r}_2) \, d\mathbf{r}_1 \, d\mathbf{r}_2$$

$$= J[\rho] + \frac{1}{2} \iint \frac{1}{r_{12}} \rho(\mathbf{r}_1) \rho_{xc}(\mathbf{r}_1, \mathbf{r}_2) \, d\mathbf{r}_1 \, d\mathbf{r}_2 \quad (2.4.15)$$

Sometimes it is convenient to have the spinless density matrices of (2.4.1) and (2.4.2) resolved into components arising from different spins or products of spins (McWeeney 1960, McWeeney and Sutcliffe 1969). First consider $\rho_1(\mathbf{r}'_1, \mathbf{r}_1)$ of (2.4.1). For any values of $\mathbf{r}'_1$ and $\mathbf{r}_1$, this is the sum over spin of the diagonal parts of γ_1; that is,

$$\rho_1(\mathbf{r}'_1, \mathbf{r}_1) = \gamma_1(\mathbf{r}'_1\alpha, \mathbf{r}_1\alpha) + \gamma_1(\mathbf{r}'_1\beta, \mathbf{r}_1\beta)$$
$$= \rho_1^{\alpha\alpha}(\mathbf{r}'_1, \mathbf{r}_1) + \rho_1^{\beta\beta}(\mathbf{r}'_1, \mathbf{r}_1) \quad (2.4.16)$$

where the second equality just defines the notation. Also, as a byproduct, the electron density itself is a sum of two components,

$$\rho(\mathbf{r}) = \rho^\alpha(\mathbf{r}) + \rho^\beta(\mathbf{r}); \quad \rho^\sigma(\mathbf{r}) = \rho_1^{\sigma\sigma}(\mathbf{r}, \mathbf{r}), \quad \sigma = \alpha, \beta \quad (2.4.17)$$

When ρ^α and ρ^β are not equal, there is spin polarization, and the *spin density*,

$$Q(\mathbf{r}) = \rho^\alpha(\mathbf{r}) - \rho^\beta(\mathbf{r}) \quad (2.4.18)$$

is not zero. Similarly, we find for ρ_2

$$\rho_2(\mathbf{r}'_1\mathbf{r}'_2, \mathbf{r}_1\mathbf{r}_2) = \rho_2^{\alpha\alpha,\alpha\alpha}(\mathbf{r}'_1\mathbf{r}'_2, \mathbf{r}_1\mathbf{r}_2) + \rho_2^{\beta\beta,\beta\beta}(\mathbf{r}'_1\mathbf{r}'_2, \mathbf{r}_1\mathbf{r}_2)$$
$$+ \rho_2^{\alpha\beta,\alpha\beta}(\mathbf{r}'_1\mathbf{r}'_2, \mathbf{r}_1\mathbf{r}_2) + \rho_2^{\beta\alpha,\beta\alpha}(\mathbf{r}'_1\mathbf{r}'_2, \mathbf{r}_1\mathbf{r}_2) \quad (2.4.19)$$

where notation analogous to that of (2.4.16) has been used.

Cusp conditions on ρ_1 and ρ_2 are given on pp. 42–44 and 103–104 of Davidson (1976).

2.5 Hartree–Fock theory in density-matrix form

The trial wave function for the Hartree–Fock method is the single determinant of spin orbitals, (1.3.1). We now rework the Hartree–Fock theory in density-matrix language (Blaizot and Ripka 1986, p. 177, Löwdin 1955a,b). The result will be a formulation in which the dependent variable is the first-order density matrix itself.

Density matrices assume very simple forms when they are derived from a single determinant. The first-order reduced density matrix, called the *Fock–Dirac density matrix*, is

$$\gamma_1(\mathbf{x}'_1, \mathbf{x}_1) = \sum_{i=1}^{N} \psi_i(\mathbf{x}'_1) \psi_i^*(\mathbf{x}_1) \quad (2.5.1)$$

where the ψ_i are orthonormal spin orbitals. To prove this formula, expand the determinant (1.3.1) in the first row and use (2.3.3), noting that the integration over $\mathbf{x}_2, \ldots, \mathbf{x}_N$ of a product of two $(N-1)$-electron Slater determinants gives $(N-1)!$ if the orbitals are the same in both, and gives zero otherwise. The second-order reduced density matrix can be calculated in a similar way by expanding the determinant in the first two rows (Laplace expansion). The result is

$$\gamma_2(\mathbf{x}_1'\mathbf{x}_2', \mathbf{x}_1\mathbf{x}_2) = \frac{1}{2} \begin{vmatrix} \gamma_1(\mathbf{x}_1', \mathbf{x}_1) & \gamma_1(\mathbf{x}_2', \mathbf{x}_1) \\ \gamma_1(\mathbf{x}_1', \mathbf{x}_2) & \gamma_1(\mathbf{x}_2', \mathbf{x}_2) \end{vmatrix}$$

$$= \tfrac{1}{2}[\gamma_1(\mathbf{x}_1', \mathbf{x}_1)\gamma_1(\mathbf{x}_2', \mathbf{x}_2) - \gamma_1(\mathbf{x}_1', \mathbf{x}_2)\gamma_1(\mathbf{x}_2', \mathbf{x}_1)] \quad (2.5.2))$$

More generally, one finds (Löwdin 1954b)

$$\gamma_p(\mathbf{x}_1'\mathbf{x}_2' \cdots \mathbf{x}_p', \mathbf{x}_1\mathbf{x}_2 \cdots \mathbf{x}_p)$$

$$= \frac{1}{p!} \begin{vmatrix} \gamma_1(\mathbf{x}_1', \mathbf{x}_1) & \gamma_1(\mathbf{x}_1', \mathbf{x}_2) & \cdots & \gamma_1(\mathbf{x}_1', \mathbf{x}_p) \\ \gamma_1(\mathbf{x}_2', \mathbf{x}_1) & \gamma_1(\mathbf{x}_2', \mathbf{x}_2) & \cdots & \gamma_1(\mathbf{x}_2', \mathbf{x}_p) \\ \vdots & \vdots & & \vdots \\ \gamma_1(\mathbf{x}_p', \mathbf{x}_1) & \gamma_1(\mathbf{x}_p', \mathbf{x}_2) & \cdots & \gamma_1(\mathbf{x}_p', \mathbf{x}_p) \end{vmatrix} \quad (2.5.3)$$

so that the density matrix of any order is calculable from first-order density matrices.

In operator form, (2.5.1) reads

$$\hat{\gamma}_1 = \sum_{i=1}^{N} |\psi_i\rangle\langle\psi_i| \quad (2.5.4)$$

and can be regarded as the projector onto the space spanned by the N occupied spin orbitals. Not only is this form for $\hat{\gamma}_1$ a consequence of the wave function being a single determinant, but, conversely, if $\hat{\gamma}_1$ is of this form, the wave function must be a single determinant. The proof has two parts. First, it follows from (2.5.4) that $\hat{\gamma}_1$ has N eigenvectors with eigenvalues all equal to 1, the "occupied orbitals," and an infinite number of other eigenvectors with eigenvalues zero [compare (2.3.29)]. Construct an N-electron Slater determinant D with N eigenfunctions of $\hat{\gamma}_1$ with all eigenvalues equal to 1. This is a determinant giving $\hat{\gamma}_1$. The occupied-orbital set is not unique, because the determinant is invariant up to a phase factor to a unitary transformation [see the discussion following (1.3.26)]. Second, we can show that there can be no other determinant giving $\hat{\gamma}_1$. To see this, expand the wave function in a complete set of determinants built from the whole set of natural orbitals of $\hat{\gamma}_1$, calculate $\hat{\gamma}_1$, and compare with (2.5.4). All coefficients will be zero except the one associated with the original determinant built from the N occupied orbitals. There is then a one-to-one mapping between a Slater determinant and a density matrix of the form (2.5.1).

Equivalently, one may say that a necessary and sufficient condition for the N-electron wave function to be a single determinant (unique up to a unitary transformation) is that $\hat{\gamma}_1$ be idempotent with trace N:

$$\hat{\gamma}_1 \hat{\gamma}_1 = \hat{\gamma}_1 \tag{2.5.5}$$

$$\text{Tr}\,\hat{\gamma}_1 = N \tag{2.5.6}$$

These formulas follow immediately from (2.5.4); the argument that they imply (2.5.4) is that (2.5.5) requires the eigenvalues of $\hat{\gamma}_1$ to be either 1 or 0 and there are N eigenvectors with eigenvalues all equal to 1 [from (2.5.6)]. In coordinate representation, (2.5.5) and (2.5.6) read

$$\int \gamma_1(\mathbf{x}_1', \mathbf{x}_1'')\gamma_1(\mathbf{x}_1'', \mathbf{x}_1)\,d\mathbf{x}_1'' = \gamma_1(\mathbf{x}_1', \mathbf{x}_1) \tag{2.5.7}$$

and

$$\int \gamma_1(\mathbf{x}_1, \mathbf{x}_1)\,d\mathbf{x}_1 = N \tag{2.5.8}$$

Inserting (2.5.2) in (2.3.27), the Hartree–Fock energy becomes [compare (1.3.2)]

$$E_{\text{HF}}[\gamma_1] = \int [(-\tfrac{1}{2}\nabla_1^2 + v(\mathbf{x}_1))\gamma_1(\mathbf{x}_1', \mathbf{x}_1)]_{\mathbf{x}_1' = \mathbf{x}_1}\,d\mathbf{x}_1$$

$$+ \frac{1}{2}\iint \frac{1}{r_{12}}[\gamma_1(\mathbf{x}_1, \mathbf{x}_1)\gamma_1(\mathbf{x}_2, \mathbf{x}_2) - \gamma_1(\mathbf{x}_1, \mathbf{x}_2)\gamma_1(\mathbf{x}_2, \mathbf{x}_1)]\,d\mathbf{x}_1\,d\mathbf{x}_2 \tag{2.5.9}$$

In the Hartree–Fock method, one seeks to minimize this functional of γ_1, over the set of all γ_1 of the form (2.5.1), or, equivalently, over all γ_1 satisfying (2.5.7) and (2.5.8).

This constrained minimization can be implemented by using Lagrange undetermined multipliers (see Appendix A). The condition (2.5.7) depends on $\mathbf{x}_1'$ and $\mathbf{x}_1$, so that the multiplier associated with it must depend on $\mathbf{x}_1'$ and $\mathbf{x}_1$; call it $\alpha(\mathbf{x}_1, \mathbf{x}_1')$. Let β be the multiplier for (2.5.8). The variational problem may then be written as

$$\delta\bigg\{E_{\text{HF}}[\gamma_1] - \iint d\mathbf{x}_1'\,d\mathbf{x}_1\,\alpha(\mathbf{x}_1, \mathbf{x}_1')\bigg[\int \gamma_1(\mathbf{x}_1', \mathbf{x}_1'')\gamma_1(\mathbf{x}_1'', \mathbf{x}_1)\,d\mathbf{x}_1'' - \gamma_1(\mathbf{x}_1', \mathbf{x}_1)\bigg]$$

$$- \beta\bigg[\iint \delta(\mathbf{x}_1' - \mathbf{x}_1)\gamma_1(\mathbf{x}_1', \mathbf{x}_1)\,d\mathbf{x}_1'\,d\mathbf{x}_1 - N\bigg]\bigg\} = 0 \tag{2.5.10}$$

Taking the functional derivatives (see Appendix A), one gets the

Euler–Lagrange equation for the problem:

$$F(\mathbf{x}_1, \mathbf{x}_1') - \int d\mathbf{y}_1 \alpha(\mathbf{y}_1, \mathbf{x}_1') \gamma_1(\mathbf{x}_1, \mathbf{y}_1)$$

$$- \int d\mathbf{z}_1 \alpha(\mathbf{x}_1, \mathbf{z}_1) \gamma_1(\mathbf{z}_1, \mathbf{x}_1') + \alpha(\mathbf{x}_1, \mathbf{x}_1') - \beta \, \delta(\mathbf{x}_1' - \mathbf{x}_1) = 0 \quad (2.5.11)$$

where $F(\mathbf{x}_1, \mathbf{x}_1')$ is the matrix for the Fock operator $\hat{F}$ in the coordinate representation,

$$F(\mathbf{x}_1, \mathbf{x}_1') = \frac{\delta E_{\text{HF}}[\gamma]}{\delta \gamma(\mathbf{x}_1', \mathbf{x}_1)}$$

$$= (-\tfrac{1}{2}\nabla_1^2 + v(\mathbf{x}_1)) \, \delta(\mathbf{x}_1' - \mathbf{x}_1) + \delta(\mathbf{x}_1' - \mathbf{x}_1) \int \frac{1}{r_{12}} \gamma_1(\mathbf{x}_2, \mathbf{x}_2) \, d\mathbf{x}_2$$

$$- \frac{1}{r_{11'}} \gamma_1(\mathbf{x}_1, \mathbf{x}_1') \quad (2.5.12)$$

The equivalence of this definition of the Fock operator with the previous (1.3.9) becomes clear if we calculate

$$\langle \mathbf{x}_1' | \hat{F} | \psi \rangle = \int d\mathbf{x}_1 \, \langle \mathbf{x}_1' | \hat{F} | \mathbf{x}_1 \rangle \langle \mathbf{x}_1 | \psi \rangle$$

$$= [-\tfrac{1}{2}\nabla_1^2 + v(\mathbf{x}_1') + \hat{j}(\mathbf{x}_1') - \hat{k}(\mathbf{x}_1')] \psi(\mathbf{x}_1') \quad (2.5.13)$$

where $\hat{j}$ and $\hat{k}$ are as defined in (1.3.11) and (1.3.12). Note the simple form of the last, the nonlocal exchange, term in (2.5.12).

Equation (2.5.11) can be regarded as the coordinate representation of an operator equation in one-particle space:

$$\hat{F} - \hat{\gamma}_1 \hat{\alpha} - \hat{\alpha} \hat{\gamma}_1 + \hat{\alpha} - \beta \hat{I} = 0 \quad (2.5.14)$$

where $\hat{I}$ is the identity operator. Multiplying this equation on the right by $\hat{\gamma}_1$, and then on the left, and subtracting the results, we obtain

$$\hat{F} \hat{\gamma}_1 - \hat{\gamma}_1 \hat{F} = 0 \quad (2.5.15)$$

Thus the operators $\hat{F}$ and $\hat{\gamma}_1$ commute and have common eigenfunctions. These common eigenfunctions are the Hartree–Fock orbitals of (1.3.31). The $\hat{\gamma}_1$ that is a solution of (2.5.15) can be constructed from these orbitals via (2.5.4). This γ_1, denoted by γ_1^{HF}, thus minimizes the Hartree–Fock energy functional $E_{\text{HF}}[\gamma_1]$ of (2.5.9); that is,

$$E_{\text{HF}}[\gamma_1^{\text{HF}}] \leq E_{\text{HF}}[\gamma_1] \quad (2.5.16)$$

for any idempotent γ_1 of trace N. It is obvious that $E_{\text{HF}}[\gamma_1^{\text{HF}}]$ is above the true ground-state energy E (by the correlation energy), since $E_{\text{HF}}[\gamma_1^{\text{HF}}]$ is the minimum over only the class of all determinantal wave functions. Lieb (1981) in fact proved a more general result than (2.5.16): γ_1 on the right-hand side of (2.5.16) can be replaced by any N-representable Γ_1 (satisfying (2.3.30)). In summary,

$$E_0 \leq E_{\mathrm{HF}}[\gamma_1^{\mathrm{HF}}] \leq E_{\mathrm{HF}}[\Gamma_1] \tag{2.5.17}$$

Finally, we can express the Hartree–Fock energy functional in terms of the spinless first-order density matrix $\rho_1(\mathbf{r}_1', \mathbf{r}_1)$ of the previous section, and exhibit the pair-correlation function and exchange-correlation hole for Hartree–Fock theory. From (2.5.2), we have, for the components of the diagonal part of ρ_2,

$$\rho_2^{\alpha\alpha,\alpha\alpha}(\mathbf{r}_1\mathbf{r}_2, \mathbf{r}_1\mathbf{r}_2) = \tfrac{1}{2}[\rho^\alpha(\mathbf{r}_1)\rho^\alpha(\mathbf{r}_2) - \rho_1^\alpha(\mathbf{r}_1, \mathbf{r}_2)\rho_1^\alpha(\mathbf{r}_2, \mathbf{r}_1)]$$
$$\rho_2^{\alpha\beta,\alpha\beta}(\mathbf{r}_1\mathbf{r}_2, \mathbf{r}_1\mathbf{r}_2) = \tfrac{1}{2}\rho^\alpha(\mathbf{r}_1)\rho^\beta(\mathbf{r}_2) \tag{2.5.18}$$

where the notation of the last section has been used. Two other similar components can be written down if we exchange α and β spin labels in the above formulas. [Since the function $\rho_2^{\sigma_1\sigma_2,\sigma_1\sigma_2}(\mathbf{r}_1\mathbf{r}_2, \mathbf{r}_1\mathbf{r}_2)$ is proportional to the probability of finding an electron at $\mathbf{r}_1$ with spin σ_1 and another electron at $\mathbf{r}_2$ with spin σ_2, note from (2.5.18) that the determinantal wave function only describes the correlation of like-spin electrons—the "exchange effect" in Hartree–Fock theory.] The spinless second-order density matrix, by (2.4.19) and (2.5.18), becomes

$$\rho_2(\mathbf{r}_1\mathbf{r}_2, \mathbf{r}_1\mathbf{r}_2) = \tfrac{1}{2}\rho(\mathbf{r}_1)\rho(\mathbf{r}_2)$$
$$- \tfrac{1}{2}[\rho_1^{\alpha\alpha}(\mathbf{r}_1, \mathbf{r}_2)\rho_1^{\alpha\alpha}(\mathbf{r}_1, \mathbf{r}_2) + \rho_1^{\beta\beta}(\mathbf{r}_1, \mathbf{r}_2)\rho_1^{\beta\beta}(\mathbf{r}_2, \mathbf{r}_1)] \tag{2.5.19}$$

Inserting (2.5.19) into (2.4.9), one gets the total energy formula

$$E_{\mathrm{HF}}[\rho_1] = \int [(-\tfrac{1}{2}\nabla_1^2 + v(\mathbf{r}_1))\rho_1(\mathbf{r}_1', \mathbf{r}_1)]_{\mathbf{r}_1'=\mathbf{r}_1} \, d\mathbf{r}_1$$
$$+ \frac{1}{2}\iint \frac{1}{r_{12}}\rho(\mathbf{r}_1)\rho(\mathbf{r}_2) \, d\mathbf{r}_1 \, d\mathbf{r}_2 - \frac{1}{2}\iint \frac{1}{r_{12}}[\rho_1^{\alpha\alpha}(\mathbf{r}_1, \mathbf{r}_2)\rho_1^{\alpha\alpha}(\mathbf{r}_2, \mathbf{r}_1)$$
$$+ \rho_1^{\beta\beta}(\mathbf{r}_1, \mathbf{r}_2)\rho_1^{\beta\beta}(\mathbf{r}_2, \mathbf{r}_1)] \, d\mathbf{r}_1 \, d\mathbf{r}_2$$
$$= T[\rho_1] + V_{ne}[\rho] + J[\rho] - K[\rho_1] \tag{2.5.20}$$

where

$$T[\rho_1] = \int [-\tfrac{1}{2}\nabla_1^2 \rho_1(\mathbf{r}_1', \mathbf{r}_1)]_{\mathbf{r}_1'=\mathbf{r}_1} \, d\mathbf{r}_1 \tag{2.5.21}$$

$$V_{ne}[\rho] = \int v(\mathbf{r})\rho(\mathbf{r}) \, d\mathbf{r} \tag{2.5.22}$$

$$J[\rho] = \frac{1}{2}\iint \frac{1}{r_{12}}\rho(\mathbf{r}_1)\rho(\mathbf{r}_2) \, d\mathbf{r}_1 \, d\mathbf{r}_2 \tag{2.5.23}$$

and

$$K[\rho_1] = \frac{1}{2} \int\int \frac{1}{r_{12}} [\rho_1^{\alpha\alpha}(\mathbf{r}_1, \mathbf{r}_2)\rho_1^{\alpha\alpha}(\mathbf{r}_2, \mathbf{r}_1) + \rho_1^{\beta\beta}(\mathbf{r}_1, \mathbf{r}_2)\rho_1^{\beta\beta}(\mathbf{r}_2, \mathbf{r}_1)] \, d\mathbf{r}_1 \, d\mathbf{r}_2 \quad (2.5.24)$$

For the closed-shell case of an even number of electrons pairwise occupying $N/2$ spatial orbitals,

$$\rho_1^{\alpha\alpha}(\mathbf{r}_1, \mathbf{r}_2) = \rho_1^{\beta\beta}(\mathbf{r}_1, \mathbf{r}_2) = \tfrac{1}{2}\rho(\mathbf{r}_1, \mathbf{r}_2) \quad \text{[closed shell]} \quad (2.5.25)$$

and (2.5.24) becomes

$$K[\rho_1] = \frac{1}{4} \int\int \frac{1}{r_{12}} \rho_1(\mathbf{r}_1, \mathbf{r}_2)\rho_1(\mathbf{r}_2, \mathbf{r}_1) \, d\mathbf{r}_1 \, d\mathbf{r}_2 \quad \text{[closed shell]} \quad (2.5.26)$$

This is equivalent to the last term of (1.3.19).

Comparing the electron–electron repulsion in (2.5.20) with (2.4.15), we see that in the Hartree–Fock approximation the exchange-correlation hole is given by

$$\rho_{xc}^{HF}(\mathbf{r}_1, \mathbf{r}_2) = \rho_x^{HF}(\mathbf{r}_1, \mathbf{r}_2) = -\frac{1}{2}\frac{|\rho_1(\mathbf{r}_1, \mathbf{r}_2)|^2}{\rho(\mathbf{r}_1)} \quad \text{[closed shell]} \quad (2.5.27)$$

From (2.4.13) the pair correlation function is

$$h^{HF}(\mathbf{r}_1, \mathbf{r}_2) = -\frac{1}{4}\frac{|\rho_1(\mathbf{r}_1, \mathbf{r}_2)|^2}{\rho(\mathbf{r}_1)\rho(\mathbf{r}_2)} \quad \text{[closed shell]} \quad (2.5.28)$$

The "correlation" included here is among electrons of the same spin only, as again can be seen by noting in (2.5.9) that the spin integrations over the $\gamma_1(\mathbf{x}_1, \mathbf{x}_2)\gamma_1(\mathbf{x}_2, \mathbf{x}_1)$ factor give zero for contributions from different spin states for particles 1 and 2. The term *correlation* as defined in §1.4 is over and above this correlation already included in the single-determinant description, so we sometimes use the notation ρ_x^{HF} for the *exchange hole* of (2.5.28). Note that the sum rules of (2.4.12) and (2.4.14) are satisfied in Hartree–Fock theory:

$$\int \rho(\mathbf{r}_2) h^{HF}(\mathbf{r}_1, \mathbf{r}_2) \, d\mathbf{r}_2 = \int \rho_x^{HF}(\mathbf{r}_2) \, d\mathbf{r}_2 = -1 \quad (2.5.29)$$

This property is preserved for any approximate ρ_2 that is N-representable.

2.6 The N-representability of reduced density matrices

The necessary and sufficient conditions for a given first-order density matrix to be derivable from a mixed-state density operator were stated in

(2.3.30). In this section, we give the proof of this theorem and also briefly discuss the corresponding problem for the second-order density matrix.

Necessary conditions on Γ_1 and Γ_2 are conditions that they satisfy when they satisfy (2.3.19) for a proper Γ_N. Sufficient conditions are those that guarantee the existence of a Γ_N that reduces to this Γ_1 and/or Γ_2. The set of Γ_1 or Γ_2 that simultaneously satisfies both necessary and sufficient conditions is called the set of N-representable Γ_1 or Γ_2. Thus, if the energy (2.4.9) is minimized over sets Γ_1 and Γ_2 satisfying only necessary conditions (such sets are larger than the corresponding N-representable sets), an energy lower than the true energy—lower bound to the energy—can be obtained; if the energy is minimized over sets satisfying only sufficient conditions (such sets are smaller than the corresponding N-representable sets), an energy higher than the true energy is delivered—an upper bound to the energy. In the Hartree–Fock method, for example, the density matrix γ_1 satisfies the sufficients (2.5.5) and (2.5.6); Hartree–Fock energies are consequently upper bounds for true energies. Lower-bound studies are rarer; they include pioneering works by Garrod and Percus (1964) and Garrod and Fusco (1976). Note that if one minimizes over *all* sets satisfying the sufficient conditions, the ground-state energy is obtained.

We now derive some necessary conditions on γ_1 and γ_2 imposed by N-representability—the so-called Pauli conditions (Coleman 1981). These are that if $|\psi_i\rangle$ is some normalized spin-orbital state and $|\psi_i\psi_j\rangle$ is a normalized 2×2 Slater determinantal state built from orthonormal ψ_i and ψ_j, then

$$0 \leq \langle \psi_i | \hat{\gamma}_1 | \psi_i \rangle \leq 1 \tag{2.6.1}$$

and

$$0 \leq \langle \psi_i\psi_j | \hat{\gamma}_2 | \psi_i\psi_j \rangle \leq 1 \tag{2.6.2}$$

or, in the coordinate representation,

$$0 \leq \iint d\mathbf{x}_1 \, d\mathbf{x}'_1 \psi_i^*(\mathbf{x}'_1) \gamma_1(\mathbf{x}'_1, \mathbf{x}_1) \psi_i(\mathbf{x}_1) \leq 1 \tag{2.6.3}$$

and

$$0 \leq \tfrac{1}{2} \iiiint d\mathbf{x}_1 \, d\mathbf{x}'_1 \, d\mathbf{x}_2 \, d\mathbf{x}'_2 \det[\psi_i^*(\mathbf{x}'_1)\psi_j^*(\mathbf{x}'_2)]$$
$$\gamma_2(\mathbf{x}'_1\mathbf{x}'_2, \mathbf{x}_1\mathbf{x}_2) \det[\psi_i(\mathbf{x}_1)\psi_j(\mathbf{x}_2)] \leq 1 \tag{2.6.4}$$

These results can be derived in various ways.

We follow Coleman (1981) and use the method of second quantization as described in Appendix C. For each member of an arbitrary orthonormal set $\{\psi_i\}$, there is a *creation operator* $\hat{a}_j^+$ and an *annihilation operator* $\hat{a}_j$. The corresponding creation and annihilation *field operators* are

defined as
$$\hat{\psi}^+(\mathbf{x}) = \sum_i \psi_i^*(\mathbf{x})\hat{a}_i^+ \qquad (2.6.5)$$
and
$$\hat{\psi}(\mathbf{x}) = \sum_i \psi_i(\mathbf{x})\hat{a}_i \qquad (2.6.6)$$

In terms of these, for a state $|\Psi\rangle$ the first- and second-order reduced density matrices γ_1 and γ_2 of (2.3.2) and (2.3.3) are given, respectively, by (see Appendix C)

$$\gamma_1(\mathbf{x}_1', \mathbf{x}_1) = \langle\Psi|\,\hat{\psi}^+(\mathbf{x}_1)\hat{\psi}(\mathbf{x}_1')\,|\Psi\rangle \qquad (2.6.7)$$

$$\gamma_2(\mathbf{x}_1'\mathbf{x}_2', \mathbf{x}_1\mathbf{x}_2) = \tfrac{1}{2}\langle\Psi|\,\hat{\psi}^+(\mathbf{x}_2)\hat{\psi}^+(\mathbf{x}_1)\hat{\psi}(\mathbf{x}_1')\hat{\psi}(\mathbf{x}_2')\,|\Psi\rangle \qquad (2.6.8)$$

We use these to establish, in turn, (2.6.1) and (2.6.2).

Inserting (2.6.7) into the integral in (2.6.3) and using (2.6.5) and (2.6.6), we find, if we take the set $\{\psi_j\}$ to include ψ_i as one of its members (which we can always do),

$$\langle\psi_i|\,\hat{\gamma}_1\,|\psi_i\rangle = \iint d\mathbf{x}_1\,d\mathbf{x}_1'\psi_i^*(\mathbf{x}_1')\langle\Psi|\sum_i \psi_i^*(\mathbf{x}_1)\hat{a}_i^+ \sum_j \psi_j(\mathbf{x}_1')\hat{a}_j|\Psi\rangle\,\psi_i(\mathbf{x}_1)$$
$$= \langle\Psi|\,\hat{a}_i^+\hat{a}_i\,|\Psi\rangle \qquad (2.6.9)$$

Here, the operator
$$\hat{N}_i = \hat{a}_i^+\hat{a}_i \qquad (2.6.10)$$
is the occupation number operator for the ith orbital. It is idempotent:
$$\hat{N}_i^2 = \hat{a}_i^+\hat{a}_i\hat{a}_i^+\hat{a}_i = \hat{a}_i^+(1 - \hat{a}_i^+\hat{a}_i)\hat{a}_i = \hat{N}_i \qquad (2.6.11)$$

Thus $\hat{N}_i$ is a projection operator. But the expectation value of any projection operator is always nonnegative and not greater than 1:

$$\langle\Psi|\,\hat{P}^2\,|\Psi\rangle = \sum_j |\langle\Psi|\,\hat{P}\,|\Psi_j\rangle|^2 \leq \sum_j |\langle\Psi\,|\,\Psi_j\rangle|^2 = 1 \qquad (2.6.12)$$

where $\{\Psi_j\}$ is any complete set.

Therefore the right-hand side of (2.6.9) is a nonnegative number less than or equal to 1, proving (2.6.1). We note in passing that the total number operator is
$$\hat{N} = \sum_i \hat{N}_i = \sum_i \hat{a}_i^+\hat{a}_i \qquad (2.6.13)$$

For the proof of (2.6.2), we need the quantity

$$\hat{\psi}(\mathbf{x}_1)\hat{\psi}(\mathbf{x}_2) = \sum_i \psi_i(\mathbf{x}_1)\hat{a}_i \sum_i \psi_j(\mathbf{x}_2)\hat{a}_j$$
$$= \sum_j \sum_i \psi_i(\mathbf{x}_1)\psi_j(\mathbf{x}_2)\hat{a}_i\hat{a}_j$$
$$= \sum_{i<j} [\psi_i(\mathbf{x}_1)\psi_j(\mathbf{x}_2) - \psi_j(\mathbf{x}_2)\psi_i(\mathbf{x}_1)]\hat{a}_i\hat{a}_j$$
$$= \sum_{i<j} \det[\psi_i(\mathbf{x}_1)\psi_j(\mathbf{x}_2)]\hat{a}_i\hat{a}_j \qquad (2.6.14)$$

where the fact has been used that $\hat{a}_i$ and $\hat{a}_j$ anticommute (see Appendix C). Inserting (2.6.14) and its adjoint into (2.6.8) and then (2.6.4), and using the orthogonality and normalization property of the 2×2 determinants, we find

$$\langle \phi_i \phi_j | \hat{\gamma}_2 | \phi_i \phi_j \rangle = \langle \Psi | \hat{a}_j^+ \hat{a}_i^+ \hat{a}_i \hat{a}_j | \Psi \rangle$$
$$= \langle \Psi | \hat{N}_j \hat{N}_i | \Psi \rangle \qquad (2.6.15)$$

From the fact that the product of two projection operators is still a projection operator, we then see that $\langle \Psi | \hat{N}_j \hat{N}_i | \Psi \rangle$ in (2.6.15) is a nonnegative number less than or equal to 1. Equation (2.6.2) follows.

Equation (2.6.1) is equivalent to the requirement that the eigenvalues of $\hat{\gamma}_1$, the occupation numbers n_i of (2.3.12), fall in the range 0 to 1:

$$0 \leq n_i \leq 1 \qquad (2.6.16)$$

Equation (2.6.3) is not a condition on the eigenvalues of $\hat{\gamma}_2$, however, since the eigenfunctions of $\hat{\gamma}_2$ are not in general 2×2 Slater determinants.

Leaving the analysis of the N-representability of $\hat{\gamma}_2$ aside, we complete the basic story of $\hat{\gamma}_1$ by asserting and proving that conditions (2.6.16) are not only necessary but also sufficient conditions for the ensemble-N-representability of a first-order reduced density matrix (Coleman 1963, 1981). Necessity has just been proved; it remains to prove sufficiency.

We need a simple lemma about vectors and convex sets. Recall from §2.3 that a set is convex if an arbitrary positively weighted average of any two elements in the set also belongs to the set. Define an *extreme element* of a convex set as an element E such that $E = p_1 Y_1 + p_2 Y_2$ implies that Y_1 and Y_2 are both multiples of E. Then the lemma states that the set $\mathcal{L}$ of vectors $\mathbf{v} = (v_1, v_2, \ldots)$ in a space of arbitrary but fixed dimension with $0 \leq v_i \leq 1$ and $\sum v_i = N$ is convex and its extreme elements are the vectors with N components equal to 1 and all other components equal to zero. That $\mathcal{L}$ is convex is obvious. The condition on $\mathcal{L}$ requires that each element has at least N positive components. Thus, a mean of any two vectors has more than N components unless the nonzero components of the original vectors are the same and all equal to 1. The vectors with N components equal to 1 are therefore extreme. No other type of vector is extreme, as any vector with more than N positive components can always be resolved into an average of several extreme elements.

Given this lemma, it is clear that any $\hat{\gamma}_1$ or $\hat{\Gamma}_1$ satisfying (2.6.16) is an element of a convex set whose extreme elements are those γ_1^0 (or Γ_1^0) that have N eigenvalues equal to 1 and the rest equal to zero. Each of these $\hat{\gamma}_1^0$ (or $\hat{\Gamma}_1^0$), according to the discussion in §2.5, determines up to a phase a determinantal N-electron wave function and a unique corresponding pure-state density operator $\hat{\gamma}_N^0$. Some positively weighted sum of these γ_N^0

will be the Γ_N that reduces to the given $\hat{\gamma}_1$ (or $\hat{\Gamma}_1$) through (3.3.20). Thereby, sufficiency is proved.

2.7 Statistical mechanics

As was shown in §2.2, when a system is in a mixed state, rather than a pure state, it is mandatory to describe it with density operators. Such is the case for a system at some finite temperature, for the ignorance of the details of the interaction between system and surroundings makes defining an appropriate complete Hamiltonian impossible. There will be a probability distribution over accessible pure states of the form (2.2.7), with probability p_i for the accessible state $|\Psi_i\rangle$, in the ensemble density operator $\hat{\Gamma}$. The probabilities must be determined by a statistical-mechanical argument appropriate for each particular case (Gibbs 1931, Feynman 1972, McQuarrie 1976).

The key quantity to consider is the *entropy* of the probability distribution,

$$S = -k_B \sum_i p_i \ln p_i \tag{2.7.1}$$

$$= -k_B \, \text{Tr} \, (\hat{\Gamma} \ln \hat{\Gamma}) \tag{2.7.2}$$

The first formula is the definition, the second follows from the fact that the trace of an operator is the sum of its eigenvalues. The constant k_B is Boltzmann's constant.

First we consider the *canonical ensemble*, which is a mixture of pure states all having the same particle number N. Consequently

$$\hat{\Gamma}_N = \sum_i p_{Ni} |\Psi_{Ni}\rangle \langle \Psi_{Ni}| \tag{2.7.3}$$

We seek to determine the p_{Ni} by the maximum-entropy principle. At equilibrium, S will be a maximum subject to two constraints. The probabilities must sum to unity, to which we attach a Lagrange multiplier λ, and the expectation value of $\hat{H}_N$ must equal the observed energy,

$$E = \text{tr} \, (\hat{\Gamma}_N \hat{H}) \tag{2.7.4}$$

to which we attach another Lagrange multiplier α. (The notation tr indicates a trace over N constant states only.) The variational principle thus is

$$\delta\{-k_B \, \text{tr} \, (\hat{\Gamma}_N \ln \hat{\Gamma}_N) + \alpha(\text{tr} \, \hat{\Gamma}_N \hat{H} - E) + \lambda(\text{tr} \, \hat{\Gamma}_N - 1)\} = 0 \tag{2.7.5}$$

One can carry out the variation either by varying both the P_{Ni} and $|\psi_{Ni}\rangle$ in (2.7.3), or more elegantly, by working with $\hat{\Gamma}_N$ itself. Doing the latter, use $\hat{\Gamma}_N = \hat{\Gamma}_N^0 + \delta\hat{\Gamma}_N$ in (2.7.5) and find

$$\text{tr} \, [\delta\hat{\Gamma}_N(-k_B \ln \hat{\Gamma}_N^0 - k_B + \alpha\hat{H} + \lambda)] = 0 \tag{2.7.6}$$

2.7 DENSITY MATRICES

This gives, after determination of λ from the normalization condition on $\hat{\Gamma}_N^0$,

$$\hat{\Gamma}_N^0 = \frac{e^{-\beta \hat{H}}}{\text{tr}(e^{-\beta \hat{H}})} \tag{2.7.7}$$

where

$$\beta = \frac{-\alpha}{k_B} = \frac{1}{k_B \theta} \tag{2.7.8}$$

in which θ is the temperature. Note that the final $\hat{\Gamma}_N^0$ commutes with $\hat{H}_N$, in agreement with (2.2.18). That (2.7.7) corresponds to a maximum S can be checked readily.

These results can be reformulated as follows. Define the *Helmholtz free energy* A for the density operator $\hat{\Gamma}_N$ by

$$A[\hat{\Gamma}_N] \equiv \text{tr}\, \hat{\Gamma}_N \left(\frac{1}{\beta} \ln \hat{\Gamma}_N + \hat{H}\right) = E - \theta S \tag{2.7.9}$$

Then for all positive and unit trace $\hat{\Gamma}_N$,

$$A[\Gamma_N^0] \leq A[\Gamma_N] \tag{2.7.10}$$

where

$$A[\hat{\Gamma}_N^0] = -\frac{1}{\beta} \ln Z_N \tag{2.7.11}$$

with Z_N the *partition function*

$$Z_N = \text{tr}\, e^{-\beta \hat{H}} \tag{2.7.12}$$

One proof of (2.7.10) is given by Feynman (1972). Another is presented in §3.5.

Now we turn to the *grand-canonical-ensemble* approach. The system is allowed to have a more general $\hat{\Gamma}$ in which there are nonzero probabilities associated with different particle numbers, but an average number of particles equal to some observed number

$$\text{Tr}(\hat{\Gamma}\hat{N}) = N \tag{2.7.13}$$

The entropy must be maximized subject to constant energy (2.7.4) and constant particle number (2.7.13). The result is the formula for the equilibrium Γ^0,

$$\hat{\Gamma}^0 = \frac{e^{-\beta(\hat{H}-\mu\hat{N})}}{\text{Tr}\, e^{-\beta(\hat{H}-\mu\hat{N})}} \tag{2.7.14}$$

where μ is the Lagrange multiplier for the constraint of (2.7.13) and is called the *chemical potential*.

Again there is a reformulation. Define the *grand potential* by

$$\Omega[\hat{\Gamma}] \equiv \text{Tr}\, \hat{\Gamma}\left(\frac{1}{\beta} \ln \hat{\Gamma} + \hat{H} - \mu\hat{N}\right) = E - \theta S - \mu N \tag{2.7.15}$$

Then for all positive and unit trace $\hat{\Gamma}$,

$$\Omega[\hat{\Gamma}^0] \leq \Omega[\hat{\Gamma}] \qquad (2.7.16)$$

where

$$\Omega[\hat{\Gamma}^0] = -\frac{1}{\beta} \ln Z \qquad (2.7.17)$$

and Z is the *grand partition function*

$$Z = \text{Tr}\, e^{-\beta(\hat{H} - \mu\hat{N})} \qquad (2.7.18)$$

A somewhat involved proof of (2.7.16) may be found in a paper by Mermin (1965), who followed Gibbs (1931, p. 131). A much simpler proof is given in §3.6.

It is instructive to compare the three minimum principles we have now established: (1.2.3) for the ground-state energy, (2.7.10) for the equilibrium Helmholtz free energy, and (2.7.16) for the grand potential. The second extends the first to finite temperature within the same N-particle Hilbert space. The third also applies to a finite-temperature system but to the larger Fock space that is the product space of all the different N-constant Hilbert spaces. The density-functional theory described in the next chapter will be structured similarly, proceeding from the ground-state theory based on (1.2.3) to finite-temperature extensions based on (2.7.10) and (2.7.16). The zero-temperature limits of the finite-temperature theories are of special interest.

3
DENSITY-FUNCTIONAL THEORY

3.1 The original idea: the Thomas–Fermi model

We are now ready to begin to expound the density-functional theory of electronic structure, the principal subject of this book. This is a remarkable theory that allows one to replace the complicated N-electron wave function $\Psi(\mathbf{x}_1, \mathbf{x}_2, \ldots, \mathbf{x}_N)$ and the associated Schrödinger equation by the much simpler electron density $\rho(\mathbf{r})$ and its associated calculational scheme. Remarkable indeed!

There is a long history of such theories, which until 1964 only had status as models. The history begins with the works of Thomas and Fermi in the 1920s (Thomas 1927; Fermi 1927, 1928a, 1928b; March 1975). What these authors realized was that statistical considerations can be used to approximate the distribution of electrons in an atom. The assumptions stated by Thomas (1927) are that: "Electrons are distributed uniformly in the six-dimensional phase space for the motion of an electron at the rate of two for each h^3 of volume," and that there is an effective potential field that "is itself determined by the nuclear charge and this distribution of electrons." The Thomas–Fermi formula for electron density can be derived from these assumptions. We here give a slightly different, but equivalent, derivation of the Thomas–Fermi theory; see Chapter 6 for additional viewpoints.

We divide the space into many small cubes (cells), each of side l and volume $\Delta V = l^3$, each containing some fixed number of electrons ΔN (which may have different values for different cells), and we assume that the electrons in each cell behave like independent fermions at the temperature 0 K, with the cells independent of one another.

The energy levels of a particle in a three-dimensional infinite well are given by the formula

$$\varepsilon(n_x, n_y, n_z) = \frac{h^2}{8ml^2}(n_x^2 + n_y^2 + n_z^2)$$

$$= \frac{h^2}{8ml^2} R^2 \qquad (3.1.1)$$

where $n_x, n_y, n_z = 1, 2, 3, \ldots$, and the second equality defines the quantity R. For high quantum numbers, that is, for large R, the number of distinct energy levels with energy smaller than ε can be approximated by the volume of one octant of a sphere with radius R in the space (n_x, n_y, n_z). This number is

$$\Phi(\varepsilon) = \frac{1}{8}\left(\frac{4\pi R^3}{3}\right) = \frac{\pi}{6}\left(\frac{8ml^2\varepsilon}{h^2}\right)^{3/2} \tag{3.1.2}$$

The number of energy levels between ε and $\varepsilon + \delta\varepsilon$ is accordingly

$$g(\varepsilon)\Delta\varepsilon = \Phi(\varepsilon + \delta\varepsilon) - \Phi(\varepsilon)$$

$$= \frac{\pi}{4}\left(\frac{8ml^2}{h^2}\right)^{3/2} \varepsilon^{1/2}\,\delta\varepsilon + O((\delta\varepsilon)^2) \tag{3.1.3}$$

where the function $g(\varepsilon)$ is the *density of states at energy* ε.

To compute the total energy for the cell with ΔN electrons, we need the probability for the state with energy ε, to be occupied, which we call $f(\varepsilon)$. This is the Fermi–Dirac distribution,

$$f(\varepsilon) = \frac{1}{1 + e^{\beta(\varepsilon - \mu)}} \tag{3.1.4}$$

which at 0 K reduces to a step function:

$$f(\varepsilon) = \begin{cases} 1, & \varepsilon < \varepsilon_F \\ 0, & \varepsilon > \varepsilon_F \end{cases} \quad \text{as } \beta \to \infty \tag{3.1.5}$$

where ε_F is the so-called Fermi energy. All the states with energy smaller than ε_F are occupied and those with energy greater than ε_F are unoccupied. The Fermi energy ε_F is the zero-temperature limit of the chemical potential μ.

Now we find the total energy of the electrons in this cell by summing the contributions from the different energy states:

$$\Delta E = 2 \int \varepsilon f(\varepsilon) g(\varepsilon)\,d\varepsilon$$

$$= 4\pi \left(\frac{2m}{h^2}\right)^{3/2} l^3 \int_0^{\varepsilon_F} \varepsilon^{3/2}\,d\varepsilon$$

$$= \frac{8\pi}{5}\left(\frac{2m}{h^2}\right)^{3/2} l^3 \varepsilon_F^{5/2} \tag{3.1.6}$$

where the factor 2 enters because each energy level is doubly occupied, by one electron with spin α and another with spin β. The Fermi energy ε_F

is related to the number of electrons ΔN in the cell, through the formula

$$\Delta N = 2 \int f(\varepsilon) g(\varepsilon) \, d\varepsilon$$

$$= \frac{8\pi}{3} \left(\frac{2m}{h^2}\right)^{3/2} l^3 \varepsilon_F^{3/2} \tag{3.1.7}$$

Eliminating ε_F from (3.1.6) by (3.1.7), we obtain

$$\Delta E = \frac{3}{5} \Delta N \, \varepsilon_F$$

$$= \frac{3h^2}{10m} \left(\frac{3}{8\pi}\right)^{2/3} l^3 \left(\frac{\Delta N}{l^3}\right)^{5/3} \tag{3.1.8}$$

This derivation can be found, for example, in McQuarrie (1976, pp. 164–166).

Equation (3.1.8) is a relation between total kinetic energy and the electron density $\rho = \Delta N/l^3 = \Delta N/\Delta V$ for each cell in the space. (Note that different cells can have different values of ρ.) Adding the contributions from all cells, we find the total kinetic energy to be, now reverting to atomic units,

$$T_{TF}[\rho] = C_F \int \rho^{5/3}(\mathbf{r}) \, d\mathbf{r}, \qquad C_F = \tfrac{3}{10}(3\pi^2)^{2/3} = 2.871 \tag{3.1.9}$$

where the limit $\Delta V \to 0$, with $\rho = \Delta N/\Delta V = \rho(\mathbf{r})$ finite, has been taken to give an integration instead of a summation. This is the famous Thomas–Fermi kinetic energy functional, which Thomas and Fermi dared to apply to electrons in atoms, in the manner we are about to describe. [We here first encounter one of the most important ideas in modern density-functional theory, the *local density approximation* (LDA). In this approximation, electronic properties are determined as functionals of the electron density by applying *locally* relations appropriate for a homogeneous electronic system. In later chapters the LDA is employed for properties other than the kinetic energy.]

What (3.1.9) accomplishes is approximation of the electronic kinetic energy in terms of the density $\rho(\mathbf{r})$, whereas the rigorous energy formula of (2.4.9) gives the kinetic energy in terms of the first-order density matrix. If we further neglect the exchange and correlation terms in (2.4.9), thus only taking into consideration the classical electrostatic energies of electron–nucleus attraction and electron–electron repulsion, we get, using (2.4.10), an energy formula for an atom in terms of electron density alone:

$$E_{TF}[\rho(\mathbf{r})] = C_F \int \rho^{5/3}(\mathbf{r}) \, d\mathbf{r} - Z \int \frac{\rho(\mathbf{r})}{r} \, d\mathbf{r} + \frac{1}{2} \iint \frac{\rho(\mathbf{r}_1)\rho(\mathbf{r}_2)}{|\mathbf{r}_1 - \mathbf{r}_2|} \, d\mathbf{r}_1 \, d\mathbf{r}_2 \tag{3.1.10}$$

This is the energy functional of the Thomas–Fermi theory of atoms. For molecules, the second term is modified appropriately.

We now *assume* that for the ground state of an atom of interest the electron density minimizes the energy functional $E_{TF}[\rho(\mathbf{r})]$, under the constraint

$$N = N[\rho(\mathbf{r})] = \int \rho(\mathbf{r}) \, d\mathbf{r} \tag{3.1.11}$$

where N is the total number of electrons in the atom. One may incorporate this constraint by the method of Lagrange multipliers (see Appendix A). The ground-state electron density must satisfy the variational principle

$$\delta\left\{E_{TF}[\rho] - \mu_{TF}\left(\int \rho(\mathbf{r}) \, d\mathbf{r} - N\right)\right\} = 0 \tag{3.1.12}$$

which yields the Euler–Lagrange equation

$$\mu_{TF} = \frac{\delta E_{TF}[\rho]}{\delta \rho(\mathbf{r})} = \tfrac{5}{3} C_F \rho^{2/3}(\mathbf{r}) - \phi(\mathbf{r}) \tag{3.1.13}$$

where $\phi(\mathbf{r})$ is the *electrostatic potential* at point $\mathbf{r}$ due to the nucleus and the entire electron distribution:

$$\phi(\mathbf{r}) = \frac{Z}{r} - \int \frac{\rho(\mathbf{r}_2)}{|\mathbf{r} - \mathbf{r}_2|} \, d\mathbf{r}_2 \tag{3.1.14}$$

Equation (3.1.13) can be solved in conjunction with the constraint (3.1.11), and the resulting electron density then inserted in (3.1.10) to give the total energy. This is the Thomas–Fermi theory of the atom, an exquisitely simple model.

Countless modifications and improvements of the Thomas–Fermi theory have been made over the years. Some of them will be discussed in Chapter 6, where the underlying approximations will also be examined in some detail. Unfortunately, the primitive method just described founders when one comes to molecules. As will be shown in Chapter 6, no molecular binding whatever is predicted in the method (Teller 1962). This, plus the fact that the accuracy for atoms is not high as that with other methods, caused the method to come to be viewed as an oversimplified model of not much real importance for quantitative predictions in atomic or molecular or solid-state physics.

However, the situation changed with the publication of the landmark paper by Hohenberg and Kohn (1964). They provided the fundamental theorems showing that for ground states the Thomas–Fermi model may be regarded as an approximation to an exact theory, the *density-functional theory*. There exists an exact energy functional $E[\rho]$, and there exists also an exact variational principle of the form of (3.1–12). This exact theory will now be described, first in its original form, then in more mature versions.

3.2 The Hohenberg–Kohn theorems

Recall that for an electronic system described by the Hamiltonian (1.1.2), both the ground-state energy and the ground-state wave function are determined by the minimization of the energy functional $E[\Psi]$ of (1.2.1) and (1.2.3). But for an N-electron system, the external potential $v(\mathbf{r})$ completely fixes the Hamiltonian; thus N and $v(\mathbf{r})$ determine all properties for the ground state. (Only nondegenerate ground states are considered in this section; degeneracy presents no difficulty, as will be discussed in §3.4.) This of course is not surprising since $v(\mathbf{r})$ defines the whole nuclear frame for a molecule, which together with the number of electrons determines all the electronic properties.

In place of N and $v(\mathbf{r})$, the first Hohenberg–Kohn theorem (Hohenberg and Kohn 1964) legitimizes the use of electron density $\rho(\mathbf{r})$ as basic variable. It states: *The external potential $v(\mathbf{r})$ is determined, within a trivial additive constant, by the electron density $\rho(\mathbf{r})$.* Since ρ determines the number of electrons, it follows that $\rho(\mathbf{r})$ also determines the ground-state wave function Ψ and all other electronic properties of the system. Note that $v(\mathbf{r})$ is not restricted to Coulomb potentials.

The proof of this theorem of Hohenberg and Kohn is disarmingly simple. All that is employed is the minimum-energy principle for the ground state. Consider the electron density $\rho(\mathbf{r})$ for the nondegenerate ground state of some N-electron system. It determines N by simple quadrature [(1.5.2)]. It also determines $v(\mathbf{r})$, and hence all properties. For if there were two external potentials v and v' differing by more than a constant, each giving the same ρ for its ground state, we would have two Hamiltonians H and H' whose ground-state densities were the same although the normalized wave functions Ψ and Ψ' would be different. Taking Ψ' as a trial function for the $\hat{H}$ problem, we would then have, using (1.2.3),

$$E_0 < \langle \Psi' | \hat{H} | \Psi' \rangle = \langle \Psi' | \hat{H}' | \Psi' \rangle + \langle \Psi' | \hat{H} - \hat{H}' | \Psi' \rangle$$
$$= E_0' + \int \rho(\mathbf{r})[v(\mathbf{r}) - v'(\mathbf{r})] \, d\mathbf{r} \qquad (3.2.1)$$

where E_0 and E_0' are the ground-state energies for $\hat{H}$ and $\hat{H}'$, respectively. Similarly, taking Ψ as a trial function for the $\hat{H}'$ problem,

$$E_0' < \langle \Psi | \hat{H}' | \Psi \rangle = \langle \Psi | \hat{H} | \Psi \rangle + \langle \Psi | \hat{H}' - \hat{H} | \Psi \rangle$$
$$= E_0 - \int \rho(\mathbf{r})[v(\mathbf{r}) - v'(\mathbf{r})] \, d\mathbf{r}. \qquad (3.2.2)$$

Adding (3.2.1) and (3.2.2), we would obtain $E_0 + E_0' < E_0' + E_0$, a contradiction, and so there cannot be two different v that give the same ρ for their ground states.

Thus, ρ determines N and v and hence all properties of the ground state, for example the kinetic energy $T[\rho]$, the potential energy $V[\rho]$, and the total energy $E[\rho]$. In place of (3.1.10) we have, writing E_v for E to make explicit the dependence on v,

$$E_v[\rho] = T[\rho] + V_{ne}[\rho] + V_{ee}[\rho]$$
$$= \int \rho(\mathbf{r})v(\mathbf{r})\,d\mathbf{r} + F_{HK}[\rho] \quad (3.2.3)$$

where

$$F_{HK}[\rho] = T[\rho] + V_{ee}[\rho] \quad (3.2.4)$$

We may write

$$V_{ee}[\rho] = J[\rho] + \text{nonclassical term} \quad (3.2.5)$$

where $J[\rho]$ is the classical repulsion of (2.4.10). The nonclassical term is a very elusive, very important quantity; it is the major part of the "exchange-correlation energy" defined and discussed at length in Chapters 7 and 8 below.

The second Hohenberg–Kohn theorem (Hohenberg and Kohn 1964) provides the energy variational principle. It reads: *For a trial density $\tilde{\rho}(\mathbf{r})$, such that $\tilde{\rho}(\mathbf{r}) \geq 0$ and $\int \tilde{\rho}(\mathbf{r})\,d\mathbf{r} = N$,*

$$E_0 \leq E_v[\tilde{\rho}] \quad (3.2.6)$$

where $E_v[\tilde{\rho}]$ is the energy functional of (3.2.3). This is analogous to the variational principle for wave functions, (1.2.3). It provides the justification for the variational principle in Thomas–Fermi theory in that $E_{TF}[\rho]$ is an approximation to $E[\rho]$. To prove this theorem, note that the previous theorem assures that $\tilde{\rho}$ determines its own $\tilde{v}$, Hamiltonian $\hat{H}$, and wave $\tilde{\Psi}$, which can be taken as a trial function for the problem of interest having external potential v. Thus,

$$\langle \tilde{\Psi}| \hat{H} |\tilde{\Psi}\rangle = \int \tilde{\rho}(\mathbf{r})v(\mathbf{r})\,d\mathbf{r} + F_{HK}[\tilde{\rho}] = E_v[\tilde{\rho}] \geq E_v[\rho] \quad (3.2.7)$$

Assuming differentiability of $E_v[\rho]$, the variational principle (3.2.6) requires that the ground-state density satisfy the stationary principle

$$\delta\left\{E_v[\rho] - \mu\left[\int \rho(\mathbf{r})\,d\mathbf{r} - N\right]\right\} = 0 \quad (3.2.8)$$

which gives the Euler–Lagrange equation

$$\mu = \frac{\delta E_v[\rho]}{\delta \rho(\mathbf{r})} = v(\mathbf{r}) + \frac{\delta F_{HK}[\rho]}{\delta \rho(\mathbf{r})} \quad (3.2.9)$$

The quantity μ is the *chemical potential*; it is discussed in detail in Chapters 4 and 5.

If we knew the exact $F_{HK}[\rho]$, (3.2.8) would be an exact equation for the ground-state electron density. Note that $F_{HK}[\rho]$ of (3.2.4) is defined independently of the external potential $v(\mathbf{r})$; this means that $F_{HK}[\rho]$ is a *universal functional* of $\rho(\mathbf{r})$. Once we have an explicit form (approximate or accurate) for $F_{HK}[\rho]$, we can apply this method to any system. Equation (3.2.9) is the basic working equation of density-functional theory.

Accurate calculational implementations of the density-functional theory are far from easy to achieve, because of the unfortunate (but challenging) fact that the functional $F_{HK}[\rho]$ is hard to come by in explicit form. We will say a great deal more about these matters in subsequent chapters. Suffice it here to emphasize that the very existence of the exact theory provides impetus both to work to advance the calculational procedures to higher and higher accuracy and also to strive to develop the conceptual consequences. In this reformulation of wave mechanics, the electron density, and only the electron density, plays the key role, and that emphatically bodes well for simple descriptive consequences.

Various mathematical questions can be put to the derivation just given of density-functional theory; the theory stands up very well. We go into these questions in some detail in the next section. In later sections, we develop the whole theory, and its extensions to variable N and finite temperature, from scratch, by methods that are more transparent and more helpful for subsequent development.

3.3 The v- and N-representability of an electron density

It is extraordinary that, as shown in the previous section, the ground-state electron density uniquely determines the properties of a ground state, particularly the ground-state energy. We now discuss some subtle aspects of this relationship.

Noting the close association of electron density with ground state in the Hohenberg–Kohn theorems, we define a density to be *v-representable* if it is the density associated with the antisymmetric ground-state wave function of a Hamiltonian of the form (1.1.2) with *some* external potential $v(\mathbf{r})$ (not necessarily a Coulomb potential). A given density may or may not be v-representable. We then can restate the first Hohenberg–Kohn theorem as the fact that there is a one-to-one mapping between ground-state wave functions and v-representable electron densities. It is through this unique mapping that a v-representable density determines the properties of its associated ground state. Thus, when we say that all ground-state properties are functionals of the electron density, we need to understand that these functionals are defined only for v-representable densities. Of particular importance is the functional $F_{HK}[\rho]$ of (3.2.4),

$$F_{HK}[\rho] = \langle \Psi | \hat{T} + \hat{V}_{ee} | \Psi \rangle \tag{3.3.1}$$

where Ψ is the ground-state wave function associated with ρ, which has to be v-representable.

The second Hohenberg–Kohn theorem simply states that for all v-representable densities,

$$E_v[\rho] \equiv F_{HK}[\rho] + \int v(\mathbf{r})\rho(\mathbf{r})\,d\mathbf{r} \geq E_v[\rho_0] \tag{3.3.2}$$

where $E_v[\rho_0]$ is the ground-state energy of the Hamiltonian with $v(\mathbf{r})$ as external potential, and ρ_0 is its ground-state density.

It is clear that both the functional $F[\rho]$ and the variational principle of (3.2.2) hinge on the v-representability of trial densities. What if a trial density is not v-representable? This is a serious difficulty, since many "reasonable" densities have been shown to be non-v-representable (Levy 1982, Lieb 1982). We mention one example. Given a Hamiltonian that has q degenerate ground-state wave functions $\{\Psi_i, i = 1, \ldots, q\}$, the density $\rho(\mathbf{r}) = \sum_{i=1}^{q} C_i \rho_i(\mathbf{r})$ is generally not v-representable for $q > 2$, where $0 \leq C_i \leq 1$, $\sum_{i=1}^{q} C_i = 1$, and $\rho_i(\mathbf{r})$ is the density for Ψ_i. Englisch and Englisch (1983) showed that even for a single-particle system there are densities that do not come from a ground-state wave function of any $v(\mathbf{r})$.

The variational principle (3.3.2) ceases to apply in a practical calculation if the trial densities are not v-representable, which is entirely possible from the foregoing discussion. The conditions for a density to be v-representable are as yet unknown. In footnote 2 of their paper, Hohenberg and Kohn (1964) proved v-representability for the case of a density that is nearly uniform and, more recently, Kohn (1983) has established v-representability in a lattice version of the Schrödinger equation, for all densities near to a nondegenerate ground-state density. For further discussion on v-representability in lattice systems, see Chayes, Chayes, and Ruskai (1985).

As we shall show in the next section, fortunately it turns out that density-functional theory can be formulated in a way that only requires the density both in functionals and in variational principle to satisfy a weaker condition, the N-representability condition. A density is N-representable if it can be obtained from some antisymmetric wave function (compare §2.6). The N-representability condition is weaker than the v-representability condition, since the former is necessary for the latter.

The N-representability condition is satisfied for any reasonable density.

More mathematically stated, a density $\rho(\mathbf{r})$ is N-representable if

$$\rho(\mathbf{r}) \geq 0, \quad \int \rho(\mathbf{r}) \, d\mathbf{r} = N, \quad \text{and} \quad \int |\nabla \rho(\mathbf{r})^{1/2}|^2 \, d\mathbf{r} < \infty. \quad (3.3.3)$$

This was first shown by Gilbert (1975), who exhibited how one can represent such ρ in terms of N orthonormal orbitals based on a space partitioning (and hence generate ρ from a single-determinantal wave function); see also the proof of Lieb (1982).

An explicit construction due to Harriman (1980) [following Macke (1955a,b)], uses N smooth, continuous, and orthonormal orbitals for any density satisfying (3.3.3). For simplicity, consider a one-dimensional case, $x_1 \leq x \leq x_2$. Then the required orbitals are

$$\phi_k(x) = [\sigma(x)]^{1/2} \exp[i 2\pi k q(x)] \quad (3.3.4)$$

where

$$\sigma(x) = \rho(x)/N \quad (3.3.5)$$

$$q(x) = \int_{x_1}^{x} \sigma(x) \, dx \quad (3.3.6)$$

and either $k = 0, \pm 1, \pm 2, \ldots$, or $k = \pm 1/2, \pm 3/2, \pm 5/2, \ldots$ (Ghosh and Parr 1985). All orbitals have the same orbital density,

$$|\phi_k(x)|^2 = \sigma(x) \quad (3.3.7)$$

Also $0 = q(x_1) < q(x) < q(x_2) = 1$, and

$$\frac{dq(x)}{dx} = \sigma(x) \quad (3.3.8)$$

The orbitals are orthonormal:

$$\int_{x_1}^{x_2} \phi_k^*(x) \phi_l(x) \, dx = \int_{x_1}^{x_2} \sigma(x) \exp[2\pi q(x)(l-k)] \, dx$$

$$= \int_{x_1}^{x_2} \exp[2\pi(l-k) q(x)] \frac{dq(x)}{dx} \, dx$$

$$= \int_0^1 \exp[2\pi(l-k) q] \, dq = \delta_{kl} \quad (3.3.9)$$

Thus, any number of such orbitals can be constructed and any total density can be represented by

$$\rho(x) = \sum_k^M \lambda_k |\phi_k(x)|^2 \quad (3.3.10)$$

where $0 \leq \lambda_k \leq 1$, $\sum_k^M \lambda_k = N$ for any integer $M \geq N$. For $M = N$, we have the density $\rho(x)$ associated with an $N \times N$ determinant. Thus, construction suffices to prove the sufficiency of the conditions (3.3.3) for N-representability. For the three-dimensional case, see Zumbach and Maschke (1983), and especially Cioslowski (1988a).

Some additional terminology will be helpful later (Levy 1982). If ρ is the density from the ground-state wave function of a Hamiltonian of the form (1.1.2) with some $v(r)$, then ρ is said to be *pure-state v-representable* (so far we have referred to this simply as v-representable and will continue to do so if there is no ambiguity). If the Hamiltonian giving ρ includes no interactions between electrons, that is, if the Hamiltonian has the form

$$\hat{H} = \hat{T} + \sum_{i=1}^{N} v(\mathbf{r}_i) \qquad (3.3.11)$$

then ρ is said to be *noninteracting pure-state v-representable*. Correspondingly, for an interacting or noninteracting Hamiltonian, if ρ is derived from an ensemble density matrix composed entirely of degenerate ground-state wave functions, then ρ is said to be *ensemble v-representable* or *noninteracting ensemble v-representable*. The previously mentioned example of a non-v-representable density is in fact ensemble v-representable, interacting or noninteracting.

3.4 The Levy constrained-search formulation

Having established in the last two sections the one-to-one correspondence between ground-state electron density $\rho_0(\mathbf{r})$ and the ground-state wave function Ψ_0, we now proceed to show how one in fact can determine Ψ_0 from a given $\rho_0(\mathbf{r})$. (Here subscripts have been given to Ψ and ρ to emphasize their ground-state nature.)

The inverse of this problem is trivial: Ψ_0 gives $\rho_0(\mathbf{r})$ by quadrature. But there exist an infinite number of antisymmetric wave functions (not necessarily from ground states) that all give the same density. Given one of these functions that integrates to ρ_0, say Ψ_{ρ_0}, how do we distinguish it from the true ground-state Ψ_0?

The answer is simple (Levy 1979, 1982; Levy and Perdew 1985b). The minimum-energy principle for the ground state gives

$$\langle \Psi_{\rho_0} | \hat{H} | \Psi_{\rho_0} \rangle \geq \langle \Psi_0 | \hat{H} | \Psi_0 \rangle = E_0 \qquad (3.4.1)$$

where $\hat{H} = \hat{T} + \hat{V}_{ee} + \sum_{i}^{N} v(\mathbf{r}_i)$, the Hamiltonian for the N-electron system. Since the potential energy due to the external field $v(\mathbf{r})$ is a simple functional of density, we therefore have

$$\langle \Psi_{\rho_0} | \hat{T} + \hat{V}_{ee} | \Psi_{\rho_0} \rangle + \int v(\mathbf{r}) \rho_0(\mathbf{r}) \, d\mathbf{r} \geq \langle \Psi_0 | \hat{T} + \hat{V}_{ee} | \Psi_0 \rangle + \int v(\mathbf{r}) \rho_0(r) \, d\mathbf{r}$$
$$(3.4.2)$$

Thus,
$$\langle \Psi_{\rho_0}| \hat{T} + \hat{V}_{ee} |\Psi_{\rho_0}\rangle \geq \langle \Psi_0| \hat{T} + \hat{V}_{ee} |\Psi_0\rangle \qquad (3.4.3)$$

and we conclude that among all the wave functions giving the same ground-state density ρ_0, the ground state Ψ_0 minimizes the expectation value $\langle \hat{T} + \hat{V}_{ee}\rangle$.

If we compare (3.4.3) with (3.3.1), we see that the right-hand side of (3.4.3) is equal to $F_{HK}[\rho_0]$, the universal functional defined for any v-representable density. That is to say,

$$\begin{aligned} F_{HK}[\rho_0] &= \langle \Psi_0| \hat{T} + \hat{V}_{ee} |\Psi_0\rangle \\ &= \underset{\Psi \to \rho_0}{\text{Min}} \langle \Psi| \hat{T} + \hat{V}_{ee} |\Psi\rangle \end{aligned} \qquad (3.4.4)$$

This is a *constrained-search* definition for the density functional $F_{HK}[\rho_0]$. Searching over all the antisymmetric wave functions that yield the input density ρ_0, $F_{HK}[\rho_0]$ delivers the minimum expectation value of $\langle \hat{T} + \hat{V}_{ee}\rangle$.

Defining $F_{HK}[\rho_0]$ by (3.4.4) not only provides a new proof for the first theorem of Hohenberg and Kohn, but also eliminates the original Hohenberg–Kohn limitation that there be no degeneracy in the ground state. For in the constrained search, only one of a set of degenerate wave functions is selected, the one corresponding to ρ_0.

The variational search in (3.4.4) is "constrained" because the space of trial wave functions comprises only those that give the density $\rho_0(\mathbf{r})$, in contrast to the search for the minimum of $\langle \Psi| \hat{T} + \hat{V}_{ee} + \Sigma_i^N v(\mathbf{r}_i) |\Psi\rangle$ in (1.2.3), which is unconstrained (except for normalization) because the space of trial wave functions is the whole N-particle Hilbert space.

In the constrained-search formula for $F_{HK}[\rho_0]$, that is, the second equality in (3.4.4), there is no need to make reference to the fact that ρ_0 is a v-representable ground-state density, so long as it comes from an antisymmetric wave function. This permits one to extend the domain of definition for $F_{HK}[\rho_0]$ from v-representable densities to N-representable densities. Define

$$F[\rho] = \underset{\Psi \to \rho}{\text{Min}} \langle \Psi| \hat{T} + \hat{V}_{ee} |\Psi\rangle \qquad (3.4.5)$$

for any ρ that is N-representable. The functional $F[\rho]$ searches all Ψ that yield the input density $\rho(\mathbf{r})$ and then delivers the minimum of $\langle \hat{T} + \hat{V}_{ee}\rangle$. It follows from (3.4.4) and (3.4.5) that

$$F_{HK}[\rho_0] = F[\rho_0] \qquad (3.4.6)$$

for any ρ_0 that is v-representable.

The constrained-search formula (3.4.5) eliminates the v-representability constraint on the domain of variation in the Hohenberg–Kohn variational principle (3.3.2). This becomes clear if one divides the ground-state energy minimization procedure of (1.2.3) into two steps:

$$E_0 = \underset{\Psi}{\text{Min}} \langle \Psi | \hat{T} + \hat{V}_{ee} + \sum_i^N v(\mathbf{r}_i) | \Psi \rangle$$

$$= \underset{\rho}{\text{Min}} \left\{ \underset{\Psi \to \rho}{\text{Min}} \langle \Psi | \hat{T} + \hat{V}_{ee} + \sum_i^N v(\mathbf{r}_i) | \Psi \rangle \right\}$$

$$= \underset{\rho}{\text{Min}} \left\{ \underset{\Psi \to \rho}{\text{Min}} \left[\langle \Psi | \hat{T} + \hat{V}_{ee} | \Psi \rangle + \int v(\mathbf{r}) \rho(\mathbf{r}) \, d\mathbf{r} \right] \right\} \quad (3.4.7)$$

In the second line here, the inner minimization is constrained to all wave functions that give $\rho(\mathbf{r})$, while the outer minimization releases this constraint by searching all $\rho(\mathbf{r})$. The first term within the square brackets in the third line is precisely the functional $F[\rho]$ defined in (3.4.5). Thus,

$$E_0 = \underset{\rho}{\text{Min}} \left\{ F[\rho] + \int v(\mathbf{r}) \rho(\mathbf{r}) \, d\mathbf{r} \right\}$$

$$= \underset{\rho}{\text{Min}} E[\rho] \quad (3.4.8)$$

where

$$E[\rho] = F[\rho] + \int v(\mathbf{r}) \rho(\mathbf{r}) \, d\mathbf{r} \quad (3.4.9)$$

The prescription of (3.4.5) and (3.4.7) to (3.4.9) dispels any aura of mystery that may appear to surround the original Hohenberg–Kohn $F[\rho]$ (see Figure 3.1.)

The variation in (3.4.8) is over all N-representable densities; it requires no more than the nonnegativity, proper normalization, and continuity of the trial densities [cf. (3.3.3)]. This makes the minimization in (3.4.7) easier to carry out than the original Hohenberg–Kohn minimization of (3.3.2). The v-representability problem in the original approach has been eliminated.

The foregoing arguments are from the work of Levy (1979) [see also Lieb (1982)], who devised this constrained-search formulation in the spirit of the Percus (1978) definition of a universal kinetic-energy functional for noninteracting fermion systems. The existence of the minimum in (3.4.5) has been proved by Lieb (1982), in an elegant and rigorous discussion of the whole theory. If we call the minimizing wave function ψ_ρ^{min}, then we have

$$F[\rho] = \langle \psi_\rho^{\text{min}} | \hat{T} + \hat{V}_{ee} | \psi_\rho^{\text{min}} \rangle$$

$$= T[\rho] + V_{ee}[\rho] \quad (3.4.10)$$

where

$$T[\rho] = \langle \psi_\rho^{\text{min}} | \hat{T} | \psi_\rho^{\text{min}} \rangle \quad (3.4.11)$$

3.4 DENSITY-FUNCTIONAL THEORY

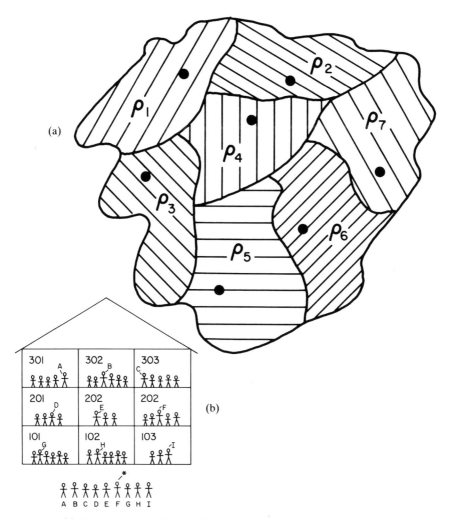

Figure 3.1(a) Percus–Levy partition of the N-electron Hilbert space. Each shaded area is the set of all Ψ that integrate to a particular ρ. The minimization in (3.4.5) for a particular ρ is constrained to the shaded area associated with this ρ, and is realized by only one point (denoted by ●) in this shaded area. The minimization in (3.4.8) is over all such points.

Figure 3.1(b) An analogous constrained-search problem. To identify the tallest child in a school, one does not have to line up all of the children in the schoolyard. Instead, one can simply ask the tallest child from each room to come to the schoolyard.

and

$$V_{ee}[\rho] = \langle \psi_\rho^{\min} | \hat{V}_{ee} | \psi_\rho^{\min} \rangle \tag{3.4.12}$$

The functional $F[\rho]$ in (3.4.5) is universal [as is $F_{HK}[\rho]$ of (3.2.4)] in the sense that it is independent of the external field $v(\mathbf{r})$. The same $F[\rho]$ applies to both a hydrogen atom and a giant protein molecule; the part of $E[\rho]$ that depends on the structure of an individual system is $\int v(\mathbf{r})\rho(\mathbf{r})\,d\mathbf{r}$, as indicated in (3.4.9). This shows the promise, as well as the challenge, of looking for an accurate method to deal with $F[\rho]$. What we have at this stage is a rigorously defined $F[\rho]$, for any N-representable density ρ characterized by (3.3.3). However, as we shall be discussing at length, what we have is a functional $F[\rho]$ that is highly difficult to approximate, which is not surprising because what we have done is to change the variational unknown from the complicated many-variable function Ψ to a single three-variable function ρ.

This situation may be contrasted with the state of the theory of reduced density matrices outlined in Chapter 2. The electronic energy as a functional of the second-order reduced density matrix Γ_2 is explicit and simple, but the domain of energy variation is the set of N-representable Γ_2, the practical characterization of which is, unfortunately, unknown.

The constrained-search definition of (3.4.5) allows one to approximate $F[\rho]$ by bounds; namely,

$$F[\rho] \leq \underset{\substack{\Psi \to \rho \\ \Psi \in S}}{\text{Min}} \langle \Psi | \hat{T} + \hat{V}_{ee} | \Psi \rangle \tag{3.4.13}$$

or, more simply,

$$F[\rho] \leq \langle \Psi_\rho | \hat{T} + \hat{V}_{ee} | \Psi_\rho \rangle \tag{3.4.14}$$

where S is any subspace of the N-electron Hilbert space and Ψ_ρ is any N-electron wave function that gives $\rho(\mathbf{r})$. This idea has been exploited by Nyden and Parr (Nyden and Parr 1983, Nyden 1983), Zumbach and Maschke (1983), Ghosh and Parr (1984), Schönhammer and Gunnarsson (1987), and Cioslowski (1988a).

In summary, there now has been proved the existence of the universal density functional $F[\rho]$ for any N-representable density, (3.4.5), and there has now been established the energy-functional minimum principle, (3.4.8). A rigorous density-functional method for determining the ground-state density and ground-state energy for any electronic system is thereby provided.

3.5 Finite-temperature canonical-ensemble theory

For macroscopic systems at finite temperature, the equilibrium state is analogous to the ground state at zero temperature. In place of the variational principle of (1.2.3), we have the principle (2.7.10) for a canonical ensemble, or the principle (2.7.16) for a grand canonical ensemble. The constrained-search ground-state density-functional theory can readily be generalized to the equilibrium states in ensembles. We begin with the canonical ensemble.

First, we provide a proof of (2.7.10). We may express a general N-particle density operator $\hat{\Gamma}_N$ in terms of its eigenvalues and normalized eigenvectors,

$$\hat{\Gamma}_N = \sum_i p_{Ni} |\Psi_{Ni}\rangle\langle\Psi_{Ni}| \qquad (3.5.1)$$

which is possible by (2.1.25). Then the Helmholtz free energy $A[\hat{\Gamma}_N]$ of (2.7.9) becomes

$$A[\{p_{Ni}, \Psi_{Ni}\}] = A[\hat{\Gamma}_N]$$
$$= \sum_i p_{Ni}\left(\frac{1}{\beta}\ln p_{Ni} + \langle\Psi_{Ni}|\hat{H}|\Psi_{Ni}\rangle\right) \qquad (3.5.2)$$

The free energy thus depends on two sets of variables $\{p_{Ni}\}$ and $\{\Psi_{Ni}\}$. Leaving $\{\Psi_{Ni}\}$ fixed for the moment, one can find the set of $\{p_{Ni}\}$ that minimizes A for a given set $\{\Psi_i\}$ by solving

$$\frac{\partial}{\partial p_i}\left\{A[\{p_{Ni}, \Psi_{Ni}\}] + \lambda\left(\sum_i p_{Ni} - 1\right)\right\} = 0 \qquad (3.5.3)$$

or equivalently,

$$\frac{1}{\beta}(\ln p_{Ni} + 1) + \langle\Psi_{Ni}|\hat{H}|\Psi_{Ni}\rangle + \lambda = 0 \qquad (3.5.4)$$

where λ is the Lagrange multiplier for the constraint

$$\sum_i p_{Ni} = 1 \qquad (3.5.5)$$

which is equivalent to $\operatorname{tr}\hat{\Gamma}_N = 1$. Solving (3.5.4) and (3.5.5) gives

$$p_{Ni}^0 = \frac{\exp\{-\beta\langle\Psi_{Ni}|\hat{H}|\Psi_{Ni}\rangle\}}{\sum_i \exp\{-\beta\langle\Psi_{Ni}|\hat{H}|\Psi_{Ni}\rangle\}} \qquad (3.5.6)$$

Inserting (3.5.6) into (3.5.2), we obtain

$$A[\{p_{Ni}^0, \Psi_{Ni}\}] = -\frac{1}{\beta}\ln\sum_i \exp\{-\beta\langle\Psi_{Ni}|\hat{H}|\Psi_{Ni}\rangle\} \qquad (3.5.7)$$

Using (3.5.2) and (3.5.6), we therefore find

$$A[\{p_{Ni}, \Psi_{Ni}\}] - A[\{p_{Ni}^0, \Psi_{Ni}\}] = \frac{1}{\beta}\sum_i p_{Ni}(\ln p_{Ni} - \ln p_{Ni}^0) \geq 0 \qquad (3.5.8)$$

The last step is from the Gibb's inequality, Equation (B.11) of Appendix B.

To complete the proof, we need an inequality for a general operator $\hat{O}$,

$$\exp(\langle\hat{O}\rangle) \leq \langle\exp(\hat{O})\rangle \tag{3.5.9}$$

where $\langle \ \rangle$ denotes the expectation value for any state. This comes from Jensen's inequality for convex functions of an operator, Equation (B.17) of Appendix B. Thus,

$$A[\{p_{Ni}^0, \Psi_{Ni}\}] = -\frac{1}{\beta}\ln\sum_i \exp(-\beta\langle\Psi_{Ni}|\hat{H}|\Psi_{Ni}\rangle)$$

$$\geq -\frac{1}{\beta}\ln\sum_i \langle\Psi_{Ni}|e^{-\beta\hat{H}}|\Psi_{Ni}\rangle$$

$$= -\frac{1}{\beta}\ln \text{tr } e^{-\beta\hat{H}}$$

$$= A[\{p_{Ni}^0, \Psi_{Ni}^0\}]$$

$$= A[\hat{\Gamma}_N^0] \tag{3.5.10}$$

where the Ψ_{Ni}^0 are the eigenstates of $\hat{H}$.

The inequality becomes an equality only when $\Psi_{Ni} = \Psi_{Ni}^0$ for all i. We have here used the fact that $\{\Psi_{Ni}\}$ is a complete set. Combining (3.5.8) and (3.5.10) leads to the variational principle (2.7.10).

Having the variational principle in hand, one can now construct the density-functional theory for the canonical ensemble by dividing the minimization of the Helmhotz free energy into two steps:

$$A^0 = A[\Gamma_N^0] = \underset{\Gamma_N}{\text{Min tr}}\left[\hat{\Gamma}_N\left(\hat{H} + \frac{1}{\beta}\ln\hat{\Gamma}_N\right)\right]$$

$$= \underset{\rho}{\text{Min}}\left\{\underset{\hat{\Gamma}_N\to\rho}{\text{Min tr}}\left[\hat{\Gamma}_N\left(\hat{T} + \hat{U} + \frac{1}{\beta}\ln\hat{\Gamma}_N\right)\right] + \int\rho(\mathbf{r})v(\mathbf{r})\,d\mathbf{r}\right\} \tag{3.5.11}$$

where

$$\hat{H} = \hat{T} + \hat{U} + \sum_i^N v(\mathbf{r}_i) \tag{3.5.12}$$

with $\hat{T}$ the total kinetic-energy operator, $\hat{U}$ the particle–particle interaction energy (not necessarily Coulomb repulsion), and $v(\mathbf{r})$ the external field. Defining the universal density functional by the constrained search

$$F[\rho(\mathbf{r})] = \underset{\hat{\Gamma}_N\to\rho}{\text{Min tr}}\left[\hat{\Gamma}_N\left(\hat{T} + \hat{U} + \frac{1}{\beta}\ln\hat{\Gamma}_N\right)\right] \tag{3.5.13}$$

one obtains the free-energy variational principle:

$$A^0 = \min_\rho \left\{ F[\rho(\mathbf{r})] + \int v(\mathbf{r})\rho(\mathbf{r})\, d\mathbf{r} \right\}$$
$$= \min_\rho A[\rho(\mathbf{r})] \quad (3.5.14)$$

where

$$A[\rho] = F[\rho] + \int v(\mathbf{r})\rho(\mathbf{r})\, d\mathbf{r} \quad (3.5.15)$$

Note that $F[\rho]$ is a universal functional of ρ: while it depends on the statistics of the particles (which determines the symmetry of $\hat{\Gamma}_N$) and the particle–particle interaction potential U, it does not depend on the external field. We use the same notation $F[\rho]$ to denote different functionals in (3.5.13) and (3.4.5) in order to emphasize the similarity of the different density-functional theories. This should not cause confusion, because one rarely uses two types of density-functional theories at one time. We shall also use F for the universal functionals constructed in the next two sections.

The functional $F[\rho]$ defined in (3.5.13) can be decomposed into three pieces, namely

$$F[\rho] = T[\rho] + U[\rho] - \theta S[\rho] \quad (3.5.16)$$

with the kinetic-energy functional being

$$T[\rho] = \mathrm{tr}\,(\hat{\Gamma}_{N,\rho}^{\min} \hat{T}) \quad (3.5.17)$$

the particle–particle interaction potential energy functional being

$$U[\rho] = \mathrm{tr}\,(\hat{\Gamma}_{N,\rho}^{\min} \hat{U}) \quad (3.5.18)$$

and the entropy functional being

$$S[\rho] = -k_B\, \mathrm{tr}\,(\hat{\Gamma}_{N,\rho}^{\min} \ln \hat{\Gamma}_{N,\rho}^{\min}) \quad (3.5.19)$$

where $\hat{\Gamma}_{N,\rho}^{\min}$ is the density operator with which the minimum in (3.5.13) is achieved, k_B is the Boltzmann constant, and θ is the absolute temperature.

If the density $\rho_0(\mathbf{r})$ for an equilibrium state of some $v(\mathbf{r})$ is given, the constrained-search of (3.5.13) reaches the minimum at the equilibrium density operator $\hat{\Gamma}_N$ for this $v(\mathbf{r})$, in analogy with (3.4.6). This follows from (3.5.11) and (3.5.14):

$$F[\rho_0] = A[\rho_0] - \int v(\mathbf{r})\rho_0(\mathbf{r})\, d\mathbf{r}$$
$$= \mathrm{tr}\left[\hat{\Gamma}_N^0 \left(\hat{T} + \hat{U} + \frac{1}{\beta} \ln \hat{\Gamma}_N^0 \right) \right] \quad (3.5.20)$$

In this way, $\rho_0(\mathbf{r})$ determines $\hat{\Gamma}_N^0$.

In summary, we have shown that in a canonical ensemble at a given temperature: (1) The equilibrium-state particle density $\rho_0(\mathbf{r})$ uniquely determines the equilibrium density operator $\hat{\Gamma}_N^0$ and hence all the properties of the equilibrium state. (2) There holds a free-energy variational principle (3.5.14) in terms of particle density into which there enters the universal functional (3.5.13).

3.6 Finite-temperature grand-canonical-ensemble theory

We now turn to the grand canonical ensemble, in which the state of a system is described by a density operator in Fock space, that is, with nonzero probabilities for different particle numbers. Such a general density operator is

$$\hat{\Gamma} = \sum_N \sum_i p_{Ni} |\Psi_{Ni}\rangle \langle \Psi_{Ni}| \qquad (3.6.1)$$

and the grand potential $\Omega[\Gamma]$ of (2.7.15) is

$$\Omega[\{p_{Ni}, \Psi_{Ni}\}] = \Omega[\hat{\Gamma}] = \operatorname{Tr} \hat{\Gamma}\left(\frac{1}{\beta} \ln \hat{\Gamma} + \hat{H} - \mu \hat{N}\right)$$

$$= \sum_N \sum_i p_{Ni} \left(\frac{1}{\beta} \ln p_{Ni} + \langle \Psi_{Ni}| \hat{H} - \mu \hat{N} |\Psi_{Ni}\rangle\right)$$

$$= \sum_N \sum_i p_{Ni} \left(\frac{1}{\beta} \ln p_{Ni} + \langle \Psi_{Ni}| \hat{H} |\Psi_{Ni}\rangle - \mu N\right)$$

(3.6.2)

where μ is the chemical potential of the system (a parameter describing the whole ensemble). The proof of the variational principle for the grand potential, (2.7.16), proceeds similarly to the proof in the last section of the variational principle for the Helmholtz free energy in the canonical ensemble. One first holds the set of variables $\{\Psi_{Ni}\}$ fixed, and looks for the minimum of Ω with respect to $\{p_{Ni}\}$. Thus, we solve

$$\frac{\partial}{\partial p_{Ni}} \left[\Omega[\{p_{Ni}, \Psi_{Ni}\}] + \lambda \left(\sum_N \sum_i p_{Ni} - 1\right)\right] = 0 \qquad (3.6.3)$$

or equivalently

$$\frac{1}{\beta}(\ln p_{Ni} + 1) + \langle \Psi_{Ni}| \hat{H} - \mu \hat{N} |\Psi_{Ni}\rangle + \lambda = 0 \qquad (3.6.4)$$

where λ is the Lagrange multiplier for the constraint that probability be normalized. There results

$$p_{Ni}^0 = \frac{\exp(-\beta \langle \Psi_{Ni}| \hat{H} - \mu \hat{N} |\Psi_{Ni}\rangle)}{\sum_N \sum_i \exp(-\beta \langle \Psi_{Ni}| \hat{H} - \mu \hat{N} |\Psi_{Ni}\rangle)} \qquad (3.6.5)$$

Hence
$$\Omega[\{p_{Ni}^0, \Psi_{Ni}\}] = -\frac{1}{\beta} \ln \sum_N \sum_i \exp(-\beta \langle \Psi_{Ni}| \hat{H} - \mu \hat{N} |\Psi_{Ni}\rangle) \quad (3.6.6)$$

We now prove that $\Omega[\{p_{Ni}^0, \Psi_{Ni}\}]$ is a minimum for fixed $\{\Psi_{Ni}\}$:

$$\Omega[\{p_{Ni}, \Psi_{Ni}\}] - \Omega[\{p_{Ni}^0, \Psi_{Ni}\}]$$
$$= \sum_N \sum_i p_{Ni} \left(\frac{1}{\beta} \ln p_{Ni} + \langle \Psi_{Ni}| \hat{H} - \mu \hat{N} |\Psi_{Ni}\rangle\right)$$
$$+ \frac{1}{\beta} \ln \sum_N \sum_i \exp(-\beta \langle \Psi_{Ni}| \hat{H} - \mu \hat{N} |\Psi_{Ni}\rangle)$$
$$= \frac{1}{\beta} \sum_N \sum_i p_{Ni} (\ln p_{Ni} - \ln p_{Ni}^0) \geq 0 \quad (3.6.7)$$

Here we have used (3.6.5) and the Gibbs' inequality (B.11). Consequently, again using (B.17),

$$\Omega[\{p_{Ni}^0, \Psi_{Ni}\}] = -\frac{1}{\beta} \ln \sum_N \sum_i \exp(-\beta \langle \Psi_{Ni}| \hat{H} - \mu \hat{N} |\Psi_{Ni}\rangle)$$
$$\geq -\frac{1}{\beta} \ln \sum_N \sum_i \langle \Psi_{Ni}| e^{-\beta(H - \mu \hat{N})} |\Psi_{Ni}\rangle$$
$$= -\frac{1}{\beta} \ln \text{Tr}\, e^{-\beta(\hat{H} - \mu \hat{N})}$$
$$= \Omega[\{p_{Ni}^0, \Psi_{Ni}^0\}] \quad (3.6.8)$$

The inequality in this last formula becomes an equality when each Ψ_{Ni} is Ψ_{Ni}^0, an eigenstate of $(\hat{H} - \mu \hat{N})$. The density operator determined by the sets $\{p_{Ni}^0, \Psi_{Ni}^0\}$ is the equilibrium one of (2.7.14). Adding (3.6.7) and (3.6.8), we obtain the desired variational principle, (2.7.16).

Construction of a density-functional theory by constrained search is now straightforward:

$$\Omega^0 = \Omega[\hat{\Gamma}^0] = \underset{\Gamma}{\text{Min}} \,\text{Tr}\left[\hat{\Gamma}\left(\hat{H} - \mu \hat{N} + \frac{1}{\beta} \ln \hat{\Gamma}\right)\right]$$
$$= \underset{\rho}{\text{Min}} \left\{\underset{\Gamma \to \rho}{\text{Min}} \,\text{Tr}\left[\hat{\Gamma}\left(\hat{T} + \hat{U} + \frac{1}{\beta} \ln \hat{\Gamma}\right)\right] + \int (v(\mathbf{r}) - \mu)\rho(\mathbf{r})\, d\mathbf{r}\right\}$$
$$(3.6.9)$$

Define
$$F[\rho(\mathbf{r})] = \underset{\Gamma \to \rho}{\text{Min}} \,\text{Tr}\left[\hat{\Gamma}\left(\hat{T} + \hat{U} + \frac{1}{\beta} \ln \hat{\Gamma}\right)\right] \quad (3.6.10)$$

Then
$$\Omega^0 = \underset{\rho(\mathbf{r})}{\text{Min}}\, \Omega[\rho(\mathbf{r})] \quad (3.6.11)$$

where the grand potential functional is

$$\Omega[\rho(\mathbf{r})] = F[\rho(\mathbf{r})] + \int \rho(\mathbf{r})(v(\mathbf{r}) - \mu)\,d\mathbf{r} \quad (3.6.12)$$

One can split $F[\rho(\mathbf{r})]$ into its components,

$$F[\rho(\mathbf{r})] = T[\rho(\mathbf{r})] + U[\rho(\mathbf{r})] - \theta S[\rho(\mathbf{r})]$$

$$= \mathrm{Tr}\left[\hat{\Gamma}_\rho^{\min}\left(\hat{T} + \hat{U} + \frac{1}{\beta}\ln\hat{\Gamma}_\rho^{\min}\right)\right] \quad (3.6.13)$$

where $\hat{\Gamma}_\rho^{\min}$ is the density operator that minimizes $\mathrm{Tr}\,\hat{\Gamma}(\hat{T} + \hat{U} + (1/\beta)\ln\hat{\Gamma})$ and simultaneously gives $\rho(\mathbf{r})$. If the density $\rho_0(\mathbf{r})$ for an equilibrium state of some $v(\mathbf{r})$ is given, the constrained search of (3.6.10) reaches the minimum at the equilibrium density operator $\hat{\Gamma}^0$ for the same $v(\mathbf{r})$. This follows from (3.6.11) and (3.6.2):

$$F[\rho_0(\mathbf{r})] = \Omega[\Gamma^0] - \int \rho_0^0(\mathbf{r})(v(\mathbf{r}) - \mu)\,d\mathbf{r}$$

$$= \mathrm{Tr}\,\hat{\Gamma}^0\left(\hat{T} + \hat{U} + \frac{1}{\beta}\ln\hat{\Gamma}^0\right) \quad (3.6.14)$$

Note that the chemical potential μ only appears in the combination $v(\mathbf{r}) - \mu$. Note also that because we have here described a grand canonical ensemble, $F[\rho]$ is a universal functional of density $\rho(\mathbf{r})$ that may integrate to a noninteger, the average particle number N.

In summary, we have shown that in a grand canonical ensemble at a given temperature and chemical potential μ: (1) The equilibrium-state particle density $\rho_0(\mathbf{r})$ uniquely determines the equilibrium density operator $\hat{\Gamma}^0$ and hence all the properties of the equilibrium state. (2) There holds the grand potential variational principle in terms of particle density, (3.6.11), with the universal functional $F[\rho]$ defined by (3.6.10).

The density-functional theory of the grand canonical ensemble is due in the first instance to Mermin (1965), whose proof was a proof by contradiction patterned after the Hohenberg–Kohn proof of §3.2.

3.7 Finite-temperature ensemble theory of classical systems

For classical systems at finite temperature, a theory may also be formulated through constrained search. We do that here, choosing the grand-canonical-ensemble approach. Following the work of Mermin (1965) for quantum electronic systems, Ebner and Saam (1975) and Ebner, Saam and Stroud (1976) developed the density-functional theory for classical systems through a proof by contradiction. For subsequent developments, see Evans (1979) and Rowlinson and Widom (1982). For earlier related work, see, for example, Stillinger and Buff (1962) or Lebowitz and Percus (1963).

In classical statistical mechanics, the equilibrium state of a system in a finite-temperature grand canonical ensemble is described by the probability distribution function

$$f^0(\mathbf{r}_1, \ldots, \mathbf{r}_N, \mathbf{p}_1, \ldots, \mathbf{p}_N) = \frac{e^{-\beta(H_N - \mu N)}}{Z} \qquad (3.7.1)$$

where H_N is the Hamiltonian when there are N particles present,

$$H_N = \sum_i \frac{p_i^2}{2m} + U(\mathbf{r}_1, \ldots, \mathbf{r}_N) + \sum_i^N v(\mathbf{r}_i) \qquad (3.7.2)$$

μ is the chemical potential, $\beta = 1/k_B \theta$, and Z is the grand partition function

$$Z = \mathrm{Tr}_{\mathrm{cl}} [\exp(-\beta(H_N - \mu N))] \qquad (3.7.3)$$

The trace symbol here connotes integration over phase space and then summation over all different numbers of particle; namely,

$$\mathrm{Tr}_{\mathrm{cl}} \equiv \sum_N^\infty \frac{1}{(h)^{3N} N!} \int d\mathbf{r}_1, \ldots, d\mathbf{r}_N \, d\mathbf{p}_1, \ldots, d\mathbf{p}_N \qquad (3.7.4)$$

In (3.7.1), f^0 is the probability density for the system at equilibrium in a grand canonical ensemble, in the microscopic state of N particles described by the phase-space coordinates $\{\mathbf{r}_1, \ldots, \mathbf{r}_N, \mathbf{p}_1, \ldots, \mathbf{p}_N\}$. Note the similarity with the formula (2.7.14) for a quantum system.

Like the quantum mechanical Γ^0, f^0 can be derived from the maximum-entropy principle of §2.7. Or, as in §3.6, F can be derived from the *minimum grand potential principle*

$$\Omega[f^0] \leq \Omega[f] \qquad (3.7.5)$$

for all probability densities f such that

$$\mathrm{Tr}_{\mathrm{cl}} f = 1 \qquad (3.7.6)$$

The grand potential is defined as

$$\Omega[f] = \mathrm{Tr}_{\mathrm{cl}} \left[f \left(H_N - \mu N + \frac{1}{\beta} \ln f \right) \right] \qquad (3.7.7)$$

Its value at f^0 is

$$\Omega[f^0] = -\frac{1}{\beta} \ln Z \qquad (3.7.8)$$

The proof of (3.7.5) is straightforward. The Euler–Lagrange equation associated with the variational principle of (3.7.5) is

$$\delta\{\Omega[f] - \lambda(\text{Tr}_{cl} f - 1)\} = 0 \qquad (3.7.9)$$

or

$$\frac{1}{\beta}(\ln f + 1) + H_N - \mu N - \lambda = 0 \qquad (3.7.10)$$

The solution of (3.7.10) satisfying (3.7.6) is (3.7.1). From (3.7.7) and (3.7.8) there follows

$$\Omega[f] - \Omega[f^0] = \text{Tr}_{cl}\left[f\left(H_N - \mu N + \frac{1}{\beta}\ln f\right) - \frac{1}{\beta}\ln Z\right]$$

$$= \frac{1}{\beta}\text{Tr}_{cl}\left[f(\ln f - \ln f^0)\right]$$

$$\geq 0 \qquad (3.7.11)$$

where in the second equality (3.7.1) has been used to express $H_N - \mu N$ in terms of f^0 and Z. The inequality follows from the Gibbs inequality of (B.11) for the continuous-variable case.

Given the minimum principle, constrained-search implementation of it is straightforward:

$$\Omega[f^0] = \underset{\rho(\mathbf{r})}{\text{Min}}\left\{\underset{f \to \rho(\mathbf{r})}{\text{Min}} \text{Tr}_{cl}\left[f\left(H_N - \mu N + \frac{1}{\beta}\ln f\right)\right]\right\}$$

$$= \underset{\rho(\mathbf{r})}{\text{Min}}\left\{\underset{f \to \rho(\mathbf{r})}{\text{Min}} \text{Tr}_{cl}\left[f\left(T + U + \frac{1}{\beta}\ln f\right)\right] + \int (v(\mathbf{r}) - \mu)\rho \, d\mathbf{r}\right\}$$

$$(3.7.12)$$

where $\rho(\mathbf{r})$ is the particle density

$$\rho(\mathbf{r}) = \text{Tr}_{cl}(\hat{\rho}(\mathbf{r})f) \qquad (3.7.13)$$

with

$$\hat{\rho}(\mathbf{r}) = \sum_i^N \delta(\mathbf{r}_i - \mathbf{r}) \qquad (3.7.14)$$

Define

$$F[\rho] = \underset{f \to \rho}{\text{Min}} \text{Tr}_{cl}\left[f\left(T + U + \frac{1}{\beta}\ln f\right)\right] \qquad (3.7.15)$$

Then

$$\Omega^0 = \Omega[f^0] = \underset{\rho}{\text{Min}}\, \Omega[\rho(\mathbf{r})]$$

$$= \underset{\rho}{\text{Min}}\left\{F[\rho(\mathbf{r})] + \int (v(\mathbf{r}) - \mu)\rho(\mathbf{r}) \, d\mathbf{r}\right\} \qquad (3.7.16)$$

Note that $F[\rho]$ is a universal functional of $\rho(\mathbf{r})$, independent of the external potential $v(\mathbf{r})$. Finally, split $F[\rho(\mathbf{r})]$ into its components:

$$F[\rho] = T[\rho] + U[\rho] - \theta S[\rho]$$

$$= \text{Tr}_{\text{cl}}\left[f_\rho^{\min}\left(T + U + \frac{1}{\beta}\ln f_\rho^{\min}\right)\right] \quad (3.7.17)$$

Here $f_\rho^{\min}$ is the probability density that minimizes $\text{Tr}_{\text{cl}}[f(T + U + (1/\beta)\ln f)]$ and simultaneously gives the particle density $\rho(\mathbf{r})$.

In the foregoing constrained-search formulation, the minimum principle of (3.7.15) is *not* restricted to the class of particle densities that come from an equilibrium state of some external potential (classical ensemble v-representability). This is required in the earlier formulation of Evans (1979); for further discussion see Chayes, Chayes, and Lieb (1984). The only requirement now is that trial $\rho(\mathbf{r})$ be derivable from a classical probability distribution. In case the given density $\rho(\mathbf{r})$ does come from an equilibrium state the constrained search in (3.7.15) reaches its minimum at the equilibrium probability density f^0 for the same external potential $v(\mathbf{r})$. This follows from (3.7.15) and (3.7.16):

$$F[\rho_0(\mathbf{r})] = \Omega[f^0] - \int \rho_0(\mathbf{r})(v(\mathbf{r}) - \mu)\,d\mathbf{r}$$

$$= \text{Tr}_{\text{cl}}\left[f^0\left(T + U + \frac{1}{\beta}\ln f^0\right)\right] \quad (3.7.18)$$

This section has provided the following density functional theorems. In a classical grand canonical ensemble at a given temperature and chemical potential μ: (1) The equilibrium one-particle density $\rho_0(\mathbf{r})$ uniquely determines the equilibrium probability distribution f^0, and hence all the properties of the equilibrium state. (2) There holds the grand-potential-variational principle in terms of one-particle density, (3.7.16), with the universal functional $F[\rho]$ defined by (3.7.15).

4
THE CHEMICAL POTENTIAL

4.1 Chemical potential in the grand canonical ensemble at zero temperature

In the last chapter were obtained the principal equations of the density-functional theory of electronic systems, the equations for the determination of the electron density and the energy for a ground state or equilibrium state of a system of interest. For the ground state, $E[\rho]$ reaches its minimum among densities ρ that integrate to the number of electrons N for the system of interest, where $E[\rho]$ is given by (3.2.3) or (3.4.9). For an equilibrium state at temperature θ in a canonical ensemble, $A[\rho]$ reaches its minimum among densities ρ that integrate to N, where $A[\rho]$ is given by (3.5.15). And for an equilibrium state at temperature θ and chemical potential μ in a grand canonical ensemble, $\Omega[\rho]$ reaches its minimum among all ρ, where $\Omega[\rho]$ is defined by (3.6.12).

We are particularly interested in the zero-temperature limit, for at $\theta = 0$ ($\beta = \infty$), the equilibrium state and ground state become one and the same—the state of primary interest to us. As θ goes to zero, the term $\theta S[\rho]$ disappears from $A[\rho]$ of (3.5.15) and $\Omega[\rho]$ of (3.6.12). At this limit, do the corresponding variational principles remain valid? The answer does not follow from the arguments of §3.5 and §3.6, for in the proofs of the theorems of those sections the concavity of the entropy $S[\hat{\Gamma}]$, as a functional of the density operator, plays a vital role. Detailed analysis for each case is needed. In this section and the next two, we consider the grand canonical ensemble. We take up the canonical ensemble in §4.4.

In the limit as β tends to infinity, the universal functional $F[\rho]$ defined in (3.6.10) becomes

$$F_{GC}[\rho(\mathbf{r})] = \underset{\hat{\Gamma} \to \rho}{\text{Min}} \, \text{Tr} \, [\hat{\Gamma}(\hat{T} + \hat{V}_{ee})] \qquad (4.1.1)$$

This is the universal ground-state functional proposed by Perdew, Parr, Levy, and Balduz (1982). It extends Levy's constrained search to general density operators $\hat{\Gamma}$ in Fock space (grand canonical ensemble). In (4.1.1), $\hat{V}_{ee}$ is written in place of the general particle–particle interaction potential

$\hat{U}$ because we specialize henceforth to electronic systems. The grand potential at $\beta \to \infty$ correspondingly reduces to

$$\Omega[\rho] = F_{GC}[\rho] + \int \rho(\mathbf{r})(v(\mathbf{r}) - \mu)\, d\mathbf{r}$$

$$= E_{GC}[\rho] - \mu N \qquad (4.1.2)$$

into which enters the grand-canonical-ensemble energy functional

$$E_{GC}[\rho] = F_{GC}[\rho] + \int \rho(\mathbf{r})v(\mathbf{r})\, d\mathbf{r} \qquad (4.1.3)$$

Note that both $\Omega[\rho]$ and $E_{GC}[\rho]$ are defined for densities integrating to any finite positive number, in contrast with $E[\rho]$ of (3.4.9), which is defined only for densities integrating to positive integers.

Assume now the $\beta \to \infty$ limit of the $\Omega[\hat{\Gamma}]$ variational principle of (3.6.9),

$$\Omega^0 = E^0(N) - \mu N \leq \text{Tr}\, \hat{\Gamma}(\hat{H} - \mu \hat{N}) \qquad (4.1.4)$$

where $E^0(N)$ and μ are the energy and the chemical potential of the ground state with N electrons. Following the prescription of (3.6.9), the corresponding density-functional variational principle is then

$$E^0(N) - \mu N \leq E_{GC}[\rho(\mathbf{r})] - \mu \int \rho(\mathbf{r})\, d\mathbf{r} \qquad (4.1.5)$$

from which follows the variational equation for the ground-state electron density and energy

$$\delta\left\{ E_{GC}[\rho] - \mu \int \rho(\mathbf{r})\, d\mathbf{r} \right\} = 0 \qquad (4.1.6)$$

and the Euler–Lagrange equation

$$\frac{\delta E_{GC}[\rho]}{\delta \rho(\mathbf{r})} - \mu = 0 \qquad (4.1.7)$$

Equivalently,

$$\frac{\delta F_{GC}[\rho]}{\delta \rho(\mathbf{r})} + v(\mathbf{r}) - \mu = 0 \qquad (4.1.8)$$

In the last equation, it has been assumed that $F_{GC}[\rho]$ is differentiable [see Englisch and Englisch (1984a, b) for discussion of the validity of this assumption].

We now show that (4.1.4) is not true in general [see Blaizot and Ripka 1986, p. 410)]; nevertheless, it is valid for atomic and molecular systems. Consider a trial $\hat{\Gamma}$ that describes a ground state with the average number of electrons $N + \Delta N = \text{Tr }\hat{\Gamma}\hat{N}$. Then we have from (4.1.4),

$$E^0(N) - \mu N \leq E^0(N + \Delta N) - \mu \cdot (N + \Delta N) \tag{4.1.9}$$

Similarly, from another $\hat{\Gamma}$ with $\text{Tr }\hat{\Gamma}\hat{N} = N - \Delta N$,

$$E^0(N) - \mu N \leq E^0(N - \Delta N) - \mu \cdot (N - \Delta N) \tag{4.1.10}$$

It follows from (4.1.9) that if $E^0(N)$ is a differentiable function,

$$\mu = \frac{\partial E^0(N)}{\partial N} \tag{4.1.11}$$

Adding (4.1.9) and (4.1.10), we obtain

$$E^0(N + \Delta N) + E^0(N - \Delta N) - 2E^0(N) \geq 0 \tag{4.1.12}$$

which demands that the function $E^0(N)$ be convex (see Appendix B). If $E^0(N)$ is twice differentiable, then (4.1.12) implies

$$\frac{\partial^2 E^0(N)}{\partial N^2} \geq 0 \tag{4.1.13}$$

Conditions (4.1.11) and (4.1.12) or (4.1.13) are also sufficient to guarantee the validity of (4.1.9). See the theorem before Equation (B.13) in Appendix B.

Setting $\Delta N = 1$ in (4.1.12), we see that

$$E^0(N + 1) - E^0(N) \geq E^0(N) - E^0(N - 1) \tag{4.1.14}$$

or

$$I(N + 1) \leq I(N) \tag{4.1.15}$$

where $I(N)$ is the ionization potential of the N-electron ground state. Equation (4.1.15) states that successive ionization potentials are not decreasing (for fixed external potential).

For atoms and molecules, no counterexample is known to (4.1.15), although a first-principles proof has never been given. As examples, in Table 4.1 we list all of the successive ionization potentials for the oxygen and carbon atoms; in Figure 4.1 we plot the energy of oxygen as a function of the number of electrons, determined from the formula

$$E(N) = -\sum_{M=1}^{N} I(M) \tag{4.1.16}$$

Phillips and Davidson (1984) give examples of nonconvex $E^0(N)$, though not for electronic systems.

Summarizing the foregoing, we have shown that for systems for which $E^0(N)$ is convex, including atoms and molecules, the zero-temperature limit of grand-canonical-ensemble theory exists, as manifest in (4.1.4),

4.1 THE CHEMICAL POTENTIAL

Table 4.1 Ionization Energies for Carbon and Oxygen

Species	Ionization Potential (eV)	Species	Ionization Potential (eV)
C^-	1.12	O^-	1.47
C	11.26	O	13.61
C^+	24.38	O^+	35.15
C^{2+}	47.86	O^{2+}	54.93
C^{3+}	64.48	O^{3+}	77.39
C^{4+}	391.99	O^{4+}	113.87
C^{5+}	489.84	O^{5+}	138.08
		O^{6+}	739.08
		O^{7+}	871.12

(4.1.5), and (4.1.6). The chemical potential in these equations is the zero-temperature limit of the chemical potential defined for the finite-temperature grand canonical ensemble,

$$\mu = \left(\frac{\partial A}{\partial N}\right)_{\theta, v(\mathbf{r})} = \left(\frac{\partial E}{\partial N}\right)_{\theta, v(\mathbf{r})} - \theta \left(\frac{\partial S}{\partial N}\right)_{\theta, v(\mathbf{r})} \quad (4.1.17)$$

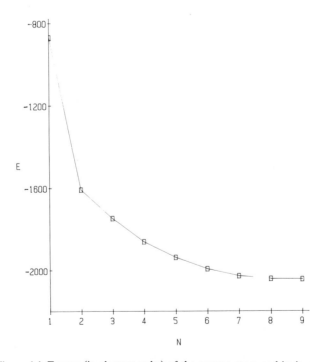

Figure 4.1 Energy (in electron volts) of the oxygen atom and its ions.

where A is the Helmholtz free energy. In this book we generally use the symbol μ to designate this zero-temperature limit, writing

$$\mu = \left(\frac{\partial E}{\partial N}\right)_v = \lim_{\theta \to 0} \left(\frac{\partial A}{\partial N}\right)_{\theta, v(\mathbf{r})} \quad (4.1.18)$$

and we simplify the notation in (4.1.6), (4.1.7), and (4.1.8) to give

$$\delta\{E[\rho] - \mu N[\rho]\} = 0 \quad (4.1.19)$$

and

$$\mu = \left[\frac{\delta E}{\delta \rho}\right]_v = v(\mathbf{r}) + \frac{\delta F[\rho]}{\delta \rho(\mathbf{r})} \quad (4.1.20)$$

Equations (4.1.19) and (4.1.20) are the basic working equations of the ground-state density-functional theory for atoms, molecules, and solids.

4.2 Physical meaning of the chemical potential

The chemical potential of density-functional theory measures the escaping tendency of an electronic cloud. It is a constant, through all space, for the ground state of an atom, molecule, or solid, and equals the slope of the E versus N curve at constant $v(\mathbf{r})$. The analogy with the chemical potential of ordinary macroscopic thermodynamics is clear and useful.

The chemical potential is the negative of the electronegativity of Pauling and Mulliken (Parr, Donnelly, Levy, and Palke 1978). For, the three-point finite-difference approximation to $\partial E/\partial N$ for a species S is

$$\mu \approx -\frac{I+A}{2} \quad (4.2.1)$$

where $I = E_S^+ - E_S$ and $A = E_S - E_S^-$ are respectively the ionization potential and electron affinity for the species. And Mulliken's formula for electronegativity (Mulliken 1934) is

$$\chi_M = \frac{I+A}{2} \quad (4.2.2)$$

Therefore

$$\mu \approx -\chi_M \quad (4.2.3)$$

and the chemical potential concept is the same as the electronegativity concept.

Mulliken's beautiful argument for (4.2.2) is as follows. Given two species, S and T, which is the more electronegative? The energy required for S to take an electron from T is $I_T - A_S$, while the energy required for T to take an electron from S is $I_S - A_T$. If the two requirements were

the same, we would have equal electronegativities:
$$I_T - A_S = I_S - A_T, \qquad I_S + A_S = I_T + A_T$$
The factor $\frac{1}{2}$ is arbitrary.

Density-functional theory is, then, a theory of ground (equilibrium) electronic states in which the electronegativity of chemistry plays in the basic variational principle (4.1.6) just the role that the energy plays in the basic variational principle (1.2.8) of wave-function theory. This result demands attention to the density-functional theory as a description of chemistry. In this description, the chemical potential is the key concept; we shall find in the next chapter that the derivatives of the chemical potential are of comparable importance.

4.3 Detailed consideration of the grand canonical ensemble near zero temperature

In order to have the variational equation (4.1.20), we have seen in §4.1 that we should take the $\theta \to 0$ limit of the grand canonical theory described in §§3.6 and 4.1. We now carefully examine this limiting process. Then we determine the ground-state energy for nonintegral N. The argument derives from Gyftopoulos and Hatsopoulos (1965) and Perdew, Parr, Levy, and Balduz (1982); see also Linderberg (1977).

To see how the analysis goes, it is most helpful first to consider a particular model system, a species that can exist in only three states: the neutral ground state with energy E_0 and number of electrons N_0, a positive-ion ground state with energy $E_0 + I$ and number of electrons $N_0 - 1$, and a negative-ion ground state with energy $E_0 - A$ and number of electrons $N_0 + 1$. All states are assumed nondegenerate. At inverse temperature β and chemical potential μ, the grand canonical partition function will be

$$Z(\beta, \mu) = e^{-\beta(E_0 - \mu N_0)}[1 + e^{-\beta(-A-\mu)} + e^{-\beta(I+\mu)}] \qquad (4.3.1)$$

Consequently we have, from standard formulas for ensemble averages,

$$\bar{N} = \frac{1}{\beta}\left[\frac{\partial \ln Z}{\partial \mu}\right]_\beta = N_0 + \left[\frac{e^{\beta(A+\mu)} - e^{-\beta(I+\mu)}}{1 + e^{\beta(A+\mu)} + e^{-\beta(I+\mu)}}\right]. \qquad (4.3.2)$$

Also

$$\bar{E} = -\left[\frac{\partial \ln Z}{\partial \beta}\right]_\mu = E_0 - \left[\frac{Ae^{\beta(A+\mu)} - Ie^{-\beta(I+\mu)}}{1 + e^{\beta(A+\mu)} + e^{-\beta(I+\mu)}}\right] \qquad (4.3.3)$$

and

$$\bar{S} = k[\ln Z + \beta\bar{E} - \beta\bar{N}\mu]$$
$$= k\left\{\ln(1 + e^{\beta(A+\mu)} + e^{-\beta(I+\mu)}) - \left[\frac{\beta(A+\mu)e^{\beta(A+\mu)} - \beta(I+\mu)e^{-\beta(I+\mu)}}{1 + e^{\beta(A+\mu)} + e^{-\beta(I+\mu)}}\right]\right\}$$
$$(4.3.4)$$

These formulas provide the various average properties for any β and μ. To obtain them for any β and $\bar{N}$, one may eliminate μ using (4.3.2).

Suppose first that $\bar{N} = N_0$—the system has an integral number of electrons. Then from (4.3.2) it immediately follows that $e^{\beta(A+\mu)} = e^{-\beta(I+\mu)}$, or $I + A + 2\mu = 0$, or

$$\mu_0 = -\frac{I+A}{2} \quad [\bar{N} = N_0] \tag{4.3.5}$$

This is just the Mulliken formula of (4.2.2), and it is valid for any temperature.

When $\bar{N} \neq N_0$, (4.3.2) can be rewritten as a quadratic equation and solved, as follows. Let

$$\mu = \mu_0 + \Delta\mu, \qquad \bar{N} = N_0 + \Delta N,$$
$$y = e^{-\beta[(I-A)/2]}, \qquad x = e^{\beta \Delta\mu} \tag{4.3.6}$$

Then (4.3.2) becomes

$$\Delta N = \frac{y(x^2 - 1)}{y(x^2 + 1) + x} \tag{4.3.7}$$

the solution of which is

$$x = \frac{\Delta N \pm \{(\Delta N)^2 + 4y^2[1 - (\Delta N)^2]\}^{1/2}}{2y(1 - \Delta N)} = e^{\beta \Delta\mu} \tag{4.3.8}$$

This gives $\Delta\mu$ and hence μ for any β and ΔN or $\bar{N}$. For $\Delta N = 0$, for example, we find $x = \pm 1 = +1$, $\Delta\mu = 0$, $\mu = \mu_0$, recovering (4.3.5) for any β. For $\Delta N \neq 0$, the zero-temperature limits are different for ΔN plus and ΔN minus. Noting that $\lim_{\beta \to \infty} y = 0$ assuming convexity of $\bar{E}(N)$, we find, in the zero-temperature limit:

$$\Delta N > 0: \qquad x = \frac{\Delta N + \Delta N\left\{1 + \frac{4y^2}{(\Delta N)^2}[1 - (\Delta N)^2]\right\}^{1/2}}{2y(1 - \Delta N)} \to \frac{\Delta N}{y(1 - \Delta N)} \tag{4.3.9}$$

$$\Delta N < 0: \qquad x = \frac{\Delta N - \Delta N\left\{1 + \frac{4y^2}{(\Delta N)^2}[1 - (\Delta N)^2]\right\}^{1/2}}{2y(1 - \Delta N)} \to \frac{-y(1 + \Delta N)}{\Delta N} \tag{4.3.10}$$

Consequently, we have for very high β or very low θ,

$$1 > \Delta N > 0: \qquad \Delta\mu = +\left(\frac{I-A}{2}\right) + \frac{1}{\beta}\ln\left(\frac{\Delta N}{1 - \Delta N}\right) \tag{4.3.11}$$

$$\mu = -A + \frac{1}{\beta}\ln\left(\frac{\Delta N}{1-\Delta N}\right) \qquad (4.3.12)$$

$$-1 < \Delta N < 0: \quad \Delta\mu = -\left(\frac{I-A}{2}\right) + \frac{1}{\beta}\ln\left[-\frac{(1+\Delta N)}{\Delta N}\right] \qquad (4.3.13)$$

$$\mu = -I + \frac{1}{\beta}\ln\left[-\frac{(1+\Delta N)}{\Delta N}\right] \qquad (4.3.14)$$

From (4.3.5), (4.3.12) and (4.3.14), we therefore infer the zero-temperature limits

$$\mu_{\theta=0} = \begin{cases} -I, & N_0 - 1 < N < N_0 \\ -\dfrac{I+A}{2}, & N = N_0 \\ -A, & N_0 + 1 > N > N_0 \end{cases} \qquad (4.3.15)$$

To obtain the ground-state energy, we insert (4.3.11) and (4.3.13) into (4.3.3), and take the limit $\beta \to \infty$. The result, writing N for $\bar{N}$, is

$$E^0(N) = \begin{cases} E^0(N_0) - I(N - N_0), & N_0 - 1 < N < N_0 \\ E^0(N_0), & N = N_0 \\ E^0(N_0) - A(N - N_0), & N_0 + 1 > N > N_0 \end{cases} \qquad (4.3.16)$$

For this three-state model, the energy at zero temperature is a continuous series of straight-line segments, as shown in Fig. 4.1. The energy for a nonintegral number of electrons is given by linear interpolation between the values for integral numbers of electrons (Perdew, Parr, Levy, and Balduz, 1982).

As can be readily verified by calculation, for typical atomic or molecular values of I and A, it requires a quite high temperature to cause much curvature to appear in $E(N)$ curves. We will return to the curvature at the end of this section. Equations (4.3.15) and (4.3.16), although derived from a three-state model, remain true in general, as will be established below. The essence of the zero-temperature grand-canonical-ensemble theory is captured in (4.3.15) and (4.3.16).

In this three-state model system, all states have been assumed to be nondegenerate. But there is little effect if degeneracy is present. If the degeneracies of positive ion, neutral species, and negative ion are g_+, g_0, and g_-, (4.3.2) is replaced by

$$\Delta N = \frac{g_- e^{\beta(A+\mu)} - g_+ e^{-\beta(I+\mu)}}{g_0 + g_- e^{\beta(A+\mu)} + g_+ e^{-\beta(I+\mu)}} \qquad (4.3.17)$$

Equation (4.3.5) is replaced by

$$\mu_0 = -\frac{I+A}{2} + \frac{1}{2\beta}\ln\frac{g_+}{g_-} \quad [\text{large } \beta, N = N_0] \qquad (4.3.18)$$

This displays a temperature dependence in μ_0, but in the $\beta \to \infty$ limit this disappears and (4.3.5) results. For N nonintegral, (4.3.12) and (4.3.14) are similarly modified by terms vanishing as $\beta \to \infty$ [Perdew 1985], so that (4.3.15) and (4.3.16) remain valid in the presence of degeneracies.

We now describe the zero-temperature limit of the full grand-canonical-ensemble theory, into which enter ground and excited states of all species, and ions of all possible positive and negative charges. The only assumption is the convexity assumption of (4.1.15). For the general case, the partition function becomes

$$Z = \sum_N \sum_j g_{Nj} e^{-\beta(E_{Nj} - \mu N)} \tag{4.3.19}$$

where g_{Nj} are degeneracies, and in place of (4.3.2)

$$\bar{N} = \frac{1}{Z} \sum_N \sum_j g_{Nj} N e^{-\beta(E_{Nj} - \mu N)} \tag{4.3.20}$$

In the $\theta \to 0$ or $\beta \to \infty$ limit, only the ground states of the various species survive. Calling the ground-state energy and degeneracy for N electrons E_N^0 and g_N^0, we get for very large β and integral N

$$\mu_0 = -\frac{E_{N-1}^0 - E_{N+1}^0}{2} + \frac{1}{2\beta} \ln \frac{g_{N-1}^0}{g_{N+1}^0} \quad [\text{large } \beta, N = N_0] \tag{4.3.21}$$

provided only that the convexity condition of (4.1.5) is satisfied.

Similarly, if $N = N_0 + \Delta N$, where N_0 is some integer and ΔN is positive, the result should be

$$\mu_{\theta=0} = E_{N_0}^0 - E_{N_0+1}^0, \quad N_0 + 1 > N > N_0 \tag{4.3.22}$$

If ΔN is negative, on the other hand,

$$\mu_{\theta=0} = E_{N_0}^0 - E_{N_0-1}^0, \quad N_0 - 1 < N < N_0 \tag{4.3.23}$$

Summarizing, we have

$$\mu_{\theta=0} = \begin{cases} -I, & N_0 - 1 < N < N_0 \\ -\dfrac{I+A}{2}, & N = N_0 \\ -A, & N_0 + 1 > N < N_0 \end{cases} \tag{4.3.24}$$

in accordance with (4.3.15). $E(N)$ as a function of N at $\beta = \infty$ is a series of straight-line segments connecting integral-N values.

To prove the foregoing, we depart from the literature by formulating the problem directly at zero temperature. Instead of using a Lagrange multiplier μ as in (4.1.4), we carry out the energy minimization with proper normalization built in at the start. Thus, we write the ground-state energy for an arbitrary number of electrons N as

$$E^0(N) = \text{Min} \left\{ \sum_M p(M) \cdot E^0(M) \right\} \quad (4.3.25)$$

with constraints

$$\sum_M p(M) = 1 \quad (4.3.26)$$

$$\sum_M M p(M) = N \quad (4.3.27)$$

$$1 \geq p(M) \geq 0 \quad \text{for all } M \quad (4.3.28)$$

where $p(M)$ and $E^0(M)$ are the probability and ground-state energy for the system in same external potential $v(\mathbf{r})$ with M electrons. M is an integer and the sum is over all positive integers. The minimization in (4.3.25) is equivalent to the one in (4.1.4), because we now use for the evaluation of $\text{Tr}\,\hat{\Gamma}\hat{H}$ the eigenstates of the Hamiltonian $\hat{H}$ in which $\hat{\Gamma}$ is diagonal with eigenvalues $p(M)$. The excited states all have zero probabilities, since $E^0(M)$ is the minimum for the expectation value of $\hat{H}$ for states with M electrons.

We now solve for $p(M)$ and $E^0(N)$. This problem is a minimization of a linear function of $p(M)$ with linear constraints; it cannot be solved by the usual Lagrange multiplier method. Instead, the minimum is reached at the boundary of the domain of admissible $p(M)$. Suppose N is between the integers $J-1$ and J. Then we can rewrite (4.3.26) and (4.3.27) as

$$p(J-1) + p(J) = 1 - \sum_L{}' p(L) \quad (4.3.29)$$

$$(J-1) \cdot p(J-1) + J \cdot p(J) = N - \sum_L{}' L \cdot p(L) \quad (4.3.30)$$

which can be solved for $p(J-1)$ and $p(J)$,

$$p(J-1) = J - N - \sum_L{}' p(L) \cdot (J - L) \quad (4.3.31)$$

$$p(J) = N - J + 1 - \sum_L{}' p(L) \cdot (L - J + 1) \quad (4.3.32)$$

where, for the primed summations,

$$\sum_L{}' = \sum_{L \neq J, J-1} \quad (4.3.33)$$

Inserting (4.3.31) and (4.3.32) into (4.3.25), we have

$$E^0(N) = (N - J + 1) \cdot E^0(J) + (J - N) E^0(J - 1)$$

$$+ \text{Min} \left\{ \sum_L{}' p(L)[-(L-J+1)E^0(J) - (J-L)E^0(J-1) + E^0(L)] \right\}$$

$$(4.3.34)$$

in which $p(L)$ needs only to satisfy (4.3.28).

For each L, the value of $p(L)$ that minimizes (4.3.34) depends on the sign of the term in square brackets in (4.3.34), the quantity

$$g(L) = -(L - J + 1)E^0(J) - (J - L)E^0(J - 1) + E^0(L) \quad (4.3.35)$$

Since the boundary values of $p(L)$ are 0 and 1, to attain the minimum of $E^0(N)$ it is necessary that

$$\left.\begin{array}{ll} p(L) = 0 & \text{if} \quad g(L) > 0 \\ p(L) = 1 & \text{if} \quad g(L) < 0 \\ 0 \leq p(L) \leq 1 & \text{if} \quad g(L) = 0 \end{array}\right\} \quad (4.3.36)$$

It turns out that for the atoms and molecules we always have

$$g(L) \geq 0 \quad (4.3.37)$$

which comes from the convexity of $E^0(M)$ in (4.1.15), or equivalently [see Equation (B.1) in Appendix B],

$$\alpha E^0(M_1) + \beta E^0(M_2) - E^0(\alpha M_1 + \beta M_2) \geq 0 \quad (4.3.38)$$

for $\alpha \geq 0$, $\beta \geq 0$, $\alpha + \beta = 1$, and $\alpha M_1 + \beta M_2$ also an integer. For $L > J - 1$,

$$g(L) = (L - J + 1)\left[\frac{E^0(L)}{L - J + 1} + \frac{(L - J)E^0(J - 1)}{L - J + 1} - E(J)\right]$$

$$\geq 0 \quad (4.3.39)$$

using (4.3.38). For $L < J - 1$, one finds

$$g(L) = (J - L)\left[\frac{E^0(L)}{J - L} + \frac{(J - L - 1)E^0(J)}{J - L} - E(J - 1)\right]$$

$$\geq 0 \quad (4.3.40)$$

using (4.3.38).

Applying (4.3.39) and (4.3.40), (4.3.36), and (4.3.34), we obtain

$$E^0(N) = (N - J + 1)E^0(J) + (J - N)E^0(J - 1) \quad (4.3.41)$$

in agreement with (4.3.16). If (4.3.38) is a strict inequality, that is, if $E^0(M)$ is strictly convex, we further have, from (4.3.31) and (4.3.32),

$$\left.\begin{array}{l} p(J - 1) = J - N, \\ p(J) = N - J + 1, \\ p(L) = 0, \, L \neq J, J - 1 \end{array}\right\} \quad (4.3.42)$$

If $E^0(M)$ is only convex, then the set of $p(M)$ that minimizes (4.2.34) is not unique. Equation (4.3.41) still holds, however, as does the chemical potential expression

$$\mu = \frac{\partial E}{\partial N} = E^0(J) - E^0(J-1) \quad \text{for } J > N > J-1 \quad (4.3.43)$$

In summary, at the limit of zero temperature the grand-canonical-ensemble theory gives $E^0(N)$ as a continuous linear interpolation among $E^0(M)$ values, where M is integral. This assumes $E^0(M)$ to be convex at all positive integers, which is the case for atoms and molecules.

To end this section, we briefly consider another quantity of interest: the curvature of E as a function of N at 0 K,

$$2\eta = \left(\frac{\partial^2 E}{\partial N^2}\right)_{\theta \to 0} = \lim_{\beta \to \infty} \left[\frac{\frac{\partial}{\partial \mu}(\partial E/\partial N)_\beta}{(\partial N/\partial \mu)_\beta}\right] \quad (4.3.44)$$

This is found to be

$$2\eta = \begin{cases} 0, & N < N_0 \\ \lim_{\beta \to \infty}[\eta_0 e^{\beta \eta_0}] = \infty, & N = N_0 \\ 0, & N > N_0 \end{cases} \quad (4.3.45)$$

where

$$\eta_0 = \frac{I-A}{2} \quad (4.3.46)$$

Alternatively, we could define

$$\eta = \left(\frac{I-A}{2}\right)\delta(N-N_0) = \frac{1}{2}\left(\frac{\partial^2 E}{\partial N^2}\right)_{T=0} \quad (4.3.47)$$

which would accord with the identity

$$\int_{N_0-\Delta}^{N_0+\Delta}\left(\frac{\partial^2 E}{\partial N^2}\right)dN = \left(\frac{\partial E}{\partial N}\right)_{N_0+\Delta} - \left(\frac{\partial E}{\partial N}\right)_{N_0-\Delta} \quad (4.3.48)$$

since $(\partial^2 E/\partial N^2)_{\theta=0}$ is zero for the straight-line segments for $N < N_0$ and $N > N_0$. The quantity $\eta_0 = (I-A)/2$ turns out to be very important; it is the *absolute hardness* of a chemical species (Parr and Pearson 1983). Hardness will be discussed at length in the next chapter.

4.4 The chemical potential for a pure state and in the canonical ensemble

We now consider the concept of chemical potential in the canonical ensemble and for a pure state, both cases in which the number of electrons is integral.

As was described in §3.5, in the canonical ensemble one has a system of interest of N_0 particles and minimizes a functional $A[\rho]$ over densities integrating to N_0. In the $\theta = 0$ limit, which is our present concern, one minimizes $E[\rho]$ subject to proper normalization of ρ; that is,

$$E^0(N) = \underset{\rho(\mathbf{r})}{\text{Min}}\, E_C[\rho] \tag{4.4.1}$$

where

$$E_C[\rho] = F_C[\rho] + \int v(\mathbf{r})\rho(\mathbf{r})\,d\mathbf{r} \tag{4.4.2}$$

with

$$F_C[\rho] = \underset{\hat{\Gamma}_N \to \rho}{\text{Min}}\, \{\text{tr}\,[\hat{\Gamma}_N(\hat{T} + \hat{V}_{ee})]\} \tag{4.4.3}$$

These are the zero-temperature limits of (3.5.13)–(3.5.15). The variational principle of (4.4.1) follows from (2.3.28), the minimizing density being the ground-state density if the ground state is nondegenerate, or an arbitrary linear combination of the ground-state densities if it is degenerate. The universal functional $F_C[\rho]$ was proposed by Valone (1980a) and Lieb (1982) to extend the Levy constrained search to all mixed states in the N-electron Hilbert space.

In the pure ground-state theory of §3.4, $F[\rho]$ of (3.4.5) is the universal functional instead of (4.4.3). The variational principle (3.4.8) is of the same form as (4.4.1). In the following discussion, these two cases are dealt with at the same time, $E[\rho]$ standing for the energy functional for both canonical ensemble and pure state.

How do we implement the variation (4.4.1) within all the densities integrating to integral N? We cannot directly use the Lagrange multiplier technique, because the constraint-free variation $\delta\{E[\rho] - \lambda[\int \rho\,d\mathbf{r} - N]\}$, where λ is the supposed Lagrange multiplier, would require $E[\rho]$ to be defined for nonintegral numbers of electron (see Appendix A), which is not the case in either the canonical ensemble (4.4.2) or the pure state (3.4.9). However, we can explicitly impose the normalization by writing

$$\rho(\mathbf{r}) = \frac{Ng(\mathbf{r})}{\int g(\mathbf{r}')\,d\mathbf{r}'} \tag{4.4.4}$$

This allows $g(\mathbf{r})$ to be any nonnegative function, while $\rho(\mathbf{r})$ always still integrates to N, in the same way that normalization of wave functions is taken care of in (1.2.1) and (1.2.3). Thus, we have the ground-state variational principle

$$\frac{\delta E[\rho]}{\delta g(\mathbf{r})} = \int \left(\frac{\delta E}{\delta \rho(\mathbf{r}')}\right)_N \frac{\delta \rho(\mathbf{r}')}{\delta g(\mathbf{r})}\,d\mathbf{r}' = 0 \tag{4.4.5}$$

where we have used the chain rule, (A.24) of Appendix A. The subscript N here indicates that the functional differentiation is performed with N fixed. Using (4.4.4), we get

$$\frac{\delta\rho(\mathbf{r}')}{\delta g(\mathbf{r})} = \frac{N}{\int g(\mathbf{r}'')\, d\mathbf{r}''}\left[\delta(\mathbf{r}' - \mathbf{r}) - \frac{g(\mathbf{r}')}{\int g(\mathbf{r}'')\, d\mathbf{r}''}\right] \quad (4.4.6)$$

so that (4.4.5) gives

$$\left(\frac{\delta E}{\delta\rho(\mathbf{r})}\right)_N = \int \left(\frac{\delta E}{\delta\rho(\mathbf{r}')}\right)_N \frac{g(\mathbf{r}')}{\int g(\mathbf{r}'')\, d\mathbf{r}''}\, d\mathbf{r}' \quad (4.4.7)$$

The right-hand side of (4.4.7) is a constant independent of $\mathbf{r}$. This constant cannot be determined by (4.4.7), because any value makes (4.4.7) an identity. Therefore,

$$\left(\frac{\delta E}{\delta\rho(\mathbf{r})}\right)_N = C \quad (4.4.8)$$

with C an arbitrary constant.

Contained in (4.4.8) is the equivalent of the Schrödinger equation for the ground state. Comparing with (4.1.7), we see that chemical potential μ does not occur in (4.4.8) and its existence in the theory at this stage is solely due to the grand-canonical ensemble extension of the ground-state theory.

The chemical potential concept can also be identified in the canonical ensemble theory, however. The simple way to do this is to follow the procedure that is used for canonical ensemble theory in statistical mechanics, the finite-difference method (see p. 41 of Ashcroft and Mermin 1976). Thus, for the system of N electrons, we take [and note that this accords with (4.3.15)]

$$\mu^- = \frac{E(N-1) - E(N)}{-1} = -I \quad \text{(slope when the system gives up an electron)} \quad (4.4.9)$$

$$\mu^+ = E(N+1) - E(N) = -A \quad \text{(slope when the system adds an electron)} \quad (4.4.10)$$

$$\mu^0 = -\frac{I+A}{2} \quad \text{(slope when an electron is neither given up nor added)} \quad (4.4.11)$$

Also, the curvature of $E(N)$ becomes

$$\eta = \tfrac{1}{2}[E(N+1) - E(N) - (E(N) - E(N-1))] = \tfrac{1}{2}(I - A) \quad (4.4.12)$$

Note the difference between this formula and (4.3.45).

Another way to recover the chemical potential is to extend the minimization (4.4.1) to *all* different *integral N*. In this case the Lagrange multiplier μ becomes fuzzy if one makes no further assumptions (Parr and Bartolotti 1983). Consider at what density the quantity $E[\rho] - \mu N[\rho]$ will in fact be a minimum, if only trial densities normalized to integers are used. If $\int \rho \, d\mathbf{r} = N_0$, the minimum value is $E[\rho_0(N_0)] - \mu N_0$, where $E[\rho_0(N_0)]$ is the ground-state energy for N_0 electrons. Similarly, if $\int \rho \, d\mathbf{r} = N_0 + 1$, the minimum value will be $E[\rho_0(N_0 + 1)] - \mu(N_0 + 1)]$; if $\int \rho \, d\mathbf{r} = N_0 - 1$, it will be $E[\rho_0(N_0 - 1)] - \mu(N_0 - 1)$. Of these three numbers, the smallest depends on the value of μ and on the values of the three energies. The smallest will be the first, $E[\rho_0(N_0)] - \mu N_0$, if μ is in the range

$$A \leq -\mu \leq I \qquad (4.4.13)$$

where I and A respectively are the ionization potential and electron affinity of the species with $N = N_0$. Note that (4.4.9) to (4.4.11) give μ values in this range. With any value of μ satisfying (4.4.13), the minimum of $E[\rho] - \mu N[\rho]$ over all ρ integrating to different integers will give $E[\rho_0(N_0)] - \mu N_0$. Equation (4.4.13) also implies that $E^0(N)$ needs to be convex [see (4.1.15)]. Trial ρ may in fact be limited to those integrating to integers.

4.5 Discussion

Although the zero-temperature limit of the grand canonical ensemble provides a natural definition for the ground state of nonintegral-N systems, the continuous straight-line segments for $E^0(N)$ need not be correct for a species imbedded in some environment, or for an atom or functional group in a molecule. In most actual situations we can reasonably imagine interpolating among values of ground-state energies for integral numbers of electrons, thereby extending the minimum search of (4.4.1) to arbitrary densities. Then we will be able to use Lagrange-multiplier techniques,

$$\delta\left\{E[\rho] - \mu \int \rho \, d\mathbf{r}\right\} = 0 \qquad (4.5.1)$$

or

$$\mu = \frac{\delta E[\rho]}{\delta \rho(\mathbf{r})} \qquad (4.5.2)$$

For example, a parabolic fit of the three points $E^0(N-1)$, $E^0(N)$, and $E^0(N+1)$ for integral N, gives $\mu = -(I+A)/2$, in agreement with (4.4.11), and $\eta = (I-A)/2$ in agreement with (4.4.12).

In fact, (4.5.2) is more convenient to handle than (4.4.8), because

4.5 THE CHEMICAL POTENTIAL

$(\delta E/\delta \rho(\mathbf{r}))_N$ is hard to evaluate. Thus, in many practical applications of density functional theory, instead of (4.4.8), (4.5.2) is used, where some interpolation or other is implicit in the treatment of $E[\rho]$.

The difference between $\delta E/\delta \rho$ and $(\delta E/\delta \rho)_N$ is itself worth some elaboration (Parr and Bartolotti 1983). One has $E = E[\rho, v]$. Writing

$$\rho = N\sigma \qquad (4.5.3)$$

where σ is a *shape factor*,

$$\int \sigma(\mathbf{r}) \, d\mathbf{r} = 1 \qquad (4.5.4)$$

one has $E = E[N, \sigma, v]$ as well. Therefore we may write

$$\delta E = \int \left[\frac{\delta E}{\delta \rho(\mathbf{r})}\right]_v \delta \rho(\mathbf{r}) \, d\mathbf{r} + \int \left[\frac{\delta E}{\delta v(\mathbf{r})}\right]_\rho \delta v(\mathbf{r}) \, d\mathbf{r}$$

$$= \int \left[\frac{\delta E}{\delta \rho(\mathbf{r})}\right]_v [\sigma(\mathbf{r}) \, \delta N + N \, \delta \sigma(\mathbf{r})] \, d\mathbf{r} + \int \left[\frac{\delta E}{\delta v(\mathbf{r})}\right]_\rho \delta v(\mathbf{r}) \, d\mathbf{r} \qquad (4.5.5)$$

and also

$$\delta E = \left(\frac{\partial E}{\partial N}\right)_{\sigma,v} \delta N + \int \left[\frac{\delta E}{\delta \sigma(\mathbf{r})}\right]_{N,v} \delta \sigma(\mathbf{r}) \, d\mathbf{r} + \int \left[\frac{\delta E}{\delta v(\mathbf{r})}\right]_\rho \delta v(\mathbf{r}) \, d\mathbf{r} \qquad (4.5.6)$$

Subtracting these equations we find, since δN and $\delta v(r)$ are independent,

$$\left(\frac{\partial E}{\partial N}\right)_{\sigma,v} = \int \left[\frac{\delta E}{\delta \rho(\mathbf{r})}\right]_v \sigma(\mathbf{r}) \, d\mathbf{r} \qquad (4.5.7)$$

and

$$\int \left\{ N\left[\frac{\delta E}{\delta \rho(\mathbf{r})}\right]_v - \left[\frac{\delta E}{\delta \sigma(\mathbf{r})}\right]_{N,v} \right\} \delta \sigma(\mathbf{r}) \, d\mathbf{r} = 0 \qquad (4.5.8)$$

Equation (4.5.7) will not here concern us further.

Does (4.5.8) mean that the term in curly brackets is identically zero? No, because (4.5.4) implies that

$$\int \delta \sigma(\mathbf{r}) \, d\mathbf{r} = 0 \qquad (4.5.9)$$

Together, (4.5.8) and (4.5.9) imply that

$$N\left[\frac{\delta E}{\delta \rho(\mathbf{r})}\right]_v - \left[\frac{\delta E}{\delta \sigma(\mathbf{r})}\right]_{N,v} = \text{constant} \qquad (4.5.10)$$

or,

$$\left[\frac{\delta E}{\delta \rho(\mathbf{r})}\right]_{N,v} = \left[\frac{\delta E}{\delta \rho(\mathbf{r})}\right]_v + \text{constant} \qquad (4.5.11)$$

where the constant cannot be evaluated without more information. The mathematical lemma used here is well known (Courant and Hilbert 1953, p. 201). Note that (4.5.11) is compatible with (4.4.8) and (4.5.2).

In this chapter we have discussed the concept of chemical potential from various angles: the zero-temperature limit of the grand-canonical-ensemble theory, the finite difference in the pure-state theory, the differential in the interpolation theory, and the fuzzy result from consideration of pure states in the Fock space. Equations (4.4.9)–(4.4.12) are clear-cut definitions for most practical purposes. The discontinuity of chemical potential at integral numbers of electrons turns out to be of major importance in calculations of band gaps of solids (Perdew and Levy 1983, Sham and Schlüter 1985. Gunnarsson and Schönhammer 1986, Schönhammer and Gunnarsson 1987, Shen, Bylander, and Kleinman 1988, Sham and Schlüter 1988).

Ensembles other than the ones we have considered are also of interest. For example, see the Tachibana (1989) proposal to incorporate constraints for an apparatus, affording the possibility of turning an excited state into a ground state and thereby making it accessible to ground-state theory.

5
CHEMICAL POTENTIAL DERIVATIVES

5.1 Change from one ground state to another

Before we proceed to the further development of density-functional theory, which will require much detailed argument, we pause to demonstrate its unique power by considering the equations governing the change from one ground state to another. A splendid relevance for chemistry will be revealed.

The fundamental differential expression of density-functional theory, for the change from one ground state to another for some electronic system, is the formula

$$dE = \mu \, dN + \int \rho(\mathbf{r}) \, dv(\mathbf{r}) \, d\mathbf{r}. \quad (5.1.1)$$

Note that this is a simple generalization of the first-order perturbation formula (1.5.6) to include change in electron number. The coefficient μ is the chemical potential in the theory, the quantity that was the subject of the previous chapter.

We now prove (5.1.1). For a change from one ground state to another (the equilibrium state in the grand canonical ensemble at 0 K), the total differential of the energy $E = E[N, v]$ will be

$$dE = \left(\frac{\partial E}{\partial N}\right)_v dN + \int \left[\frac{\delta E}{\delta v(\mathbf{r})}\right]_N dv(\mathbf{r}) \, d\mathbf{r} \quad (5.1.2)$$

This must be the same as the differential of $E = E[\rho]$ defined by (3.4.9),

$$dE = \int \left[\frac{\delta E}{\delta \rho(\mathbf{r})}\right]_v d\rho(\mathbf{r}) \, d\mathbf{r} + \int \left[\frac{\delta E}{\delta v(\mathbf{r})}\right]_\rho dv(\mathbf{r}) \, d\mathbf{r} \quad (5.1.3)$$

From (3.2.9) or (4.1.20), we find, however, that the ground-state $\rho(\mathbf{r})$ must satisfy the Euler equation

$$\left[\frac{\delta E}{\delta \rho(\mathbf{r})}\right]_v = \mu = \text{constant} \quad (5.1.4)$$

while from the first-order perturbation formula (1.5.6) we have

$$\left[\frac{\delta E}{\delta v(\mathbf{r})}\right]_\rho = \left[\frac{\delta E}{\delta v(\mathbf{r})}\right]_N = \rho(\mathbf{r}) \quad (5.1.5)$$

Inserting (5.1.4) and (5.1.5) into (5.1.3) proves (5.1.1); note that $dN = \int d\rho(\mathbf{r})\, d\mathbf{r}$. Comparing (5.1.1) and (5.1.2), we thus obtain

$$\mu = \left(\frac{\partial E}{\partial N}\right)_v \qquad (5.1.6)$$

in agreement with (4.1.11). Here, again, we have employed functional derivatives: $\delta E/\delta v$, $\delta E/\delta \rho$. For a summary of the mathematical definitions and properties of these and related quantities, see Appendix A.

Similarly, the change in chemical potential associated with a change in N and/or v is given by the formula

$$d\mu = \left(\frac{\partial \mu}{\partial N}\right)_v dN + \int \left[\frac{\delta \mu}{\delta v(\mathbf{r})}\right]_N dv(\mathbf{r})\, d\mathbf{r} \qquad (5.1.7)$$

Or, introducing the symbols

$$2\eta = \left(\frac{\partial \mu}{\partial N}\right)_v \qquad (5.1.8)$$

and

$$f(\mathbf{r}) = \left[\frac{\delta \mu}{\delta v(\mathbf{r})}\right]_N = \left[\frac{\partial \rho(\mathbf{r})}{\partial N}\right]_v \qquad (5.1.9)$$

we arrive at the important formula

$$d\mu = 2\eta\, dN + \int f(\mathbf{r})\, dv(\mathbf{r})\, d\mathbf{r} \qquad (5.1.10)$$

The second formula for $f(\mathbf{r})$ in (5.1.9) is a "Maxwell relation" following from the fact that dE is an exact differential.

The density-functional equation (5.1.10) is highly pertinent for chemical concepts; the quantities in it are the subject of the next several sections of this chapter. As already stated in §4.2, μ is the negative of the *electronegativity* of the species; first introduced in §4.3, η is the *hardness* of the species. Both μ and η are numbers, *global* properties of the species. The quantity $f(\mathbf{r})$ is a *local* property that depends on $\mathbf{r}$; we call it the *Fukui function* for the species (Parr and Yang 1984); μ and η are discussed in §5.2 and §5.3, $f(\mathbf{r})$ in §5.4 and §5.5.

Returning to the differential formula of (5.1.1), we note that understanding energy changes to second order will require understanding the dependence of μ and $\rho(\mathbf{r})$ on N and $v(\mathbf{r})$. The differential coefficients of μ are given by (5.1.8) and (5.1.9). For $\rho(\mathbf{r})$ we need (5.1.9) and one other quantity, the two-variable *linear response function*,

$$\left[\frac{\delta \rho(\mathbf{r})}{\delta v(\mathbf{r}')}\right]_N = \frac{\delta^2 E}{\delta v(\mathbf{r})\, \delta v(\mathbf{r}')} = \left[\frac{\delta \rho(\mathbf{r}')}{\delta v(\mathbf{r})}\right]_N. \qquad (5.1.11)$$

This important property has already been discussed in §1.5.

Before proceeding, we should emphasize that in this chapter, beginning with (5.1.1), the ground state energy $E[N, v]$ is *assumed* to be a continuous and differentiable function of the number of electrons N for a given external potential $v(\mathbf{r})$. The zero-temperature grand-canonical-ensemble definition of the ground state for a nonintegral number of electrons, which was discussed in detail in §4.3, provides the result that μ is equal to $-I$ or $-A$ according as $dN < 0$ or $dN > 0$, where I and A are the ionization potential and electron affinity. This definition does not fully support the use of (5.1.7) (see discussion at the end of §4.3)]. Other interpolation schemes may give a smooth E versus N curve (see §4.4 and also §7.4).

There are, however, two unambiguous ways to read (5.1.6) and (5.1.8), and other formulas involving differentiation with respect to N. One is through the finite-difference methods of (4.4.9)–(4.4.12). The other is to invoke the nonzero-temperature grand-canonical-ensemble description of §3.6, in which all quantities are continuous in the ensemble-average N.

The natural variables in (5.1.1) and (5.1.10) are N and $v(\mathbf{r})$. Just as in ordinary macroscopic thermodynamics, it is convenient to define new state functions that will be particularly appropriate for use when other variables are more pertinent for a problem at hand. Define F and Ω by

$$F = E - \int \rho(\mathbf{r})v(\mathbf{r})\,d\mathbf{r} \qquad (5.1.12)$$

$$\Omega = E - N\mu \qquad (5.1.13)$$

Then, one has, for the change of any one ground state into any other,

$$dF = \mu\,dN - \int v(\mathbf{r})\,d\rho(\mathbf{r})\,d\mathbf{r} = \int [\mu - v(\mathbf{r})]\,d\rho(\mathbf{r})\,d\mathbf{r} \qquad (5.1.14)$$

$$d\Omega = -N\,d\mu + \int \rho(\mathbf{r})\,dv(\mathbf{r})\,d\mathbf{r} \qquad (5.1.15)$$

To derive (5.1.14), take the differential of (5.1.12) and make use of (5.1.1). Equation (5.1.15) follows similarly. The function F has the $\rho(\mathbf{r})$ as its natural variables, while the functional Ω has μ and $v(\mathbf{r})$ as its natural variables.

The quantity F of (5.1.12) is precisely the Hohenberg–Kohn F of (3.2.4) or (3.4.5), since of course $E = T + V_{ee} + V_{ne} = F + V_{ne}$. The quantity Ω of (5.1.13) is the zero-temperature grand potential of (3.6.12) or (4.1.4). From (5.1.12), (5.1.13), and the Euler equation (3.2.9) it follows that the ground-state density satisfies

$$\Omega = F - \int \frac{\delta F}{\delta \rho(\mathbf{r})} \rho(\mathbf{r})\,d\mathbf{r} \qquad (5.1.16)$$

The linear differential expressions dE, dF and $d\Omega$ of (5.1.1), (5.1.14) and (5.1.15) are all exact differentials. We may therefore immediately write down a number of equalities, corresponding to the Maxwell relations of classical thermodynamics (Nalewajski and Parr 1982). From (5.1.6) we obtain, equating $\partial^2 E/\partial N\, \delta v(\mathbf{r})$ to $\partial^2 E/\delta v(\mathbf{r})\, \partial N$, just (5.1.9); by equating $\delta^2 E/\delta v(\mathbf{r})\, \delta v(\mathbf{r}')$ to $\delta^2 E/\delta v(\mathbf{r}')\, \delta v(\mathbf{r})$,

$$\left[\frac{\delta \rho(\mathbf{r})}{\delta v(\mathbf{r}')}\right]_N = \left[\frac{\delta \rho(\mathbf{r}')}{\delta v(\mathbf{r})}\right]_N. \tag{5.1.17}$$

We note that this last equality is just the formula (1.5.9) of perturbation theory, but we note also that the derivation from density-functional theory is simpler than the more ponderous route through (1.5.7) and (1.5.8).

From (5.1.14), one obtains similarly,

$$\frac{\delta(\mu - v(\mathbf{r}))}{\delta \rho(\mathbf{r}')} = \frac{\delta(\mu - v(\mathbf{r}'))}{\delta \rho(\mathbf{r})} \tag{5.1.18}$$

From (5.1.15),

$$\left[\frac{\delta N}{\delta v(\mathbf{r})}\right]_\mu = -\left[\frac{\partial \rho(\mathbf{r})}{\partial \mu}\right]_v \tag{5.1.19}$$

and

$$\left[\frac{\delta \rho(\mathbf{r})}{\delta v(\mathbf{r}')}\right]_\mu = \left[\frac{\delta \rho(\mathbf{r}')}{\delta v(\mathbf{r})}\right]_\mu \tag{5.1.20}$$

Equation (5.1.20) is a response function that is related to but essentially different from the response function of (5.1.17). In a sense, the quantities in (5.1.18) are inverse response functions; see §5.5 below.

5.2 Electronegativity and electronegativity equalization

Consider again the physical meaning of the chemical potential μ. In §§4.2 and 5.1 were obtained its most important characteristics. It is a global property of a ground state, constant from point to point in the atom or molecule (or solid), in principle calculable for each N and v in terms of the fundamental constants of physics as $\mu[N, v]$. For a change from one ground state to another, $d\mu$ is given by (5.1.10).

Whatever the precise definition is chosen for what is an atom in a molecule (an important question to which we return in Chapter 10), *when atoms (or other combining groups) of different chemical potentials unite to form a molecule with its own characteristic chemical potential, to the extent that the atoms (groups) retain their identity, their chemical potentials must equalize.* This is the Sanderson Principle of Electronegativity Equalization (Sanderson 1951, 1976).

This behavior parallels the behavior of the chemical potential in classical macroscopic thermodynamics. The reader will recall that when two phases are put in contact and come to equilibrium, the chemical potential of every component equalizes among the phases.

The flow of a substance in classical thermodynamics is from the phase of high chemical potential to the phase of low chemical potential. So too in our case: when $\mu_B^0 > \mu_A^0$, and there are no complicating factors, electrons flow from B to A in formation of AB. We must show this. We write, in accordance with the definitions of μ and η, (5.1.6) and (5.1.8),

$$E_A = E_A^0 + \mu_A^0(N_A - N_A^0) + \eta_A(N_A - N_A^0)^2 + \cdots$$
$$E_B = E_B^0 + \mu_B^0(N_B - N_B^0) + \eta_B(N_B - N_B^0)^2 + \cdots$$
(5.2.1)

Then, if one can ignore all other effects, the total energy will have the form

$$E_A + E_B = E_A^0 + E_B^0 + (\mu_A^0 - \mu_B^0)\Delta N + (\eta_A + \eta_B)(\Delta N)^2 + \cdots \quad (5.2.2)$$

where

$$\Delta N = N_B^0 - N_B = N_A - N_A^0 \quad (5.2.3)$$

For $\mu_B^0 > \mu_A^0$, therefore, a positive ΔN, flow of electrons from B to A, will stabilize the system, at least for small ΔN.

The energy stabilization due to such a charge transfer is second order in $\mu_B^0 - \mu_A^0$. To obtain the explicit formula, minimize $E_A + E_B$ with respect to ΔN. The result is

$$\mu_A = \mu_B \quad (5.2.4)$$

where

$$\mu_A = \left(\frac{\partial E_A}{\partial N_A}\right)_v = \mu_A^0 + 2\eta_A \Delta N + \cdots$$
$$\mu_B = \left(\frac{\partial E_B}{\partial N_B}\right)_v = \mu_B^0 - 2\eta_B \Delta N + \cdots$$
(5.2.5)

Consequently, to first order,

$$\Delta N = \frac{\mu_B^0 - \mu_A^0}{2(\eta_A + \eta_B)} \quad (5.2.6)$$

The charge transfer to first order is proportional to the original chemical potential difference, which is reminiscent of Malone's idea that electronegativity difference should be determinative for dipole moment (Malone 1933, Pauling 1960, p. 88). For the energy change, there results, to second order,

$$\Delta E = -\frac{(\mu_B^0 - \mu_A^0)^2}{4(\eta_A + \eta_B)} \quad (5.2.7)$$

This behaviour is analogous to the behaviour of Pauling's "extra-ionic resonance energy" (Pauling 1960, p. 91).

The foregoing procedure for obtaining the binding energy of two atoms needs justification. It can be regarded as a most primitive implementation of the density-functional theory: electron density is parameterized by atomic charges, and the energy functional is approximated by simple addition of atomic contributions, (5.2.2). Of course, accurate density-functional theory does not rely on such models. Nevertheless, the binding energy derived has useful implications. Furthermore, the argument has been further developed into more sophisticated *electronegativity equalization schemes* for the practical estimation of atomic charge for very large molecules (Mortier, Ghosh, and Shankar, 1986, and see §§10.4–10.6).

The join with classical structural chemistry is achieved with the identification of μ as the negative of the electronegativity χ. From (5.1.6), μ is $(\partial E/\partial N)_v$, but from §4.2, this is the negative of electronegativity,

$$\mu = \left(\frac{\partial E}{\partial N}\right)_v = -\chi \qquad (5.2.8)$$

From what has just been said about μ, we have the following basic properties of χ: (1) The electronegativity (chemical potential) is a property of the state of the system, calculable in terms of the constants of physics from density functional theory, or determinable from experiment using (4.2.2) or (4.4.9)–(4.4.11). (2) Electronegativity differences (chemical potential differences) drive electron transfer. Electrons tend to flow from a region of low electronegativity (high chemical potential) to a region of high electronegativity (low chemical potential). In the simplest case, as covered by the above model, the number of electrons that flow is to first order proportional to the electrongativity (chemical potential) difference: the concurrent energy stabilization is proportional to its square. (3) On formation of a molecule, electronegativities (chemical potentials) of constituent atoms or groups equalize (neutralize), all becoming equal to the electronegativity of the final molecule (Sanderson's Principle).

These properties of electronegativity are all known; indeed, they have been known for a long time. What is new is that everything follows from the density-functional theory of electronic structure. The density-functional formulation of the quantum theory leads rigorously to the concept of electronegativity and the principle of electronegativity equalization.

The complete history of the electronegativity concept will not be reviewed here, but we pinpoint the main developments. Pauling (1960) used the concept as a descriptive indicator of electron-attracting power.

He made up a relative scale based in the first instance on bond energies but implicitly more or less on dipole moments—our Equation (5.2.6). He termed the electron stabilization energy due to charge transfer "extra-ionic resonance energy" and set it proportional to the square of electronegativity difference—our Equation (5.2.7). Then came Mulliken's work, already described in §4.2 (Mulliken 1934). Mulliken apparently did not think of electronegativity as the slope of an E versus N curve. That idea is due to Pritchard and Sumner (1956) and has been followed up by many people, especially Iczkowski and Margrave (1961).

Many people also have endorsed an electronegativity-equalization principle (Klopman 1965, Baird, Sichel, and Whitehead 1968, Ray, Samuels, and Parr 1979). That principle was first proposed by Sanderson (1951) and argued for at length by him (Sanderson 1976). Sanderson also proposed a *geometric mean equalization principle*,

$$\chi_{AB} = (\chi_A^0 \chi_B^0)^{1/2} \tag{5.2.9}$$

or

$$\mu_{AB} = -(\mu_A^0 \mu_B^0)^{1/2} \tag{5.2.10}$$

This rule is roughly correct, although it cannot explain electronegativity changes that occur on homonuclear bond formation. It has been shown (Parr and Bartolotti, 1982) how (5.2.9) or (5.2.10) can arise using (5.2.4). Namely, if one has atomic energies that decay exponentially with number of electrons, with the decay parameter the same for each of the atoms being bound, (5.2.9) follows. This situation roughly conforms to nature:

$$E(N) \approx E(N_0) e^{-\gamma(N-N_0)} \quad \text{with } \gamma \approx 2.2 \pm 0.6 \tag{5.2.11}$$

It was an important part of Mulliken's paper on electronegativity, and indeed it was already implicit in Pauling's work, that electronegativity must be dependent on valence state: different atoms have different electronegativities in different circumstances. There have been many follow-up studies of this aspect (for example, Hinze and Jaffe 1962).

Valence-state dependence of electronegativity is naturally accommodated in density-functional theory. Recall from the discussion in Chapter 4 that there is an essential ambiguity in the whole E versus N curve: the curve through integral-N points can vary from circumstance to circumstance. And note also that the molecular environment can perturb the isolated ground state in specific ways; see Chapter 10 for more discussion of this point.

Not surprisingly, there have been many definitions of electronegativity other than Mulliken's. One of several due to Gordy (1946) is of particular interest. Gordy took the electronegativity to be the electrostatic potential $\phi(r)$ due to an atom [(3.1.14)], at the covalent radius of the atom. There is an intimate density-functional relation between μ and $\phi(r)$, as can be

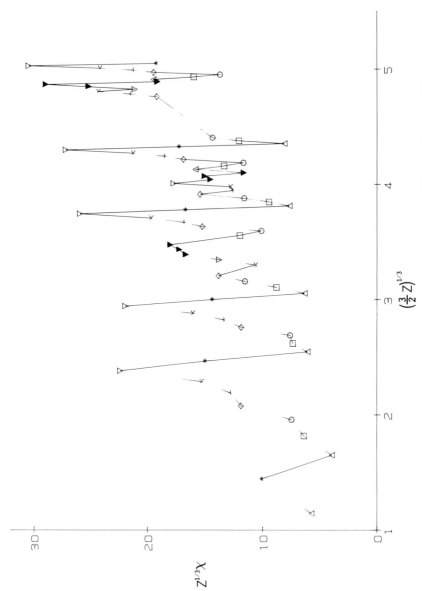

Figure 5.1 Periodic behavior of the chemical potential (after Gázquez, Vela, and Galvan, 1986).

seen from (3.2.9). If one takes μ from Mulliken's formula and calculates $\phi(r)$ from theory, one can determine for an atom the r value for which $\mu = -\phi(r)$. There results a nice set of values for atomic radii (Politzer, Parr, and Murphy 1985); see also Boyd and Markus (1981). An increase in electronegativity by loss of electron decreases diamagnetic shielding (Ray and Parr 1980).

A limiting law for electronegativity or chemical potential has been stated by March and Parr (1980) and confirmed by Gázquez, Vela, and Galvan (1986) and Gázquez, Galvan, Ortiz, and Vek (1987): As the atomic number Z of a neutral atom goes to infinity, the chemical potential is of order $Z^{-1/3}$, i.e.,

$$\mu \sim Z^{-1/3} \quad \text{for large } Z \qquad (5.2.12)$$

This law is obeyed well as one goes down a column of the periodic table, as shown in Figure 5.1.

Appendix F presents a table of electronegativity values, taken from Pearson (1988).

To summarize, the electronegativity or chemical potential of the ground state of a microscopic species (atom, ion, molecule, or solid) is a property of the system that behaves much like the chemical potential of macroscopic thermodynamics. It measures the escaping tendency of the electrons from the system and is constant through a system in equilibrium. We prefer the thermodynamically descriptive term *chemical potential* for this quantity, and generally use that term in this book.

5.3 Hardness and softness

We now turn to the derivatives of the chemical potential $\mu[N, v]$. The first is what may be called the *absolute hardness* of a species (Parr and Pearson 1983),

$$\eta = \frac{1}{2}\left(\frac{\partial \mu}{\partial N}\right)_v = \frac{1}{2}\left(\frac{\partial^2 E}{\partial N^2}\right)_v \qquad (5.3.1)$$

The inverse of hardness is *softness* (Yang and Parr 1985),

$$S = \frac{1}{2\eta} = \left(\frac{\partial N}{\partial \mu}\right)_v \qquad (5.3.2)$$

The first property we have, from the convexity assumption of (1.1.14) or (4.1.15), is positivity,

$$\eta \geq 0 \qquad (5.3.3)$$

There is no known counterexample to this rule (see discussion in §4.1).

Secondly, the finite-difference approximation for hardness is the formula,

$$\eta \approx \frac{I - A}{2} \tag{5.3.4}$$

For an insulator or semiconductor, hardness is half of the band gap. Numerical values for many species are given in Appendix F.

Figure 5.2 depicts the situation as regards η and the chemical potential (electronegativity) μ. From the fact that the absolute value of the total energy is of little importance for understanding chemical processes, one could conclude without any theoretical argument that the second derivative of E versus N curves would be next in importance after the first derivative. Both derivatives deserve names: the first derivative is the chemical potential (electronegativity); the second is hardness. Compare the discussion in §4.2 and Figure 4.1.

The physical meaning of hardness becomes clear if one considers the disproportionation reaction in which an electron is taken from S and given to S:

$$\dot{S} + \dot{S} \to S^+ + \ddot{S}^- \tag{5.3.5}$$

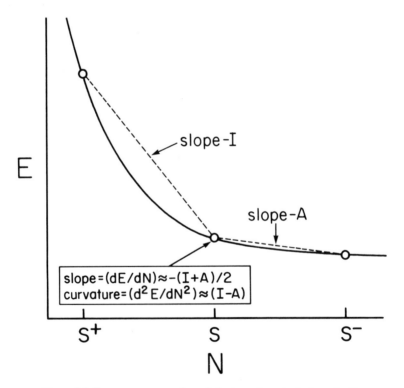

Figure 5.2 Energy versus number of electrons for a typical species S.

The energy change is

$$\Delta E_s = I_s - A_s \qquad (5.3.6)$$

This is the hardness. Small or zero hardness means that it is easy for electrons to go from S to S; that is, S is a soft species. Compare the Mulliken derivation of electronegativity at the end of §4.2. Hardness is even simpler.

It is notable that the disproportionation reaction of (5.3.5) has played an important continuing role in the development of orbital theories of electronic structure over the last 50 years. Suppose that one took the naive view that the two electrons in $\ddot{S}^-$ each had the same energy, the same as the energy of one electron in $\dot{S}$. The energy charge accompanying (5.3.6) would then be zero, a typical result of Hückel-type methods (Coulson, O'Leary, and Mallion 1978).

$$\Delta E_s = 0 \quad [\text{Hückel}] \qquad (5.3.7)$$

This is clearly wrong, as can be seen by inspection of hardness values in Appendix F, but this level of approximation nevertheless dominated the quantum theory of molecular electronic structure up to about 1945 and is much used even today.

Better than (5.3.7) would be to take ΔE_s to be the repulsion between the two valence electrons in $\ddot{S}^-$. Assume S^- is a closed-shell, even-number-of-electrons system, and call the highest occupied orbital s. Then we would have [see (1.3.44)]

$$\Delta E_s = (ss \mid ss) \quad [\text{purely theoretical, frozen core}] \qquad (5.3.8)$$

This description dominated the scene from about 1945 to about 1955. But it too is wrong, because of the "frozen core" idea in it. Better in fact, indeed much better, is the idea rooted in the work of Moffitt (1951) but first made explicit for modern theory by Pariser (1953): Use (5.3.6) as a substitute for (5.3.8); that is, set

$$(ss \mid ss) = I_s - A_s \quad [\text{semiempirical}] \qquad (5.3.9)$$

instead of computing $(ss \mid ss)$ directly from the orbital. Other things must be done too, and there result the semiempirical theories that have been useful from 1953 to the present day. In quantum chemistry, these are theories of so-called PPP, CNDO, INDO, MINDO type (Parr 1963, Segal 1977); in solid-state physics these are theories deriving from the so-called Hubbard model (Hubbard 1963) [this is a special case of the PPP model, the very special case of the PPP model that had been put forward by Longuet-Higgins and Salem (1960)]. What do these theories do that the earlier semiempirical theories did not do? They not only get electronegativities of atoms right, which the Hückel theory does through its choice of the so-called α parameters, but they also get hardnesses of atoms right, which the Hückel theory sets equal to zero and the purely theoretical valence-only theories overestimate.

Just as with electronegativity, and for the same reasons, there will be a state dependence and an environmental dependence of the hardness of a chemical species, which will have to be elucidated in any particular circumstance. The rule (5.2.11) has implications for softness (Yang, Lee, and Ghosh 1985) and hardness (Datta 1986).

Hardness and softness are terms that have been in the chemical vocabulary since the early 1950s or before (for example, see Mulliken 1952, p. 819). The definitive work introducing the concepts was the paper of Pearson (1963). Whereas until recently there has not been a consensus as to the precise meaning of the terms hard and soft (Pearson 1973), there has been agreement as to the general characteristics of hard and soft species (Pearson 1966, Klopman 1968). These are as follows.

Soft base. Donor atom has high polarizability and low electronegativity, is easily oxidized, and is associated with empty low-lying orbitals.

Hard base. Donor atom has low polarizability and high electronegativity, is hard to oxidize, and is associated with empty orbitals of high energy.

Soft acid. Acceptor atom is of low positive charge, is of large size, and has easily excited outer electrons.

Hard acid. Acceptor atom has high positive charge and small size, and does not have easily-excited outer electrons.

Our definitions in no way contravene these characteristics. Rather, they provide a quantitative scale for the most part in agreement with them. Thus, Politzer (1987) has shown that the softnesses of atoms are linearly correlated with their polarizabilities. Putting an earlier argument by Huheey (1965, 1971, 1978) in hardness-softness terms, Politzer (1987) also describes how the softness of a species can be thought of as its capacity to accept charge. As for hardness, Zhou, Parr and Garst (1988) demonstrate that it is an excellent measure of aromaticity.

Pearson found the need to introduce the terms hard and soft because of the impossibility of understanding acid–base reactions on the basis of one parameter per species. He enunciated the *HSAB Principle*: Both in their thermodynamic and kinetic properties, hard acids prefer hard bases, and soft acids prefer soft bases. Hardness or softness is a second parameter, then, with electronegativity itself being the first parameter.

Our precise definition of hardness should permit of a derivation of the HSAB Principle, or at least a rationalization of it. We defer discussion of this and other chemical applications of these ideas to Chapter 10.

5.4 Reactivity index: the Fukui function

The second important derivative of the electronic chemical potential $\mu[N, v]$ is the space-dependent (local) function

$$f(\mathbf{r}) = \left[\frac{\delta\mu}{\delta v(\mathbf{r})}\right]_N = \left[\frac{\partial\rho(\mathbf{r})}{\partial N}\right]_v \quad (5.4.1)$$

which, for reasons that will shortly become clear, is called the *Fukui function* for the system (Parr and Yang 1984). It is normalized:

$$\int f(\mathbf{r})\,d\mathbf{r} = 1 \quad (5.4.2)$$

The physical meaning of $f(\mathbf{r})$ is implied by its definition as $[\delta\mu/\delta v(\mathbf{r})]_N$: it measures how sensitive a system's chemical potential is to an external perturbation at a particular point; we shall say more about this soon.

The second formula for $f(\mathbf{r})$, written as $[\partial\rho(\mathbf{r})/\partial N]_v$, shows that it is a quantity involving the electron density of the atom or molecule in its frontier, valence regions. Suppose that in going from N electrons to $N + \delta$ electrons, for some system, the only change in the electronic structure were the addition of δ electrons to some density ρ_δ normalized to unity. Under this "frozen core" assumption, $[\partial\rho(\mathbf{r})/\partial N]_v$ would be just this ρ_δ. Note further, from the discussion in §4.3, that the derivative $\partial\rho/\partial N$ at some integral value of N will in general have one value from the right, one from the left, and an average. Consequently we have three indices:

$$f^+(\mathbf{r}) = \left[\frac{\partial\rho(\mathbf{r})}{\partial N}\right]_v^+ \quad \text{(derivative as } N \text{ increases from } N_0 \text{ to } N + \delta\text{)}$$
$$(5.4.3)$$

$$f^-(\mathbf{r}) = \left[\frac{\partial\rho(\mathbf{r})}{\partial N}\right]_v^- \quad \text{(derivative as } N \text{ increases from } N_0 - \delta \text{ to } N_0\text{)}$$
$$(5.4.4)$$

$$f^0(\mathbf{r}) = \tfrac{1}{2}[f^+(\mathbf{r}) + f^-(\mathbf{r})] \quad \text{(mean of left and right derivatives)} \quad (5.4.5)$$

with the approximate formulas

$$f^+(\mathbf{r}) \sim \rho_{\text{LUMO}}(\mathbf{r}) \quad (5.4.6)$$

$$f^-(\mathbf{r}) \approx \rho_{\text{HOMO}}(\mathbf{r}) \quad (5.4.7)$$

$$f^0(\mathbf{r}) \approx \tfrac{1}{2}[\rho_{\text{LUMO}}(\mathbf{r}) + \rho_{\text{HOMO}}(\mathbf{r})] \quad (5.4.8)$$

where in a conventional notation HOMO represents the *highest occupied*

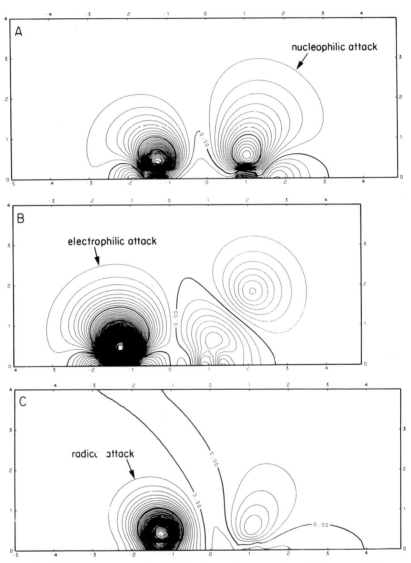

Figure 5.3 Fukui contour maps for H_2CO. The carbon atom is at 1.015 a.u. and the oxygen atom at -1.27 a.u. The increment between contour levels is 0.005 a.u.
A. $f^+(\mathbf{r})$ contours on the plane perpendicular to the H_2CO molecular plane.
B. $f^-(\mathbf{r})$ contours on the H_2CO molecular plane.
C. $f^0(\mathbf{r})$ contours on the plane perpendicular to the H_2CO molecular plane.

molecular orbital in the species in question, and LUMO the *lowest unoccupied molecular orbital*.

Returning to the physical meaning of $f(\mathbf{r})$, we see that it is a chemical reactivity index in the sense of the frontier-electron theory of reactivity as invented by Fukui and collaborators (Fukui, Yonezawa, and Shingu 1952, Fukui, Yonezawa, Nagata, and Shingu 1954, Fukui 1975, 1987), for ρ_{HOMO}, ρ_{LUMO}, and their average are the very indices of that theory, measuring the preferences for a site of a reagent of one kind or another.

$f^+(\mathbf{r}) \approx \rho_{\text{LUMO}}(\mathbf{r})$ measures reactivity toward a nucleophilic reagent

(5.4.9)

$f^-(\mathbf{r}) \approx \rho_{\text{HOMO}}(\mathbf{r})$ measures reactivity toward an electrophilic reagent

(5.4.10)

$f^0(\mathbf{r}) \approx \frac{1}{2}[\rho_{\text{HOMO}}(\mathbf{r}) + \rho_{\text{LUMO}}(\mathbf{r})]$ measures reactivity toward an

innocuous (radical) reagent (5.4.11)

More importantly, we infer from $f = \delta\mu/\delta v$ a principle that generates the frontier-electron theory. *Of two different sites with generally similar dispositions for reacting with a given reagent, the reagent prefers the one which on the reagent's approach is associated with the maximum response of the system's chemical potential* (Parr and Yang 1984). In short, $|d\mu|$ big is good.

This is perhaps an oversimplified view of chemical reactivity, but it is useful. And in any case $f(\mathbf{r})$ is established as an index of considerable importance for understanding molecular behavior—the natural reactivity index of density-functional theory. Note that $f(\mathbf{r})$ is defined independently of any model, while the concepts of classical frontier theories are couched in the language of the independent-particle model.

For particular cases, tables and contour maps of $f(\mathbf{r})$ can readily be constructed (Lee, Yang, and Parr 1987). An example is shown in Figure 5.3, which depicts f^+, f^-, and f^0 for the species H_2CO.

5.5 Local softness, local hardness, and softness and hardness kernels

A soft atom or molecule generally has a high propensity to form covalent bonds, and reactivity indices of the frontier type generally identify sites where chemical bonds are likely to form. One therefore expects the hardness–softness ideas of §5.3 to be intimately related to the reactivity index ideas of §5.4. We here explore this relationship, and in the process discover natural definitions for local softness, local hardness, and softness and hardness kernels.

Local softness is easiest to identify (Yang and Parr 1985). Equation

(5.3.2) shows that if we define *local softness* by the formula

$$s(\mathbf{r}) = \left[\frac{\partial \rho(\mathbf{r})}{\partial \mu}\right]_{v(\mathbf{r})} \quad (5.5.1)$$

we will have

$$S = \int s(\mathbf{r})\, d\mathbf{r} \quad (5.5.2)$$

This makes (5.5.1) most natural. Furthermore, the identity,

$$\left[\frac{\partial \rho(\mathbf{r})}{\partial \mu}\right]_v = \left[\frac{\partial \rho(\mathbf{r})}{\partial N}\right]_v \left[\frac{\partial N}{\partial \mu}\right]_v \quad (5.5.3)$$

and the definition (5.4.1) of the Fukui function give

$$s(\mathbf{r}) = f(\mathbf{r})S; \quad f(\mathbf{r}) = \frac{s(\mathbf{r})}{S}. \quad (5.5.4)$$

That is to say, both $f(\mathbf{r})$ and $s(\mathbf{r})$ contain the same information about relative site reactivity in a molecule; $s(\mathbf{r})$ contains additional information about the total molecular softness.

The quantities $s(\mathbf{r})$, $f(\mathbf{r})$, and S are all well known in the theory of metals. Let $g(\varepsilon_F, r)$ be the local density of states at the Fermi level and $g(\varepsilon_F)$ be the total density of states at the Fermi level, at 0 K. Then

$$s(\mathbf{r}) = g(\varepsilon_F, \mathbf{r}) \quad (5.5.5)$$
$$S = g(\varepsilon_F) \quad (5.5.6)$$

and

$$f(\mathbf{r}) = \frac{g(\varepsilon_F, \mathbf{r})}{g(\varepsilon_F)} \quad (5.5.7)$$

Another very interesting property of $s(\mathbf{r})$ is a formula for it as a fluctuation in the grand canonical ensemble (Yang and Parr 1985),

$$s(\mathbf{r}) = \frac{1}{kT}\{\langle N\rho(\mathbf{r})\rangle - \langle N\rangle\langle \rho(\mathbf{r})\rangle\} \quad (5.5.8)$$

This would appear to agree with an argument of Falicov and Somorjai (1985) that "low-energy electronic fluctuations" are determinative for catalysis.

Berkowitz and Parr (1988) have given a derivation of local softness that reveals its relation to its reciprocal property, local hardness. The idea is to define the appropriate two-variable kernels for hardness and softness, and then to generate local hardness and local softness from them in a way that accords with (5.5.1), (5.5.2), and the definition of hardness, (5.3.1).

We consider a ground state, or change of one ground state to another.

5.5 CHEMICAL POTENTIAL DERIVATIVES

Since $\rho(\mathbf{r})$ determines all properties, it determines μ and $v(\mathbf{r})$ and hence the modified potential

$$u(\mathbf{r}) = v(\mathbf{r}) - \mu = -\frac{\delta F[\rho]}{\delta \rho(\mathbf{r})} \tag{5.5.9}$$

In other words, $u(\mathbf{r})$ is a functional of $\rho(\mathbf{r})$ and the functional derivative $\delta u(\mathbf{r})/\delta \rho(\mathbf{r}')$ exists. We call this the *hardness kernel*,

$$-2\eta(\mathbf{r},\mathbf{r}') \equiv \frac{\delta u(\mathbf{r})}{\delta \rho(\mathbf{r}')} = \frac{\delta u(\mathbf{r}')}{\delta \rho(\mathbf{r})} \tag{5.5.10}$$

The equality here is just the previous Equation (5.1.18).

Similarly, $\rho(\mathbf{r})$ is a functional of $u(\mathbf{r})$ and the functional derivative $\delta \rho(\mathbf{r})/\delta u(\mathbf{r}')$ exists. This is implied by (5.5.9) or the grand-canonical-ensemble formula (3.6.12) (as is well known in statistical mechanics). We can then define

$$-s(\mathbf{r},\mathbf{r}') \equiv \frac{\delta \rho(\mathbf{r})}{\delta u(\mathbf{r}')} = \frac{\delta \rho(\mathbf{r}')}{\delta u(\mathbf{r})} \tag{5.5.11}$$

and call this quantity the *softness kernel*. We have immediately, since the two functional derivatives exist, a proper reciprocity in the sense of Equation (A.31) of Appendix A,

$$2\int s(\mathbf{r},\mathbf{r}')\eta(\mathbf{r}',\mathbf{r}'')\, d\mathbf{r}' = \delta(\mathbf{r}-\mathbf{r}'') \tag{5.5.12}$$

The hardness and softness kernels are reciprocals of each other.

At this stage we have yet to identify local softness and local hardness. Define the *local softness* by

$$s(\mathbf{r}) = \int s(\mathbf{r},\mathbf{r}')\, d\mathbf{r}' \tag{5.5.13}$$

and the *local hardness* by

$$\eta(\mathbf{r}) = \frac{1}{N}\int \eta(\mathbf{r},\mathbf{r}')\rho(\mathbf{r}')\, d\mathbf{r}' \tag{5.5.14}$$

These quantities are reciprocals in the sense that

$$2\int s(\mathbf{r})\eta(\mathbf{r})\, d\mathbf{r} = 1 \tag{5.5.15}$$

To prove this, multiply (5.5.12) by $\rho(\mathbf{r}'')$ and integrate over $\mathbf{r}''$, yielding

$$2\int s(\mathbf{r}',\mathbf{r})\eta(\mathbf{r}')\, d\mathbf{r}' = \frac{\rho(\mathbf{r})}{N} \tag{5.5.16}$$

and then integrate over $\mathbf{r}$.

To obtain an explicit formula for $\eta(\mathbf{r})$, insert (5.5.10) into (5.5.14) and use (5.5.9). There results

$$2\eta(\mathbf{r}) = \frac{1}{N}\int \frac{\delta^2 F[\rho]}{\delta\rho(\mathbf{r})\,\delta\rho(\mathbf{r}')}\rho(\mathbf{r}')\,d\mathbf{r}' \qquad (5.5.17)$$

This formula is due to Ghosh and Berkowitz (1985); see also Berkowitz, Ghosh, and Parr (1985). There are other possible definitions of local hardness indices (Berkowitz and Parr 1988).

Assuming that with these definitions we can prove (5.5.1), (5.5.4) would follow; putting this into (5.5.15) would then yield

$$\eta = \int f(\mathbf{r})\eta(\mathbf{r})\,d\mathbf{r} \qquad (5.5.18)$$

This is the way to get total hardness from local hardness.

It remains to show that (5.5.13) and (5.5.1) are synonomous. Employing (5.1.10) and (5.5.13) for the last step, we have

$$d\rho(\mathbf{r}) = \int \frac{\delta\rho(\mathbf{r})}{\delta u(\mathbf{r}')}\,du(\mathbf{r}')\,d\mathbf{r}'$$

$$= \int s(\mathbf{r},\mathbf{r}')[dv(\mathbf{r}') - d\mu]\,d\mathbf{r}'$$

$$= 2s(\mathbf{r})\eta\,dN + \int [-s(\mathbf{r},\mathbf{r}') + s(\mathbf{r})f(\mathbf{r}')]\,dv(\mathbf{r}')\,d\mathbf{r}' \qquad (5.5.19)$$

But also, $\rho = \rho[N, v]$ implies

$$d\rho(\mathbf{r}) = f(\mathbf{r})\,dN + \int \left[\frac{\delta\rho(\mathbf{r})}{\delta v(\mathbf{r}')}\right]_N dv(\mathbf{r}')\,d\mathbf{r} \qquad (5.5.20)$$

Hence we have

$$s(\mathbf{r}) = \frac{f(\mathbf{r})}{2\eta} = f(\mathbf{r})S \qquad (5.5.21)$$

as desired, and also the new result

$$\left[\frac{\delta\rho(\mathbf{r})}{\delta v(\mathbf{r}')}\right]_N = -s(\mathbf{r},\mathbf{r}') + \frac{s(\mathbf{r})s(\mathbf{r}')}{S} \qquad (5.5.22)$$

This last relation is an exact formula relating the conventional linear response function of (5.1.11) with softness, local softness, and softness kernel.

From (5.5.9) and (5.5.10), one finds

$$2\eta(\mathbf{r},\mathbf{r}') = \frac{\delta^2 F}{\delta\rho(\mathbf{r})\,\delta\rho(\mathbf{r}')} \qquad (5.5.23)$$

This formula may be found in Nalewajski (1985). Equation (5.5.22) for a model linear response case may be found in Handler and March (1975).

6
THOMAS–FERMI AND RELATED MODELS

6.1 The traditional TF and TFD models

Given an N-electron system of interest with Hamiltonian (1.1.2), ground-state energy E, wave function Ψ, and electron density $\rho(\mathbf{r})$, we have proved in Chapter 3, and developed at some length there and in Chapter 4, the fact that $\rho(\mathbf{r})$ determines $v(\mathbf{r})$, N, and the energy $E[\rho]$, and the fact that $\rho(\mathbf{r})$ is a solution of the stationary principle

$$\delta\{E[\rho] - \mu N[\rho]\} = 0 \qquad (6.1.1)$$

where μ is a Lagrange multiplier and $N[\rho] = \int \rho(\mathbf{r}) \, d\mathbf{r}$. $E[\rho]$ is the functional

$$E[\rho] = T[\rho] + \int \rho(\mathbf{r})v(\mathbf{r}) \, d\mathbf{r} + V_{ee}[\rho] \qquad (6.1.2)$$

where $T[\rho]$ is the kinetic energy and $V_{ee}[\rho]$ is the electron–electron interaction energy. The classical part of $V_{ee}[\rho]$ is the Coulomb potential energy

$$J[\rho] = \frac{1}{2} \iint \frac{\rho(\mathbf{r}_1)\rho(\mathbf{r}_2)}{|\mathbf{r}_2 - \mathbf{r}_1|} d\mathbf{r}_1 \, d\mathbf{r}_2 \qquad (6.1.3)$$

The problem remains, and this problem is of surpassing difficulty, of how to calculate $T[\rho]$, and how to calculate the nonclassical part of $V_{ee}[\rho]$. We will be dealing with these two questions in this and the next two chapters.

The traditional Thomas–Fermi (TF) model is a first approximation: replace $V_{ee}[\rho]$ by $J[\rho]$ of (6.1.3) and take $T[\rho]$ from the theory of a noninteracting uniform electron gas (Fermi 1927, Thomas 1927). We have already introduced this model in §3.1, but we will rederive it in this section. The traditional Thomas–Fermi–Dirac (TFD) model is a reasonable next guess: accept the $T[\rho]$ of Thomas–Fermi theory and for approximation to $V_{ee}[\rho]$ add to $J[\rho]$ the exchange-energy formula for a uniform electron gas (Dirac 1930). Never mind that an atomic or molecular electron cloud obviously is not a uniform gas; let us see what these assumptions bring. The literature of this subject is vast; for reviews see Gombas (1949), March (1957, 1975), and Lieb (1981b). Our summary in this chapter is far from complete.

We will derive TF and TFD theories together, since TF is obtained straightforwardly from TFD by ignoring the exchange contribution. We take as our starting point the Hartree–Fock model of §1.3 and §2.5. For simplicity, consider a nondegenerate closed-shell ground state described by a single-determinantal wave function of the form (1.3.1), a first-order density matrix of the form (2.5.1), and a spinless first-order density matrix of the form

$$\rho_1(\mathbf{r}_1, \mathbf{r}'_1) = 2 \sum_i^{N/2} \phi_i(\mathbf{r}_1)\phi_i^*(\mathbf{r}'_1) \qquad (6.1.4)$$

where the ϕ_i are the doubly occupied spatial orbitals. The energy then is given by (2.5.20),

$$E_{\text{HF}}[\rho_1] = \int \left[-\frac{1}{2} \nabla_1^2 \rho_1(\mathbf{r}_1, \mathbf{r}_2) \right]_{\mathbf{r}_2=\mathbf{r}_1} d\mathbf{r}_1 + \int \rho(\mathbf{r})v(\mathbf{r})\, d\mathbf{r}$$

$$+ J[\rho] - \frac{1}{4} \iint \frac{1}{r_{12}} \rho_1(\mathbf{r}_1, \mathbf{r}_2)\rho_1(\mathbf{r}_2, \mathbf{r}_1)\, d\mathbf{r}_1\, d\mathbf{r}_2 \qquad (6.1.5)$$

Comparing with (6.1.2), we identify

$$T[\rho] = \int [-\tfrac{1}{2}\nabla_1^2 \rho_1(\mathbf{r}_1, \mathbf{r}_2)]_{\mathbf{r}_2=\mathbf{r}_1}\, d\mathbf{r}_1 \qquad (6.1.6)$$

and

$$V_{ee}[\rho] = J[\rho] - K[\rho] \qquad (6.1.7)$$

where $K[\rho]$ is the HF exchange-energy functional

$$K[\rho] = \frac{1}{4} \int \frac{1}{r_{12}} |\rho_1(\mathbf{r}_1, \mathbf{r}_2)|^2\, d\mathbf{r}_1\, d\mathbf{r}_2 \qquad (6.1.8)$$

As expressed in (6.1.7), our first assumption then is that correlation effects can be ignored in $V_{ee}[\rho]$. We are left with the problem of expressing $T[\rho]$ and $K[\rho]$ in terms of the diagonal element of ρ_1, the electron density ρ. For an approximate solution to this problem, we turn to the electron gas.

In §3.1 we built the uniform-gas description from particle-in-a-box states for which the boundary conditions had the form $\psi(x=0) = \psi(x=l) = 0$. Equivalently, for a large number of particles (equivalence requires N to be large; see Acharya 1983), one can employ periodic boundary conditions, of the type $\psi(x+l) = \psi(x)$. These lead to the orbitals

$$\psi(k_x, k_y, k_z) = \frac{1}{l^{3/2}} e^{i(k_x x + k_y y + k_z z)}$$

$$= \frac{1}{V^{1/2}} e^{i\mathbf{k}\cdot\mathbf{r}} \qquad (6.1.9)$$

where

$$k_x = \frac{2\pi}{l} n_x, \qquad k_y = \frac{2\pi}{l} n_y, \qquad k_z = \frac{2\pi}{l} n_z,$$

with $n_x, n_y, n_z = 0, \pm 1, \pm 2 \cdots$. The energy levels are

$$E(n_x, n_y, n_z) = \frac{h^2}{8ml^2}[(2n_x)^2 + (2n_y)^2 + (2n_z)^2]$$

which for large quantum numbers correspond to just the number per unit energy range given by (3.1.3). With (6.1.9) for the orbitals, the density matrix of (6.1.4) becomes

$$\rho_1(\mathbf{r}_1, \mathbf{r}_2) = \frac{2}{V} \sum_{\text{all occupied } \mathbf{k}} e^{i\mathbf{k} \cdot (\mathbf{r}_1 - \mathbf{r}_2)} \tag{6.1.10}$$

If there are very many occupied states, the sum can be replaced by an integral, giving

$$\rho_1(\mathbf{r}_1, \mathbf{r}_2) = \frac{1}{4\pi^3} \int e^{i\mathbf{k} \cdot (\mathbf{r}_1 - \mathbf{r}_2)} \, d\mathbf{k}$$

$$= \frac{1}{4\pi^3} \int_0^{k_F} k^2 \, dk \iint e^{i\mathbf{k} \cdot \mathbf{r}_{12}} \sin \theta \, d\theta \, d\phi \tag{6.1.11}$$

In going from (6.1.10) to (6.1.11), we have used $dn = (l/2\pi)^3 \, d\mathbf{k} = (V/8\pi^3) \, d\mathbf{k}$. The quantity k_F, which may be a function of position, is still to be determined. Since $\rho_1(\mathbf{r}, \mathbf{r}) = \rho(\mathbf{r})$, (6.1.11) gives

$$\rho(\mathbf{r}) = \frac{k_F^3}{3\pi^2} \qquad \text{or} \qquad k_F(\mathbf{r}) = [3\pi^2 \rho(\mathbf{r})]^{1/3} \tag{6.1.12}$$

For an inhomogeneous system, the natural choice of the argument of $k_F(\mathbf{r})$ to be used in (6.1.11) is the average of $\mathbf{r}_1$ and $\mathbf{r}_2$. We therefore introduce the coordinates

$$\mathbf{r} = \tfrac{1}{2}(\mathbf{r}_1 + \mathbf{r}_2), \qquad \mathbf{s} = \mathbf{r}_1 - \mathbf{r}_2 \tag{6.1.13}$$

and proceed to carry out the integrations in (6.1.11) and to put the result into (6.1.6) and (6.1.8).

Choose $\mathbf{s}$ to lie along the k_z axis. Then (6.1.11) can be evaluated as follows:

$$\rho_1(\mathbf{r}_1, \mathbf{r}_2) = \frac{1}{4\pi^3} \int_0^{k_F} k^2 \, dk \int_{\theta=0}^{\pi} \sin \theta \, e^{ikr_{12}\cos\theta} \, d\theta \int_0^{2\pi} d\phi$$

$$= 3\rho(\mathbf{r}) \left[\frac{\sin t - t \cos t}{t^3} \right] = \rho_1(\mathbf{r}, \mathbf{s}) \tag{6.1.14}$$

where

$$t = k_F(\mathbf{r}) s \tag{6.1.15}$$

Equation (6.1.14) is the important exact formula for the first-order spinless density matrix for a uniform gas, expressed in the corrdinates **r** and s. Note that only the magnitude of **s** enters.

To evaluate the kinetic energy, we need

$$\nabla_{\mathbf{r}_1}^2 = \tfrac{1}{4}\nabla_{\mathbf{r}}^2 + \nabla_s^2 + \nabla_{\mathbf{r}}\nabla_s, \qquad \nabla_{\mathbf{r}_2}^2 = \tfrac{1}{4}\nabla_{\mathbf{r}}^2 + \nabla_s^2 - \nabla_{\mathbf{r}}\nabla_s \qquad (6.1.16)$$

Thus

$$[\nabla_1^2 \rho_1(\mathbf{r}_1, \mathbf{r}_2)]_{\mathbf{r}_2=\mathbf{r}_1} = [(\tfrac{1}{4}\nabla_{\mathbf{r}}^2 + \nabla_s^2 + \nabla_{\mathbf{r}}\nabla_s)\rho_1(\mathbf{r}, s)]_{s=0}$$
$$= \tfrac{1}{4}\nabla_{\mathbf{r}}^2 \rho(\mathbf{r}) - \tfrac{3}{5}(3\pi^2)^{2/3}\rho(\mathbf{r})^{5/3} \qquad (6.1.17)$$

Now for any well-behaved $\rho(\mathbf{r})$,

$$\int \nabla^2 \rho(\mathbf{r})\, d\mathbf{r} = 0 \qquad (6.1.18)$$

so that the kinetic energy of (6.1.6) becomes

$$T_{\mathrm{TF}}[\rho] = C_F \int \rho(\mathbf{r})^{5/3}\, d\mathbf{r}, \quad \text{with} \quad C_F = \tfrac{3}{10}(3\pi^2)^{2/3} = 2.8712 \qquad (6.1.19)$$

This is the Thomas–Fermi kinetic-energy formula, (3.1.9).

The exchange energy may be obtained similarly, by substituting (6.1.14) into (6.1.8):

$$K_D[\rho] = \frac{1}{4}\iint \frac{[\rho_1(\mathbf{r}, s)]^2}{s}\, d\mathbf{r}\, ds$$
$$= 9\pi \int \rho^2(\mathbf{r}) \frac{1}{k_F^2}\, d\mathbf{r} \left[\int_0^\infty \frac{(\sin t - t\cos t)^2}{t^5}\, dt \right]$$
$$= C_x \int \rho^{4/3}(\mathbf{r})\, d\mathbf{r}, \quad \text{with} \quad C_x = \frac{3}{4}\left(\frac{3}{\pi}\right)^{1/3} = 0.7386$$
$$(6.1.20)$$

This is the famous exchange-energy formula of Dirac (1930). The integral in the second line of this derivation can be evaluated as follows. Let $q = (\sin t/t)$. Then

$$\frac{dq}{dt} = -\left[\frac{\sin t - t\cos t}{t^2}\right] \quad \text{and} \quad \frac{d^2 q}{dt^2} = -\frac{2}{t}\frac{dq}{dt} - q$$

Therefore

$$\int_0^\infty \frac{(\sin t - t\cos t)^2}{t^5}\, dt = \int_0^\infty \left(\frac{dq}{dt}\right)\left(\frac{1}{t}\frac{dq}{dt}\right) dt$$
$$= \int_0^\infty \left(\frac{dq}{dt}\right)\left(-\frac{1}{2}q - \frac{1}{2}\frac{d^2 q}{dt^2}\right) dt$$
$$= -\frac{1}{4}\int_0^\infty \frac{d}{dt}\left[q^2 + \left(\frac{dq}{dt}\right)^2\right] dt = \frac{1}{4} \qquad (6.1.21)$$

Insertion of (6.1.19) and (6.1.20) into (6.1.2) gives, finally,

$$E_{\text{TFD}}[\rho] = C_F \int \rho(\mathbf{r})^{5/3} \, d\mathbf{r} + \int \rho(\mathbf{r}) v(\mathbf{r}) \, d\mathbf{r} + J[\rho] - C_x \int \rho(\mathbf{r})^{4/3} \, d\mathbf{r}$$
(6.1.22)

This is the Thomas–Fermi–Dirac energy functional, and is labeled TFD accordingly. The Thomas–Fermi energy functional of (3.1.10) is obtained by setting $C_x = 0$. The corresponding Euler–Lagrange equation is, from (6.1.1),

$$\mu_{\text{TFD}} = \tfrac{5}{3} C_F \rho^{2/3}(\mathbf{r}) - \tfrac{4}{3} C_x \rho^{1/3}(\mathbf{r}) - \phi(\mathbf{r})$$
(6.1.23)

where $\phi(\mathbf{r})$ is the classical electrostatic potential—see (3.1.14). This equation constitutes a generalization of the Thomas–Fermi formula (3.1.13). In the next section we will discuss solution of the TF and TFD equations.

The form of the energy functional (6.1.22) is not especially surprising, as can be seen from the following quick derivation of it up to the values of the coefficients. Suppose that the kinetic energy had the form of a local functional (see definition in Appendix A) $T[\rho] = \int t(\rho) \, d\mathbf{r}$, where $t(\rho)$ is a *function* of $\rho(\mathbf{r})$. Let the wave function be $\psi(\mathbf{r}_1, \mathbf{r}_2, \ldots)$. Then if we scale the wave function as in (1.6.16), the density becomes the scaled density

$$\rho_\lambda(\mathbf{r}) = \lambda^3 \rho(\lambda \mathbf{r})$$
(6.1.24)

and the kinetic energy becomes, *naively assuming* the scaling of (1.6.17),

$$T[\rho_\lambda] = \int t(\lambda^3 \rho(\lambda \mathbf{r})) \, d\mathbf{r} = \lambda^{-3} \int t(\lambda^3 \rho(\mathbf{r})) \, d\mathbf{r}$$

$$= \lambda^2 \int t(\rho(\mathbf{r})) \, d\mathbf{r} = \lambda^2 T[\rho]$$
(6.1.25)

Accordingly,

$$t(\lambda^3 \rho(\mathbf{r})) = \lambda^5 t(\rho(\mathbf{r}))$$
(6.1.26)

or

$$t(\lambda \rho) = \lambda^{5/3} t(\rho)$$
(6.1.27)

That is to say, $t(\rho)$ is homogeneous of degree 5/3 in ρ; $T[\rho] = A \int \rho^{5/3} \, d\mathbf{r}$, the Thomas–Fermi form. Similarly, since $K[\rho]$ is an electron–electron repulsion term, using (1.6.18) if we had $K[\rho] = \int k(\rho) \, d\mathbf{r}$, we would naively expect

$$k(\rho_\lambda) = \lambda k[\rho]$$
(6.1.28)

so that

$$k(\lambda \rho) = \lambda^{4/3} k(\rho)$$
(6.1.29)

This is to say, $k(\rho)$ is homogeneous of degree 4/3 in ρ; $K[\rho] = B \int \rho^{4/3} d\mathbf{r}$, the Dirac form. In summary, an energy functional of the form

$$E[\rho] = A \int \rho(\mathbf{r})^{5/3} d\mathbf{r} + \int \rho(\mathbf{r}) v(\mathbf{r}) d\mathbf{r} + J[\rho] - B \int \rho(\mathbf{r})^{4/3} d\mathbf{r} \quad (6.1.30)$$

follows from assumptions of locality (in the sense indicated above) of the kinetic energy and exchange energy kernels and scaling arguments; "statistical" ideas as such need not be invoked. As we shall see in §6.5, the Thomas–Fermi–Dirac values of A and B are not sacrosanct. For a discussion of the naiveté of this scaling procedure, see §11.1.

6.2 Implementation

The Euler equations for Thomas–Fermi and Thomas–Fermi–Dirac theory are (3.1.13) and (6.1.23), respectively. In each case the chemical potential μ is to be determined so that $\rho(\mathbf{r})$ is normalized to N.

Without going into a full mathematical analysis, for which see Lieb (1981b), we note that the existence of solutions of these equations, for a particular atomic or molecular system, should not be taken for granted. Further, if a solution for the density exists it may or may not have all of the qualitative properties the exact Schrödinger density should possess. For example, consider the cusp condition (1.5.3) for an atom. From (3.1.14), $\phi(0) = \infty$ for an atom, and in either TF or TFD theory we will have at the nucleus of an atom (or at any nucleus in a molecule)

$$\rho(0) = \infty \quad (6.2.1)$$

This is a basic defect in TF or TFD theory, to which we shall return in §6.4.

For a neutral atom or molecule,

$$\phi(\mathbf{r}) \to 0 \quad \text{as } r \to \infty \quad (6.2.2)$$

Since this must also be the behavior of $\rho(\mathbf{r})$, we obtain from (3.1.13)

$$\mu_{\text{TF}}(N = Z) = 0 \quad (6.2.3)$$

This also holds for TFD theory with smooth $\rho(\mathbf{r})$; from (6.1.23)

$$\mu_{\text{TFD}}(N = Z) = 0 \quad (6.2.4)$$

However, since $\rho^{2/3}$ decays faster than $\rho^{1/3}$, and at large distances $\phi \geq 0$, (6.1.23) gives an incorrect result at large r for a neutral species unless all the electronic charge is contained within some critical radius $r = r_c$; the Thomas–Fermi–Dirac neutral atom has a finite size. See Gombas (1949) and March (1957). Negative ions exist in neither model.

Explicit solution for ρ for the Thomas–Fermi neutral atom proceeds as follows. Solve (3.1.13) and (3.1.14) each for ϕ, equate, and use (6.2.3). There results

$$C\rho^{2/3}(\mathbf{r}) = \frac{Z}{r} - \int \frac{\rho(\mathbf{r}')}{|\mathbf{r}' - \mathbf{r}|} d\mathbf{r}' = \phi(\mathbf{r}) \tag{6.2.5}$$

Alternatively, $\rho(\mathbf{r})$ is a function of the electrostatic potential $\phi(\mathbf{r})$,

$$\rho(\mathbf{r}) = \left(\frac{1}{C}\right)^{3/2} [\phi(\mathbf{r})]^{3/2} \tag{6.2.6}$$

Now use the Poisson equation of classical electrostatics as applied to the atom,

$$\nabla^2 \phi(\mathbf{r}) = 4\pi \rho(\mathbf{r}) - 4\pi Z \, \delta(\mathbf{r}) \tag{6.2.7}$$

to give

$$\nabla^2 \phi(\mathbf{r}) = 4\pi \left(\frac{1}{C}\right)^{3/2} [\phi(\mathbf{r})]^{3/2} - 4\pi Z \, \delta(\mathbf{r}) \tag{6.2.8}$$

This is a differential equation for $\phi(\mathbf{r})$. It, like $\rho(\mathbf{r})$, depends only on the scalar r, since the atom must be spherically symmetric. Write

$$\phi(\mathbf{r}) = \phi(r) = \frac{Z}{r} \chi(r) \tag{6.2.9}$$

so that

$$\nabla^2 \phi(\mathbf{r}) = \frac{Z}{r} \frac{d^2 \chi(r)}{dr^2} - 4\pi Z \chi(0) \, \delta(\mathbf{r}) \tag{6.2.10}$$

(6.2.8) then becomes

$$\frac{d^2 \chi(r)}{dr^2} = 4\pi \left(\frac{1}{C}\right)^{3/2} \frac{Z^{1/2}}{r^{1/2}} [\chi(r)]^{3/2} \tag{6.2.11}$$

with the boundary condition $\chi(0) = 1$.

Finally, if we let

$$x = \alpha r, \quad \alpha = Z^{1/3}(4\pi)^{2/3}\left(\frac{1}{C}\right) = (128/9\pi^2)^{1/3} Z^{1/3} = 1.1295 Z^{1/3} \tag{6.2.12}$$

we obtain a universal differential equation for $\chi(x)$,

$$\frac{d^2 \chi(x)}{dx^2} = \frac{1}{x^{1/2}} [\chi(x)]^{3/2} \tag{6.2.13}$$

with the boundary conditions

$$\chi(0) = 1, \quad \chi(\infty) = 0 \tag{6.2.14}$$

This problem can be solved numerically. For a table of values of χ and some good discussion, see Landau and Lifshitz (1958), pp. 235–240. The derivatives of $\chi(x)$ at $x = 0$ and $x = \infty$ are of special interest; for them,

one finds
$$\chi'(0) = -1.5881, \qquad \chi'(\infty) = 0 \qquad (6.2.15)$$

From (6.2.6) and (6.2.9), the electron density as a function of the variable x is given by

$$\rho(x) = \frac{32}{9\pi^3}\left[\frac{\chi(x)}{x}\right]^{3/2} Z^2 = \frac{32Z^2}{9\pi^3 x}\left[\frac{d^2\chi(x)}{dx^2}\right] \qquad (6.2.16)$$

Since $x = \alpha r$, we infer the behavior near the nucleus to be

$$\rho(\text{small } r) \sim \frac{1}{r^{3/2}} \to \infty \qquad \text{as } r \to 0 \qquad (6.2.17)$$

For large r, the simple analytic solution of (6.2.13) is the one we want,

$$\chi(x) = \frac{144}{x^3} \qquad \text{for very large } x \qquad (6.2.18)$$

From this we infer that

$$\rho(\text{large } r) \sim \frac{1}{r^6} \to 0 \qquad \text{as } r \to \infty \qquad (6.2.19)$$

Neither (6.2.17) nor (6.2.19) reproduces the correct behavior of exact atomic densities. The radial distribution function $r^2\rho(r)$ is zero at the origin, peaks at $x = 0.42$, and falls off to zero without showing any shell structure. Half the charge is within $x = 1.50$ or $r = 1.33Z^{-1/3}$. The agreement with real atomic densities is rough at best.

The total energy of a neutral atom in Thomas–Fermi theory is given by the formula

$$E_{\text{TF}} = -0.7687 Z^{7/3} \qquad (6.2.20)$$

One can establish this result by substituting (6.2.16) into (6.1.22) and performing the integrations numerically. Somewhat more cleanly, one may make use of

$$\left(\frac{\partial E}{\partial N}\right)_{N=Z} = \left(\frac{\partial E}{\partial N}\right)_Z + \left(\frac{\partial E}{\partial Z}\right)_N \left(\frac{\partial Z}{\partial N}\right)_{N=Z}$$

$$= \mu + \left(\frac{\partial E}{\partial Z}\right)_N \qquad (6.2.21)$$

the fact that $\mu = 0$ in Thomas–Fermi theory for neutral atoms, and the fact that (from the Hellman–Feynmann theorem, which was shown by Harris, Jones and Miller (1981) to hold in density-functional theory)

$$\left(\frac{\partial E}{\partial Z}\right)_N = -\int \frac{\rho(\mathbf{r})}{r} d\mathbf{r} \qquad (6.2.22)$$

Consequently, for a neutral atom in Thomas–Fermi theory,

$$\left(\frac{\partial E}{\partial Z}\right)_{N=Z} = \left(\frac{\partial E}{\partial Z}\right)_N = -\int \frac{\rho(\mathbf{r}, z)}{r} d\mathbf{r} \qquad (6.2.23)$$

and

$$E = -\int_0^Z dZ \left[\int \frac{\rho(\mathbf{r})}{r} d\mathbf{r}\right] \qquad (6.2.24)$$

which can be evaluated employing (6.2.12), (6.2.15), and the second formula in (6.2.16):

$$E = -4\pi \int_0^Z dZ \int \frac{1}{\alpha^2} x\rho(x)\, dx$$
$$= -3/7(1.1295)[\chi'(\infty) - \chi'(0)]Z^{7/3} = -0.7687 Z^{7/3} \qquad (6.2.25)$$

This is the energy of the Thomas–Fermi neutral atom. For real atoms we have, in contrast, the Hartree–Fock values exhibited in Table 6.1. Equation (6.2.25) gives too low an energy by 54% for H, by 35% for He, by 20% for Kr, and by 15% for Rn.

Thomas–Fermi–Dirac theory does not constitute an improvement; indeed, it is worse. One can see this without any calculation at all. $K[\rho]$ is positive, so E_{TFD} of (6.1.22) is more negative than E_{TF} of (3.1.10) for a given ρ. Putting the Thomas–Fermi density ρ_{TF} into (6.1.20) gives

$$K_D[\rho_{TF}] = 0.221 Z^{5/3} \qquad (6.2.26)$$

This is negligible for large enough Z, but it pulls the energy down. And a full minimization of $E_{TFD}[\rho]$ with respect to ρ will give a lower energy still.

In both TFD and TF theories, it is readily seen that positive ions have finite size. For the TF theory for molecules, with numerical results for N_2 and C_6H_6, see Parker (1988).

Table 6.1 Energies of Neutral Atoms ($-E/Z^{7/3}$)

Atom (Z)	Hartree–Fock Energy[a]	Modified Thomas–Fermi Model[b]
He (2)	0.5678	0.4397
Ne (10)	0.5967	0.5763
Ar (18)	0.6204	0.6110
Kr (36)	0.6431	0.6439
Xe (54)	0.6562	0.6599
Rn (86)	0.6698	0.6745

[a] In conventional Thomas–Fermi theory, the energy is given by (6.2.25) of the text: $-E/Z^{7/3} = 0.7687$.

[b] Model of §6.4 of text (Parr and Ghosh 1986).

6.3 Three theorems in Thomas–Fermi theory

Many theorems have been proved in Thomas–Fermi theory (Lieb 1981b). Here we consider just three. The first theorem gives formulas relating the energy quantities T, V_{ne}, V_{ee}, $N\mu$, V_{nn}, E (total electronic energy exclusive of nuclear–nuclear repulsion), and W (total energy including nuclear–nuclear repulsion). For simplicity consider a ground-state Thomas–Fermi atom or molecule at its equilibrium nuclear configuration. Then, by (3.1.13),

$$\rho^{2/3}(\mathbf{r}) = \frac{3}{5C_F}[\phi(\mathbf{r}) + \mu] \tag{6.3.1}$$

The kinetic energy of (6.1.19) therefore becomes, using the definitions of $\phi(\mathbf{r})$, $V_{ne}[\rho]$ and $V_{ee}[\rho]$,

$$T[\rho] = -\tfrac{3}{5}V_{ne}[\rho] - \tfrac{6}{5}V_{ee}[\rho] + \tfrac{3}{5}N\mu \tag{6.3.2}$$

Now the virial theorem of (1.6.10) or (1.6.11) holds for Thomas–Fermi theory, since the scaling proof at the end of §1.6 goes through for any density-functional model with appropriate scaling behavior. Hence

$$\begin{aligned} T[\rho] &= -\tfrac{1}{2}V_{ne}[\rho] - \tfrac{1}{2}V_{ee}[\rho] - \tfrac{1}{2}V_{nn} \\ &= -W[\rho] = -E[\rho] - V_{nn} \end{aligned} \tag{6.3.3}$$

These equations plus (6.3.2) imply the formulas

$$\begin{aligned} V_{ne}[\rho] &= \tfrac{7}{3}E[\rho] + \tfrac{1}{3}V_{nn} - N\mu \\ V_{ee}[\rho] &= -\tfrac{1}{3}E[\rho] + \tfrac{2}{3}V_{nn} + N\mu \end{aligned} \tag{6.3.4}$$

These may be called the *Fraga relations* (Fraga, 1964).

The second theorem is the *nonbinding theorem*. Originally asserted by Teller (1962), with a proof faulted and improved by Balazs (1967), and generalized and made rigorous by Lieb and Simon (1973, 1977), this theorem states that in Thomas–Fermi theory no molecular system is stable relative to dissociation into constituent fragments. Indeed, whenever all internuclear distances are dilated uniformly, from any starting nuclear configuration, the *total* energy decreases. (Without the nuclear–nuclear repulsion, the contrary holds: nuclei coming together enhance the electronic component of the binding energy.)

For simplicity, consider a homonuclear neutral diatomic molecule AB, with nuclear charges λZ_A, Z_B, number of electrons $\lambda Z_A + Z_B$, at fixed internuclear distance R. We wish to prove that the binding energy

$$\Delta W(\lambda) = E_{AB}(\lambda) - E_A(\lambda) - E_B + \frac{\lambda Z_A Z_B}{R} \tag{6.3.5}$$

is positive for $\lambda = 1$. To do this we shall use [much as in the derivation of

(6.2.24)] the Hellman–Feynmann theorem to determine $\partial \Delta W(\lambda)/\partial \lambda$, and then calculate $\Delta W(1) = \int_0^1 [\partial W(\lambda)/\partial \lambda]\, d\lambda$. Thus,

$$\frac{\partial \Delta W(\lambda)}{\partial \lambda} = -Z_A \int \frac{\rho_{AB}(\mathbf{r}, \lambda)}{r_A}\, d\mathbf{r} + Z_A \int \frac{\rho_A(\mathbf{r}, \lambda)}{r_A}\, d\mathbf{r} + \frac{Z_A Z_B}{R} \quad (6.3.6)$$

where $\rho_{AB}(\mathbf{r}, \lambda)$ and $\rho_A(\mathbf{r}, \lambda)$ are the (Thomas–Fermi) electron densities for the molecule AB and the atom A, respectively. This can be rewritten as

$$\frac{\partial \Delta W(\lambda)}{\partial \lambda} = Z_A[\phi_{AB}(0, \lambda) - \phi_A(0, \lambda)] \quad (6.3.7)$$

where $\phi_{AB}(0, \lambda)$ is the electrostatic potential at nucleus A in the molecule and $\phi_A(0, \lambda)$ is the corresponding potential in the atom (the terms in ϕ_{AB} and ϕ_A representing the potential due to nucleus A cancel in $\phi_{AB} - \phi_A$.) Now, there is a lemma in Thomas–Fermi theory [Lieb 1981b, Theorem 34 on p. 612], that if the external potential $v(\mathbf{r})$ is increased everywhere (as by the increase of a nuclear charge), with an accompanying electron number change keeping μ constant (at zero in the present case), ϕ increases everywhere. Hence, $\phi_{AB}(0, \lambda) > \phi_A(0, \lambda)$, $\partial \Delta W/\partial \lambda > 0$ for all λ, and since $\Delta W = 0$ for $\lambda = 0$, the theorem follows.

The implications for theoretical chemistry are, of course, devastating. There is no hope of accounting for chemical bonding by Thomas–Fermi theory. It is even worse than that. Balazs (1967) argues that the same will be true for any other model! (for example the Thomas–Fermi–Dirac model) for which the electron density turns out to be a function of the total classical electrostatic potential,

$$\rho(\mathbf{r}) = \rho(\phi) \quad \text{(precludes bonding)} \quad (6.3.8)$$

On the other hand, it has been demonstrated (Parr and Berk 1981) that binding is allowed if the electron density is a function of the bare nuclear (external) potential,

$$\rho(\mathbf{r}) = \rho(v) \quad \text{(can sustain binding)} \quad (6.3.9)$$

This last would mean contours of the electron density would be parallel to contours of the bare nuclear potential, which is usually a fair description (Parr and Berk 1981, but see Politzer and Zilles 1984).

The third theorem is rather more constructive. In the limit of high atomic numbers and electron numbers, Thomas–Fermi theory is asymptotically correct in a relative sense. Specifically, let E_Q and E_{TF} be the Schrödinger exact energy and the Thomas–Fermi energy for a system of nuclear charges Z_α, at locations R_α, having N electrons, and let the corresponding electron densities be $\rho_Q(\mathbf{r})$ and $\rho_{TF}(\mathbf{r})$. Then

$$\lim_{\lambda \to \infty} \lambda^{-7/3} E_Q(\lambda Z_\alpha, \lambda N, \lambda R_\alpha) = \lim_{\lambda \to \infty} \lambda^{-7/3} E_{TF}(\lambda Z_\alpha, \lambda N, \lambda R_\alpha) \quad (6.3.10)$$

and

$$\lim_{\lambda \to \infty} \lambda^{-2} \rho_Q(\lambda^{-1/3}r) = \lim_{\lambda \to \infty} \lambda^{-2} \rho_{TF}(\lambda^{-1/3}r) \qquad (6.3.11)$$

provided that ρ_{TF} exists. For the neutral-atom case, then, (6.2.20) is the correct formula for the energy at high enough Z. Real atoms are still far from reaching this limit, however, as may be seen from the Hartree–Fock values displayed in Table 6.1. We shall say more about this later.

This theorem was first established by Lieb and Simon (1973, 1977), to whose work one may refer for the formal proof.

6.4 Assessment and modification

The results described in the last section hardly commend the Thomas–Fermi theory as a quantitative theory for atoms and molecules. And the Thomas–Fermi–Dirac theory in most respects is worse.

There are ways to improve these functionals by adding terms involving $\nabla \rho$, $\nabla^2 \rho$, $(\nabla \rho)^2$, etc. These will be considered in §§6.7–6.9 below, but more needs to be said before we turn to detailed descriptions of them. We begin by reviewing certain evidence that indicates that (6.1.22) and (3.1.10) are not so poor as they at first seem.

First, as was initially pointed out and documented by Fraga (1964), equations like (6.3.2) and (6.3.4) are surprisingly accurate for atoms and molecules. For simplicity, take neutral atoms and let $\mu = 0$ in (6.3.2) (the Thomas–Fermi neutral-atom result), and consider the implication if we apply the result to the Hartree–Fock formula (1.3.15). It will give

$$\sum \varepsilon_i = E + V_{ee} = \tfrac{2}{3}E, \qquad E = \tfrac{3}{2} \sum \varepsilon_i \qquad (6.4.1)$$

and it is formulas of this general type that appear to have semiquantitative general validity. The extension to molecules entails complications due to V_{nn}, but even for molecules a rough proportionality of E to $\sum \varepsilon_i$ holds (Ruedenberg 1977, Cioslowski 1988b). This is related to a rough homogeneity property of the energy with respect to atomic numbers, which the Thomas–Fermi formula of (6.2.25) exemplifies (Parr and Gadre 1980). For a recent study, see Chen and Spruch (1987).

Second, considerable evidence has accumulated that the TF and TFD energy functionals (and their gradient-expansion extensions of §§6.7 and 6.8) give good energies if one restricts the admissible densities in the minimization procedure to reasonable forms. One may mention the early work of Csavinsky (1968) in this connection, or that of Gázquez and Parr (1978), where exponential forms or their extensions were used. Or there are the studies of Wang and Parr, wherein it is shown that if one requires an atomic density to be piecewise exponential, with as many pieces as principal quantum shells, one obtains both good energies and densities that exhibit shell structure (Wang and Parr 1977, Wang 1982). And probably most important of all, if one inserts into these energy functionals Hartree–Fock densities themselves, both the total energies and individual energy components come out far better than the 15% and greater errors associated with (6.2.25) (Gordon and Kim 1972, Shih, Murphy, and Wang 1980; and see numerical examples in §§6.8 and 10.2).

According to the foregoing arguments, it seems that in the TF model energy functional one has a better-quality estimate of the value of energy than of the energy functional derivatives that determine the behavior of the TF electron density via the Euler equation. This idea suggests that one should try to retain the good property of the energy functional for the evaluation of energy but to improve on the density or on the differential equation for the density. Thus, Ashby and Holzman (1970) have proposed to eliminate the divergence of the density near the nucleus by using a cut-off of the TF density in the vicinity of the nucleus and replacing this region by a wave-mechanical density. Recently, Englert, and Schwinger (1984a) have also arrived at a modified TF density that is nondivergent at the nucleus. Their density results from minimization of the energy derived from the TF functional modified by replacing the contribution from the strongly bound electrons by the correct quantum-mechanical energy of these electrons. Attempts have also been made (Englert and Schwinger 1982) to improve the long-range behavior of the TF density. Englert and Schwinger (1984b,c, 1985a,b,c) have achieved substantial improvement in Thomas–Fermi theory.

In a recent study emphasizing this weakness (Parr and Ghosh 1986, Ghosh and Parr 1987b; see also Goldstein and Rieder 1987), it was pointed out that the Thomas–Fermi atomic electron-density behavior, $\rho \sim r^{3/2}$ as $r \to 0$, implies that $\int \nabla^2 \rho(\mathbf{r}) \, d\mathbf{r} = \infty$ in Thomas–Fermi theory; the Thomas–Fermi density is discontinuous at the origin. One ought to require continuity at $r = 0$. To do this, one can impose the constraint

$$\int e^{-2kr} \nabla^2 \rho(\mathbf{r}) \, d\mathbf{r} = \text{finite} \quad (6.4.2)$$

where k is to be determined. Attaching a Lagrange multiplier λ to this gives, in place of (6.1.23), a modified Euler equation for an atom,

$$\mu = \tfrac{5}{3} C_F \rho^{2/3}(\mathbf{r}) - \frac{Z}{r} + \int \frac{\rho(\mathbf{r}')}{|\mathbf{r}' - \mathbf{r}|} \, d\mathbf{r}' - \lambda \nabla^2(e^{-2kr}) \quad (6.4.3)$$

Now the infinity at the origin can be eliminated by requiring $4k\lambda = Z$, which gives

$$\mu = \tfrac{5}{3} C_F \rho^{2/3}(\mathbf{r}) - \frac{Z}{r}(1 - e^{-2kr}) + \int \frac{\rho(\mathbf{r}')}{|\mathbf{r}' - \mathbf{r}|} \, d\mathbf{r}' - Zk \, e^{-2kr} \quad (6.4.4)$$

with the parameter k still to be determined. The value of k can be chosen so that the cusp condition (1.5.3) is satisfied. When this is done, remarkably improved values of the energy are obtained, as indicated in Table 6.1. [Note that comparison of Thomas–Fermi energy with Hartree–Fock energy is reasonable in this semiquantitative context because exchange is a small part of the Hartree–Fock energy (~5%).] The electron density at the nucleus, in fact close to proportional to Z^3 but incorrectly predicted to be infinite in Thomas–Fermi theory, turns out well too. It is straightforward to perform the minimization to get these results. Thus a Thomas–Fermi theory in which ρ is required to be continuous and to satisfy the cusp constraint can provide a more reasonable description of atoms than the original Thomas–Fermi theory.

6.5 An alternative derivation and a Gaussian model

In the derivation of Thomas–Fermi–Dirac theory in §6.1, we started from the Hartree–Fock functional of (6.1.5) and arbitrarily substituted into it the first-order density matrix for a uniform gas, (6.1.14). Here we start again with (6.1.5), but this time we avoid making approximations for as long as possible. We will again arrive at TFD theory, but we will also come to a slightly modified version of it, a model that may be called a *Gaussian model* (Lee and Parr 1987; compare Meyer, Bartel, Brack, Quentin, and Aicher 1986).

In addition to (6.1.5), (6.1.6), and (6.1.8), we need one other formula, the normalization condition for the first-order density matrix (for a closed-shell),

$$2N = \iint \rho_1(\mathbf{r}_1, \mathbf{r}_2)\rho_1(\mathbf{r}_2, \mathbf{r}_1)\, d\mathbf{r}_1\, d\mathbf{r}_2 \qquad (6.5.1)$$

This follows from (2.5.7) for the closed-shell case. The full idempotency conditions of (2.5.7) constitute an infinity of conditions that we are going to ignore except for their most important single consequence, (6.5.1).

Introducing the interparticle coordinates $\mathbf{r}$ and $\mathbf{s}$ of (6.1.13), we note first that the kinetic energy may be written

$$T = -\frac{1}{2}\int \left[\nabla_s^2\left[\rho_1\left(\mathbf{r}+\frac{\mathbf{s}}{2}, \mathbf{r}-\frac{\mathbf{s}}{2}\right)\right]_{s=0}\right] d\mathbf{r} \qquad (6.5.2)$$

Here have been used (6.1.16), symmetry between $\mathbf{r}_1$ and $\mathbf{r}_2$, and the fact that $\nabla^2\rho(\mathbf{r})$ integrates to zero. In these coordinates we also have

$$K = \frac{1}{4}\iint \frac{\left|\rho_1\left(\mathbf{r}+\frac{\mathbf{s}}{2}, \mathbf{r}-\frac{\mathbf{s}}{2}\right)\right|^2}{s}\, d\mathbf{r}\, d\mathbf{s} \qquad (6.5.3)$$

$$N = \frac{1}{2} \iint \left|\rho_1\left(\mathbf{r}+\frac{\mathbf{s}}{2}, \mathbf{r}-\frac{\mathbf{s}}{2}\right)\right|^2 d\mathbf{r}\, d\mathbf{s} \tag{6.5.4}$$

Note the similarity between these last two formulas.

Following Berkowitz (1986) and Kemister (1986), we now introduce the expansion of $\rho_1(\mathbf{r}+(\mathbf{s}/2), \mathbf{r}-(\mathbf{s}/2))$ in the Cartesian coordinates of $\mathbf{s}$ and average over the angular coordinates of $\mathbf{s}$. Define $\beta(\mathbf{r})$ by the formulas

$$\frac{1}{\beta(\mathbf{r})} = \frac{2}{3} \frac{t(\mathbf{r})}{\rho(\mathbf{r})}, \quad \text{with } T = \int t(\mathbf{r})\, d\mathbf{r} \tag{6.5.5}$$

and

$$t(\mathbf{r}) = \frac{1}{8} \sum_i^N \nabla \rho_i \cdot \frac{\nabla \rho_i}{\rho_i} - \frac{1}{8}\nabla^2 \rho \tag{6.5.6}$$

where the ρ_i are Hartree–Fock orbital densities. Define also

$$\Gamma(\mathbf{r}, s) = \frac{\overline{\left|\rho_1\left(\mathbf{r}+\frac{\mathbf{s}}{2}, \mathbf{r}-\frac{\mathbf{s}}{2}\right)\right|^2}}{\rho^2(\mathbf{r})} \tag{6.5.7}$$

where the bar denotes the indicated spherical average. Then we find the exact formal expansion

$$\Gamma(\mathbf{r}, s) = 1 - \frac{s^2}{\beta(\mathbf{r})} + \cdots \tag{6.5.8}$$

where $\beta(\mathbf{r})$ is the "temperature" parameter of (6.5.5), and the exact formulas

$$T = \frac{3}{2} \int \frac{\rho(\mathbf{r})}{\beta(\mathbf{r})}\, d\mathbf{r} \tag{6.5.9}$$

$$K = \pi \iint \rho^2(\mathbf{r}) \Gamma(\mathbf{r}, s)\, d\mathbf{r}\, ds \tag{6.5.10}$$

$$N = 2\pi \iint \rho^2(\mathbf{r}) \Gamma(\mathbf{r}, s)\, d\mathbf{r}\, s^2\, ds \tag{6.5.11}$$

It is well known that K involves only a spherical average; the same is seen from (6.5.11) to be true for N. The Taylor series for $\Gamma(\mathbf{r}, s)$ in (6.5.8) must converge in the sense required for the integrals in (6.5.10) and (6.5.11) to exist. The range for s is 0 to ∞; $\Gamma(\mathbf{r}, s)$ is everywhere positive. A local "temperature" was introduced first by Ghosh, Berkowitz, and Parr (1984), who even set $\beta(\mathbf{r}) = 1/kT(\mathbf{r})$. See §11.2.

If we can reasonably approximate the higher-order, less important terms in Γ, we will have achieved reasonable approximation to K and N.

We consider two approximations (Lee and Parr 1987). A simple approximation for Γ is a Gaussian,

$$\Gamma(\mathbf{r}, s) \approx \Gamma_G(\mathbf{r}, s) = e^{-s^2/\beta(\mathbf{r})} \tag{6.5.12}$$

With this, (6.5.10) and (6.5.11) give

$$K_G = \frac{\pi}{2} \int \rho^2(\mathbf{r})\beta(\mathbf{r}) \, d\mathbf{r} \tag{6.5.13}$$

and

$$N_G = \frac{\pi^{3/2}}{2} \int \rho^2(\mathbf{r})\beta^{3/2}(\mathbf{r}) \, d\mathbf{r} \tag{6.5.14}$$

First derived by Ghosh and Parr (1986), (6.5.13) is a useful approximate formula for exchange energy. This is demonstrated in Table 6.2. As indicated in the table, corresponding values of the particle number are reasonably good.

Alternatively, one might try to approximate Γ using spherical Bessel functions, or, in another language, some appropriate combination of sines, cosines, and powers of s. The simplest such approximation that will do involves the square (to guarantee positivity) of the first-order spherical Bessel function $j_1(x) = [\sin x - x \cos x]/x^2$, with x proportional to s. Specifically, if we take

$$\Gamma(\mathbf{r}, s) \approx \Gamma_T(\mathbf{r}, s) = \frac{9(\sin t - t \cos t)^2}{t^6} \tag{6.5.15}$$

with

$$t = \left[\frac{5}{\beta(\mathbf{r})}\right]^{1/2} s \tag{6.5.16}$$

then we again recover (6.5.8). Equations (6.5.10) and (6.5.11) become

$$K_T = \frac{9\pi}{20} \int \rho^2(\mathbf{r})\beta(\mathbf{r}) \, d\mathbf{r} \tag{6.5.17}$$

Table 6.2 Nonlocal Approximations to the Hartree–Fock Exchange Energy and Particle Number for Neutral Atoms

Atom	N	$N_G{}^a$	$N_T{}^b$	K^c	$K_G{}^a$	$K_T{}^b$
He	2	1.646	1.565	1.026	0.913	0.826
Ne	10	9.164	8.716	12.11	11.57	10.41
Ar	18	16.24	15.45	30.18	29.24	27.16
Kr	36	33.80	32.15	93.9	94.26	84.83
Xe	54	50.74	48.26	179.1	181.7	163.5

[a] Gaussian approximation of Equations (6.5.13) and (6.5.14) of the text, calculated using Hartree–Fock wave functions.

[b] Trigonometric approximation of Equations (6.5.17) and (6.5.18), calculated using Hartree–Fock wave functions.

[c] Accurate Hartree–Fock values.

and

$$N_T = \frac{3\pi^2}{5^{3/2}} \int \rho^2(\mathbf{r})\beta^{3/2}(\mathbf{r})\, d\mathbf{r} \tag{6.5.18}$$

Results of calculations with these formulas are included in Table 6.2. They are not bad, but they are inferior to those from (6.5.13) and (6.5.14).

Comparing (6.5.15) with (6.1.14), we recognize that (6.5.15) is just a uniform-gas ansatz for the dependence of the density matrix on the interparticle coordinate s. The numerical comparisons in Table 6.2 demonstrate that for atoms the uniform-gas ansatz is not as good as the Gaussian ansatz. Figure 6.1 bears this conclusion out; in it an accurate $\Gamma(\mathbf{r}, s)$ is compared with Gaussian and uniform-gas approximations to it (Lee and Parr 1987).

At this stage we have obtained interesting approximations to Hartree–Fock theory, but how about density functionals? The preceding formulas can be turned into density-functional formulas if $\beta(\mathbf{r})$ can be expressed in terms of $\rho(\mathbf{r})$. The simplest way to do this, and an appealing idea physically, is to assume that $\beta(\mathbf{r})$ is a *function* of $\rho(\mathbf{r})$:

$$\beta(\mathbf{r}) = \beta(\rho(\mathbf{r})) \tag{6.5.19}$$

Scaling arguments then immediately give

$$\beta(\mathbf{r}) = \frac{3}{2C}\rho^{-2/3}(\mathbf{r}) \tag{6.5.20}$$

where C is a constant. Equation (6.5.9) becomes

$$T^{\mathrm{LDA}}[\rho] = C \int \rho^{5/3}(\mathbf{r})\, d\mathbf{r} \tag{6.5.21}$$

with C yet to be determined, and either (6.5.13) or (6.5.17) gives $K[\rho]$ proportional to the corresponding integral of $\rho^{4/3}(\mathbf{r})$. We use LDA to denote this *local density approximation*.

The constant C may be determined to force the number of electrons to be correct, as determined from (6.5.14) or (6.5.18). Thus $N_G = N$ gives

$$C_G = \frac{3\pi}{2^{5/3}} = 2.9686 \tag{6.5.22}$$

$$\beta_G^{\mathrm{LDA}}(\mathbf{r}) = \frac{2^{2/3}}{\pi}\rho^{-2/3}(\mathbf{r}) = 0.5053\rho^{-2/3}(\mathbf{r}) \tag{6.5.23}$$

$$T_G^{\mathrm{LDA}}[\rho] = 2.9686 \int \rho^{5/3}(\mathbf{r})\, d\mathbf{r} \tag{6.5.24}$$

$$K_G^{\mathrm{LDA}}[\rho] = 0.7937 \int \rho^{4/3}(\mathbf{r})\, d\mathbf{r} \tag{6.5.25}$$

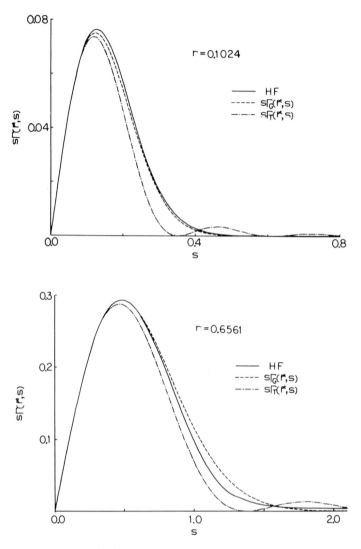

Figure 6.1 The quantity $s\Gamma(\mathbf{r}, s)$ for the neon atom. The accurate Hartree–Fock values are compared with the Gaussian approximation of (6.5.12) and the trigonometric approximation of (6.5.15) for two values of $\mathbf{r}$. Above is for the inner radial maximum, $r = 0.1024$; below for the outer maximum, $r = 0.6561$. Note the artificial, false zeros for the trigonometric approximation. (Reproduced from Lee and Parr (1987).)

Table 6.3 Approximate Kinetic Energy and Exchange Energy Density Functionals for Neutral Atoms, in Various Local-Density Approximations. Hartree–Fock Densities are Used Throughout

	He	Ne	Ar	Kr	Xe
Kinetic energy					
Exact[a]	2.862	128.55	526.81	2752.0	7232.1
Gaussian[b]	2.648	121.76	506.61	2679.3	7090.9
Trigonometric[c]	2.561	117.76	489.95	2591.2	6857.7
Exchange energy					
Exact[a]	1.026	12.11	30.18	93.85	179.1
Gaussian[b]	0.950	11.86	29.95	95.27	183.4
Trigonometric[c]	0.884	11.03	27.86	94.75	182.5

[a] Hartree–Fock values.
[b] Equations (6.5.24) and (6.5.25) of text.
[c] Equations (6.5.28) and (6.5.29) of the text (traditional Thomas–Fermi–Dirac formulas).

while $N_T = N$ gives

$$C_T = \tfrac{3}{10}(3\pi^2)^{2/3} = 2.8712 = C_F \qquad (6.5.26)$$

$$\beta_T^{\mathrm{LDA}} = 5(3\pi^2)^{-2/3}\rho^{-2/3}(\mathbf{r}) = 0.5224\rho^{-2/3}(\mathbf{r}) \qquad (6.5.27)$$

$$T_T^{\mathrm{LDA}}[\rho] = C_F \int \rho^{5/3}(\mathbf{r})\,d\mathbf{r} \qquad (6.5.28)$$

$$K_T^{\mathrm{LDA}}[\rho] = 0.7386 \int \rho^{4/3}(\mathbf{r})\,d\mathbf{r} = C_x \int \rho^{4/3}(\mathbf{r})\,d\mathbf{r} \qquad (6.5.29)$$

We have thereby recovered, with the ansatz of (6.5.15), the classical Thomas–Fermi–Dirac theory. From the alternative Gaussian ansatz of (6.5.12), we have obtained a model of precisely the same form but with modified coefficients. For atoms, the Gaussian model is better, as is shown in Table 6.3 (Lee and Parr 1987).

It is striking, to say the least, that the TFD energy functional and an improvement upon it (the Gaussian model) can be derived from two assumptions only, the resummation assumption of (6.5.12) or (6.5.15) and the local approximation of (6.5.19). The only point at which the statistics of the particles [the idempotency condition (2.5.7)] has entered is in the imposition of the gross normalization condition (6.5.1). It will be important to develop improved models by imposition of additional consequences of idempotency.

6.6 The purely local model

We pause now to show how an even simpler but still useful theory results if we make even more sweeping assumptions, not just about $K[\rho]$ but about the full $V_{ee}[\rho] = J[\rho] - K[\rho]$. After all, it is V_{ee} not K that we are really interested in, and in the Hartree–Fock approximation V_{ee}, like K, is a functional of the first-order density matrix. Furthermore, the self-repulsion (1.3.37) is a large part of both J and K but cancels between them; in modelling K one is modelling this piece that cancels anyway. This section explores how one may approximate the whole V_{ee} using approximations of the type described in the last section (Parr, Gadre, and Bartolotti 1979, Parr 1988).

From (2.4.8), the electron–electron repulsion energy, even beyond the Hartree–Fock approximation, is given by the formula

$$V_{ee} = \iint \frac{1}{r_{12}} \rho_2(\mathbf{r}_1, \mathbf{r}_2) \, d\mathbf{r}_1 \, d\mathbf{r}_2 \qquad (6.6.1)$$

where $\rho_2(\mathbf{r}_1, \mathbf{r}_2)$ is the diagonal element of the second-order spinless density matrix, (2.4.3). Furthermore, from (2.3.4) or by integration of (2.4.6),

$$\frac{N(N-1)}{2} = \iint \rho_2(\mathbf{r}_1, \mathbf{r}_2) \, d\mathbf{r}_1 \, d\mathbf{r}_2 \qquad (6.6.2)$$

We shall obtain the result we want by making an appropriate resummation assumption on $\rho_2(\mathbf{r}_1, \mathbf{r}_2)$ after expressing it in the sum and difference coordinates of (6.1.13), making a local assumption, and then imposing (6.6.2).

We write $\rho_2(\mathbf{r}_1, \mathbf{r}_2) = \rho_2(\mathbf{r} + (\mathbf{s}/2), \mathbf{r} - (\mathbf{s}/2))$ and note that ρ_2 is an even function of $\mathbf{s}$. It therefore has a formal power series expansion in the Cartesian coordinates of $\mathbf{s}$ in which linear terms are missing. Spherically averaging this, we obtain

$$\overline{\rho_2\left(\mathbf{r}+\frac{\mathbf{s}}{2}, \mathbf{r}-\frac{\mathbf{s}}{2}\right)} = \frac{1}{4\pi} \int d\Omega_s \rho_2\left(\mathbf{r}+\frac{\mathbf{s}}{2}, \mathbf{r}-\frac{\mathbf{s}}{2}\right)$$

$$= \rho_2(\mathbf{r}, \mathbf{r}) \left[1 - \frac{s^2 t_2(\mathbf{r})}{3\rho_2(\mathbf{r}, \mathbf{r})} + \cdots \right] \qquad (6.6.3)$$

where

$$t_2(\mathbf{r}) = -\tfrac{1}{2} \nabla_s^2 \rho_2\left[\left(\mathbf{r}+\frac{\mathbf{s}}{2}, \mathbf{r}-\frac{\mathbf{s}}{2}\right)\right]_{s=0} \qquad (6.6.4)$$

The equations (6.6.1) and (6.6.2) become, on performing the integrations over the angular part of $\mathbf{s}$

$$V_{ee} = 4\pi \iint \rho_2(\mathbf{r}, \mathbf{r}) \Gamma_2(\mathbf{r}, s) \, d\mathbf{r} s \, ds \qquad (6.6.5)$$

and

$$\frac{N(N-1)}{2} = 4\pi \iint \rho_2(\mathbf{r}, \mathbf{r}) \Gamma_2(\mathbf{r}, s) \, d\mathbf{r} s^2 \, ds \qquad (6.6.6)$$

where
$$\Gamma_2(\mathbf{r}, s) = 1 - \frac{s^2}{2\beta_2(\mathbf{r})} + \cdots \quad (6.6.7)$$

in which $\beta_2(\mathbf{r})$ is defined by
$$t_2(\mathbf{r}) = \frac{3}{2} \frac{\rho_2(\mathbf{r}, \mathbf{r})}{\beta_2(\mathbf{r})} \quad (6.6.8)$$

The analogy of (6.6.5) and (6.6.6) with (6.5.10) and (6.5.11) is clear.

First we make a Gaussian resummation of (6.6.7), à la Berkowitz (1986):
$$\Gamma_2(\mathbf{r}, s) \approx \exp\left(-\frac{1}{2} \frac{s^2}{\beta_2(\mathbf{r})}\right) \quad (6.6.9)$$

We get
$$V_{ee} = 4\pi \int \rho_2(\mathbf{r}, \mathbf{r}) \beta_2(\mathbf{r}) \, d\mathbf{r} \quad (6.6.10)$$

and
$$\frac{N(N-1)}{2} = (2\pi)^{3/2} \int \rho_2(\mathbf{r}, \mathbf{r}) \beta_2^{3/2}(\mathbf{r}) \, d\mathbf{r} \quad (6.6.11)$$

Now we make local density assumptions about each of the integrands, and use dimensional arguments, finding
$$\rho_2(\mathbf{r}, \mathbf{r}) \beta_2(\mathbf{r}) = A \rho^{4/3}(\mathbf{r}) \quad (6.6.12)$$

and
$$\rho_2(\mathbf{r}, \mathbf{r}) \beta_2^{3/2}(\mathbf{r}) = B \rho(\mathbf{r}) \quad (6.6.13)$$

Hence the dependence of β_2 on ρ must be (in this approximation)
$$\beta_2(\mathbf{r}) = \left(\frac{B}{A}\right)^2 \rho^{-2/3}(\mathbf{r}) \quad (6.6.14)$$

and the dependence of ρ_2 on ρ must be
$$\rho_2(\mathbf{r}, \mathbf{r}) = \frac{A^3}{B^2} \rho^2(\mathbf{r}) \quad (6.6.15)$$

But in the closed-shell Hartree–Fock case, from (2.5.19) and (2.5.25),
$$\rho_2(\mathbf{r}, \mathbf{r}) = \tfrac{1}{4} \rho^2(\mathbf{r}) \quad (6.6.16)$$

This will be a very good approximation in general, so we assume it:
$$\frac{A^3}{B^2} = \frac{1}{4} \quad (6.6.17)$$

Equation (6.6.11) then gives

$$\frac{N(N-1)}{2} = (2\pi)^{3/2}(4A^3)^{1/2}N, \qquad 4\pi A = 2^{-1/3}(N-1)^{2/3}$$

(6.6.18)

so that

$$V_{ee} = 2^{-1/3}(N-1)^{2/3} \int \rho^{4/3}(\mathbf{r})\, d\mathbf{r}$$

$$= 0.7937(N-1)^{2/3} \int \rho^{4/3}(\mathbf{r})\, d\mathbf{r} \qquad (6.6.19)$$

This is a final simplified local formula for the electron–electron repulsion energy.

Given its crude derivation, (6.6.19) is surprisingly accurate, presumably owing to the fact that forcing the normalization condition (6.6.6) on an approximation to Γ_2 makes up for errors in that approximation. Table 6.4 gives some numerical values.

In 1979 there had already been proposed (Parr, Gadre, and Bartolotti 1979) a local energy functional of the form [compare (6.1.30)]

$$E_{\text{PGB}}[\rho] = C \int \rho^{5/3}(\mathbf{r})\, d\mathbf{r} + \int \rho(\mathbf{r})v(\mathbf{r})\, d\mathbf{r} + BN^{2/3} \int \rho^{4/3}(\mathbf{r})\, d\mathbf{r} \qquad (6.6.20)$$

where $B = 0.7544$ was derived from the condition that the functional confer zero chemical potentials on neutral atoms. The agreement with (6.6.19) is extraordinary. Of course $(N-1)^{2/3}$ equals $N^{2/3}$ for large N. The former is better for small N; the atom H is nicely encompassed by (6.6.19); see also Gázquez and Robles (1981) and Pathak, Sharma, and Thakkar (1986).

Table 6.4 Test of Local Formula for V_{ee} for Atoms. Hartree–Fock Densities are Used Throughout

Atom (Z)	Hartree–Fock $V_{ee}{}^a$	$\int \rho^{4/3}(\mathbf{r})\, d\mathbf{r}^b$	V_{ee} From (6.6.19) of Text
He (2)	1.03	1.197	0.95
Ne (10)	54.0	14.94	51.3
Ar (18)	201.4	37.726	198.0
Kr (36)	1078	120.00	1019
Xe (54)	2701	230.94	2586
Rn (86)	8244	505.01	7749

[a] Values from Clementi and Roetti (1974).
[b] Values from Bartolotti (1980).

The big advantage of the energy functional of (6.6.20) is that it causes the corresponding Euler equation to have purely algebraic form. Explicit solution of the whole problem is possible, and with suitable choice of a C value (3.8738 is recommended in Parr, Gadre, and Bartolotti 1979), fairly good energies of atoms are obtained. More work on such functionals is warranted; note that they satisfy the condition (6.3.9).

6.7 Conventional gradient correction

Approximations to the kinetic energy and exchange energy functionals by the Thomas–Fermi and Dirac expressions of §6.1 are examples of the local-density approximation (LDA)—using the corresponding uniform electron gas formulas locally through the local Fermi wave vector $k_F(\mathbf{r})$ of (6.1.12). However, poor accuracy accompanies the oversimplified TF and TFD models. Systematic improvement of the TF kinetic energy functional is the subject of this section. Improvements of the Dirac exchange functional and the correlation-energy functional will be considered in §6.9 and in Chapter 8.

One would expect a better functional to exhibit effects of the inhomogeneity of the electron density, which is large in atoms and molecules but which is ignored in the LDA. This view was first taken by von Weizsacker (1935), who considered modified plane waves of the form $(1 + \mathbf{a} \cdot \mathbf{r})e^{i\mathbf{k}\cdot\mathbf{r}}$, where $\mathbf{a}$ is a constant vector and $\mathbf{k}$ is the local wave vector. There resulted the *Weizsacker correction* to the Thomas–Fermi kinetic energy; namely,

$$T_W[\rho] = \frac{1}{8}\frac{\hbar^2}{m} \int \frac{|\nabla \rho(\mathbf{r})|^2}{\rho(\mathbf{r})} d\mathbf{r} \qquad (6.7.1)$$

In order to display the dependence on $\hbar$, atomic units are not used here and throughout this section. The total kinetic energy thus becomes

$$T_{\text{TF}\lambda W}[\rho] = T_{\text{TF}}[\rho] + \lambda T_W[\rho] \qquad (6.7.2)$$

where the parameter λ is 1 in the original work of Weizsacker, but was later shown to be 1/9 from the gradient-expansion approach to be described next. There also exist other empirical values of λ, as will be considered in the next section.

The resulting TFW and TFDW models using (6.7.2) are well-defined corrections to TF and TFD theory. The behavior of the density both near and asymptotically far from the atomic nucleus is improved and the curse of the Teller nonbinding theorem of §6.3 is nullified. Various TFW and TFDW calculations for atoms and molecules will be described in the next two sections.

To generate the correction $T_W[\rho]$ and to make further improvement from first principles, formal gradient expansion techniques have been developed over the years, through the operator commutator expansion (Kirzhnits 1957, 1967, Hodges 1973, Murphy 1981), the $\hbar$ expansion of Wigner, or the Kirkwood representation of the one-particle Green's function (Jennings and Bhaduri 1975, Jennings, Bhaduri, and Brack 1975, Grammaticos and Voros 1979). We shall not trace the history, but rather simply demonstrate how $T_W[\rho]$ results from the $\hbar$ expansion of the Wigner function. The background is in Appendix D. The basic results are contained in (6.7.21) to (6.7.23).

Consider the ground state of N noninteracting electrons. The spinless first-order reduced density matrix is given by (6.1.4), or equivalently

$$\rho_1(\mathbf{r},\mathbf{r}') = 2\sum_i^\infty \phi_i(\mathbf{r})\phi_i^*(\mathbf{r}')\eta(\varepsilon_F - \varepsilon_i)$$

$$= 2\langle\mathbf{r}|\,\eta(\varepsilon_F - \hat{H})\,|\mathbf{r}'\rangle \qquad (6.7.3)$$

where $\eta(x)$ is the Heaviside step function

$$\eta(x) = \begin{cases} 1, & x > 1 \\ 0, & x < 1 \end{cases} \qquad (6.7.4)$$

and $\hat{H}$ is the one-particle Hamiltonian having ϕ_i and ε_i as its eigenstates and eigenvalues:

$$\hat{H}\phi_i(\mathbf{r}) = \{-\tfrac{1}{2}\nabla^2 + w(\mathbf{r})\}\phi_i(\mathbf{r}) = \varepsilon_i\phi_i(\mathbf{r}) \qquad (6.7.5)$$

with $w(\mathbf{r})$ the *local* potential function, that is, a local operator [see (2.1.24); note that this class of Hamiltonian rules out the Fock operator $\hat{F}$ of (2.5.12) in Hartree–Fock theory for more than two electrons] and ε_F take any value between the highest occupied and the lowest unoccupied eigenvalue. The last equality in (6.7.3) may be verified by inserting into (6.7.3) an identity operator decomposed in the eigenstates of $\hat{H}$, namely,

$$I = \sum_i^\infty |\phi_i\rangle\langle\phi_i|$$

Our goal is to express the kinetic energy T as a functional of the electron density. The idea is the following: T is determined by $\rho_1(\mathbf{r},\mathbf{r}')$ as in (6.1.6), $\rho_1(\mathbf{r},\mathbf{r}')$ in turn by $w(\mathbf{r})$ through (6.7.3); electron density $\rho(\mathbf{r})$ as the diagonal of $\rho_1(\mathbf{r},\mathbf{r}')$ is also determined by $w(\mathbf{r})$; therefore we can hope to use $w(\mathbf{r})$ as a bridge to connect T to $\rho(\mathbf{r})$.

The key problem is to find $\rho_1(\mathbf{r},\mathbf{r}')$ in terms of $w(\mathbf{r})$. Note that as expressed in (6.7.3), the N-electron quantity $\rho_1(\mathbf{r},\mathbf{r}')$ is the matrix representation of a one-particle operator $\hat{\rho}_1 = \eta(\varepsilon_F - \hat{H})$. A one-electron problem is much easier to handle than an N-electron one. Many techniques have been developed to manipulate and approximate the single-electron Green's function (propagator), defined as

$$G(\mathbf{r}, \mathbf{r}'; \beta) = \langle \mathbf{r}| e^{-\beta \hat{H}} |\mathbf{r}' \rangle$$

$$= \sum_i \phi_i(\mathbf{r})\phi_i^*(\mathbf{r}')e^{-\beta \varepsilon_i} \qquad (6.7.6)$$

where the second equality follows like (6.7.3). $G(\mathbf{r}, \mathbf{r}')$ is in turn related to $\rho_1(\mathbf{r}, \mathbf{r}')$ by an inverse Laplace transform; namely (Golden 1960),

$$\rho_1(\mathbf{r}, \mathbf{r}') = \frac{2}{2\pi i} \int_{\gamma-i\infty}^{\gamma+i\infty} \frac{d\beta}{\beta} e^{\beta \varepsilon_F} G(\mathbf{r}, \mathbf{r}'; \beta) \qquad (6.7.7)$$

where γ is any positive constant. To prove (6.7.7), we only need show the same equality in terms of eigenvalues,

$$\eta(\varepsilon_F - \varepsilon_i) = \frac{1}{2\pi i} \int_{\gamma-i\infty}^{\gamma+i\infty} \frac{d\beta}{\beta} e^{\beta \varepsilon_F} e^{-\beta \varepsilon_i} \qquad (6.7.8)$$

We may use the calculus of residues to verify (6.7.8). For $\varepsilon_F > \varepsilon_i$, the β-integration contour can be closed on the left-hand side of the complex-β plane and it becomes $C^1 + C^2$, with $R \to \infty$ (see Figure 6.2). The contribution from C^2 vanishes as $R \to \infty$. The result of closed-contour integration is given by the residue of $(1/\beta)e^{\beta(\varepsilon_F - \varepsilon_i)}$ at the pole $\beta = 0$, which is equal to 1. For $\varepsilon_F < \varepsilon_i$, the β-integration contour can be closed on the right-hand side of the complex-β plane and the result is zero because there is no singularity within the closed contour. Thereby (6.7.8) is proved, and hence (6.7.7) also.

We now invoke $\tilde{G}(\mathbf{r}, \mathbf{p}; \beta)$, the Wigner transformation of $G(\mathbf{r}, \mathbf{r}'; \beta)$ (see Appendix D for details). Following Equation (D.1), define

$$\tilde{G}(\mathbf{r}, \mathbf{p}; \beta) = \frac{1}{(2\pi\hbar)^3} \int d\mathbf{s} \left\langle \mathbf{r} - \frac{\mathbf{s}}{2} \right| e^{-\beta \hat{H}} \left| \mathbf{r} + \frac{\mathbf{s}}{2} \right\rangle e^{i\mathbf{p}\cdot\mathbf{s}/\hbar}$$

$$= \frac{1}{(2\pi\hbar)^3} \int d\mathbf{s} \, G\left(\mathbf{r} - \frac{\mathbf{s}}{2}, \mathbf{r} + \frac{\mathbf{s}}{2}; \beta\right) e^{i\mathbf{p}\cdot\mathbf{s}/\hbar} \qquad (6.7.9)$$

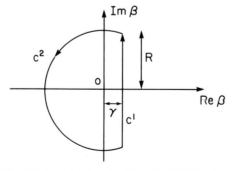

Figure 6.2 Integration contour for Equation (6.7.8) when $\varepsilon_F < \varepsilon_i$.

Then, by (D.3)

$$G(\mathbf{r}, \mathbf{r}'; \beta) = \int d\mathbf{p}\, e^{-i\mathbf{p}\cdot(\mathbf{r}-\mathbf{r}')/\hbar} \tilde{G}(\tfrac{1}{2}(\mathbf{r}+\mathbf{r}'), \mathbf{p}; \beta) \qquad (6.7.10)$$

Inserting (6.7.10) in (6.7.7), we obtain

$$\rho_1(\mathbf{r}, \mathbf{r}') = \int d\mathbf{p}\, e^{-i\mathbf{p}\cdot(\mathbf{r}-\mathbf{r}')/\hbar} \tilde{\rho}_1(\tfrac{1}{2}(\mathbf{r}+\mathbf{r}'), \mathbf{p}) \qquad (6.7.11)$$

where

$$\tilde{\rho}_1(\mathbf{r}, \mathbf{p}) = \frac{2}{2\pi i} \int_{\gamma-i\infty}^{\gamma+i\infty} \frac{d\beta}{\beta} e^{\beta \varepsilon_F} \tilde{G}(\mathbf{r}, \mathbf{p}; \beta) \qquad (6.7.12)$$

Via the foregoing equations, the $\hbar$ expansion of $\tilde{G}$ developed in Appendix D provides an explicit approximation of $\rho_1(\mathbf{r}, \mathbf{r}')$ in terms of the potential $w(\mathbf{r})$. Thus, from (D.30), to the order $\hbar^2$,

$$\tilde{G}(\mathbf{r}, \mathbf{p}; \beta) = \frac{1}{(2\pi\hbar)^3} \exp[-\beta H(\mathbf{r}, \mathbf{p})]$$
$$\times \left\{ 1 + \hbar^2 \left[-\frac{\beta^2}{8m} \nabla^2 w(\mathbf{r}) + \frac{\beta^3}{24m} |\nabla w(\mathbf{r})|^2 + \frac{\beta^3}{24m} (\mathbf{p}\cdot\nabla)(\mathbf{p}\cdot\nabla) w(\mathbf{r}) \right] \right\} \qquad (6.7.13)$$

with

$$H(\mathbf{r}, \mathbf{p}) = \frac{p^2}{2m} + w(\mathbf{r}) \qquad (6.7.14)$$

Correspondingly, by (6.7.12),

$$\tilde{\rho}_1(\mathbf{r}, \mathbf{p}) = \frac{2}{(2\pi\hbar)^3} \eta(\varepsilon_F - H(\mathbf{r}, \mathbf{p})) + \frac{2\hbar^2}{(2\pi\hbar)^3}$$
$$\times \left\{ -\frac{1}{8m} \eta^{(2)}(\varepsilon_F - H(\mathbf{r}, \mathbf{p})) \nabla^2 w(\mathbf{r}) + \eta^{(3)}(\varepsilon_F - H(\mathbf{r}, \mathbf{p})) \right.$$
$$\left. \times \left[\frac{1}{24m} |\nabla w(\mathbf{r})|^2 + \frac{1}{24m^2} (\mathbf{p}\cdot\nabla)(\mathbf{p}\cdot\nabla) w(\mathbf{r}) \right] \right\} \qquad (6.7.15)$$

where (6.7.8) has been used and

$$\eta^{(n)}(\varepsilon_F - H(\mathbf{r}, \mathbf{p})) \equiv \frac{\partial^n}{\partial \varepsilon_F^n} \eta(\varepsilon_F - H(\mathbf{r}, \mathbf{p})) \qquad (6.7.16)$$

In the limit $\hbar \to 0$, only the first term of (6.7.15)—remains. This is the step-function distribution in phase space—the classical approximation to the step-function operator (6.7.3) in Wigner transformation.

We are now ready to compute the basic quantities with which we are concerned. On inserting (6.7.15) into (6.7.11), we find for the total electron density,

$$\rho(\mathbf{r}) = \rho_1(\mathbf{r}, \mathbf{r}) = \int \tilde{\rho}_1(\mathbf{r}, \mathbf{p}) \, d\mathbf{p}$$

$$= \frac{1}{3\pi^2} \left(\frac{2m}{\hbar^2}\right)^{3/2} [\varepsilon_F - w(\mathbf{r})]^{3/2} \eta(\varepsilon_F - w(\mathbf{r}))$$

$$\times \left\{1 - \frac{1}{8} \frac{\hbar^2}{2m} [\nabla^2 w(\mathbf{r})(\varepsilon_F - w(\mathbf{r}))^{-2} + \tfrac{1}{4} |\nabla_w(\mathbf{r})|^2 (\varepsilon_F - w(\mathbf{r}))^{-3}]\right\}$$

(6.7.17)

where we have used the integral formula

$$\int \eta\left(\varepsilon_F - \frac{\mathbf{p}^2}{2m} - w(\mathbf{r})\right) d\mathbf{p} = \int \eta\left(\varepsilon_F - w(\mathbf{r}) - \frac{\mathbf{p}^2}{2m}\right) 4\pi p^2 \, dp$$

$$= \frac{4\pi}{3} p_F^3(\mathbf{r}) = \frac{4\pi}{3} [2m(\varepsilon_F - w(\mathbf{r}))]^{3/2} \eta(\varepsilon_F - w(\mathbf{r}))$$

(6.7.18)

and its derivatives with respect to ε_F. The total kinetic energy according to (D.8) is given by

$$T = \int \frac{p^2}{2m} \tilde{\rho}_1(\mathbf{r}, \mathbf{p}) \, d\mathbf{r} \, d\mathbf{p}$$

$$= \int d\mathbf{r} \frac{\hbar^2}{10\pi^2 m} \left(\frac{2m}{\hbar^2}\right)^{5/2} [\varepsilon_F - w(\mathbf{r})]^{5/2} \eta(\varepsilon_F - w(\mathbf{r}))$$

$$\times \left\{1 - \frac{5}{8} \frac{\hbar^2}{2m} [\tfrac{5}{3}\nabla^2 w(\mathbf{r})(\varepsilon_F - w(\mathbf{r}))^{-2} - \tfrac{3}{4} |\nabla w(\mathbf{r})|^2 (\varepsilon_F - w(\mathbf{r}))^{-3}]\right\}$$

(6.7.19)

where we have employed the integral formula

$$\int \eta\left(\varepsilon_F - \frac{\mathbf{p}^2}{2m} - w(\mathbf{r})\right) \mathbf{p}^2 \, d\mathbf{p} = \int \eta\left(\varepsilon_F - \frac{\mathbf{p}^2}{2m} - w(\mathbf{r})\right) 4\pi p^4 \, dp$$

$$= \frac{4\pi}{5} [2m(\varepsilon_F - w(\mathbf{r}))]^{5/2} \eta(\varepsilon_F - w(\mathbf{r})) \quad (6.7.20)$$

and its derivatives with respect to ε_F.

It is now elementary to eliminate $\varepsilon_F - w(\mathbf{r})$ from (6.7.19) by use of (6.7.17). We only need to keep the accuracy of this elimination consistently to the order $\hbar^2$. The result found is the TF-$\frac{1}{9}$W functional:

$$T^{(2)}[\rho] = T_{TF}[\rho] + \frac{1}{9}T_W[\rho]$$

$$= \frac{\hbar^2}{m}\left[\frac{3}{10}(3\pi^2)^{2/3}\rho(\mathbf{r})^{5/3}\,d\mathbf{r} + \frac{1}{9}\cdot\frac{1}{8}\int\frac{|\nabla\rho(\mathbf{r})|^2}{\rho(\mathbf{r})}\,d\mathbf{r}\right] \quad (6.7.21)$$

If we used only the zeroth-order approximation, we would obtain only the TF kinetic energy. For a derivation of this formula from a mean-path approximation to $G(\mathbf{r}, \mathbf{r}'; \beta)$, see Yang (1986).

We can also use the $\hbar$ expansion to higher order in $\hbar$ to get higher-order gradient correction; for details, see Hodges (1973), Murphy (1981), and Grammaticos and Voros (1979). The gradient expansion to fourth order is given by (Hodges 1973)

$$T^{(4)}[\rho] = T_{TF}[\rho] + \frac{1}{9}T_W[\rho] + T_H[\rho] \quad (6.7.22)$$

$$T_H[\rho] = \frac{\hbar^2}{m}\frac{1}{540(3\pi^2)^{2/3}}$$

$$\times \int\left[\frac{(\nabla^2\rho)^2}{\rho^{5/3}} - \frac{9}{8}\frac{\nabla^2\rho(\nabla\rho)^2}{\rho^{8/3}} + \frac{1}{3}\frac{(\nabla\rho)^4}{\rho^{11/3}}\right]d\mathbf{r} \quad (6.7.23)$$

The sixth-order formula may be found in Murphy (1981).

It is very satisfying that the TF kinetic energy and its systematic gradient corrections can be derived in this way from the semiclassical $\hbar$-expansion of the Wigner transform of the one-particle Green's function. Indeed in zeroth order this constitutes a derivation from first principles of the Thomas–Fermi kinetic-energy formula itself that makes no use of an ad hoc assumption of local behavior.

One might naively expect that higher accuracy would be achieved with higher-order gradient corrections. For atoms and molecules this is only true to fourth order, as will be documented in the next two sections. To one's dismay, the sixth-order gradient correction for atoms diverges!

As an approximation to the exact $T[\rho]$, $T^{(4)}[\rho]$ is still poor; only an average behavior of the electrons is described. And the road to better accuracy through gradient corrections ends at fourth order. However, the idea of employing just the electron density as in the TF theory is not dead, for recently there has been developed an integral formulation of density-functional theory that uses it. This will be the subject of §7.5.

6.8 The Thomas–Fermi–Dirac–Weizsacker model

Having developed the gradient correction for the Thomas–Fermi kinetic-energy functional, we now add it to the existing Thomas–Fermi–Dirac energy functional of (6.1.22). To second order we then obtain the Thomas–Fermi–Dirac–Weizsacker (TFDW or TFD-λW) model,

$$E_{\text{TFD-}\lambda\text{W}}[\rho] = C_F \int \rho^{5/3}(\mathbf{r})\, d\mathbf{r} + \lambda \frac{1}{8} \int \frac{|\nabla \rho(\mathbf{r})|^2}{\rho(\mathbf{r})}\, d\mathbf{r}$$
$$+ \int \rho(\mathbf{r}) v(\mathbf{r})\, d\mathbf{r} + J[\rho] - C_x \int \rho^{4/3}(\mathbf{r})\, d\mathbf{r} \qquad (6.8.1)$$

The parameter λ is 1/9 in the conventional gradient expansion of the last section. Other empirical values of λ have also been employed, however: $\lambda = 1/5$ by Yonei and Tomishima (1965), based on a fitting for hydrogenic atoms (noninteracting electrons in the Coulomb potential due to the nucleus); $\lambda = 0.186$ by Lieb (1981) from analysis of the large-atomic-number limit of atoms; and $\lambda = 1.4/9 \sim 1.5/9$ ($\sim 1/6$) by Brack (1985) as a means to effectively include the fourth-order effect in the second-order formula. Justification from a Green's-function approach has been provided for all these λ values (Yang 1986). Compare the discussion in Levy, Perdew, and Sahni (1984), which emphasizes the case $\lambda = 1$.

The Euler equation for TFD-λW atoms or molecules becomes

$$\mu_{\text{TFD-}\lambda\text{W}} = \frac{5}{3} C_F \rho^{2/3}(\mathbf{r}) + \frac{\lambda}{8} \left[\frac{|\nabla \rho(\mathbf{r})|^2}{\rho^2(\mathbf{r})} - 2 \frac{\nabla^2 \rho(\mathbf{r})}{\rho(\mathbf{r})} \right]$$
$$- \phi(\mathbf{r}) - \frac{4}{3} C_x \rho^{1/3}(\mathbf{r}) \qquad (6.8.2)$$

where the rules of Appendix A have been employed to evaluate the functional derivative of the gradient term, and $\phi(\mathbf{r})$ is the classical electrostatic potential. The solution of (6.8.2), with $\mu_{\text{TFD-}\lambda\text{W}}$ determined by proper normalization, is the electron density of the TFD-λW atom. Although this model has been established for a long time, numerical investigations of it have not been available until recently (Tomishima and Yonei 1966, Gross and Dreizler 1979, Stich, Gross, Malzacher, and Dreizler 1982, Yang 1986).

The TFDW model provides considerable improvement over the TF or TFD model (Gross and Dreizler 1979). The electron density is finite at the nucleus instead of diverging as it does in the TF or TFD model—Equation (6.2.17). The electron density decreases exponentially far from the nucleus,

$$\rho_{\text{TFD-}\lambda\text{W}} \sim \left(\frac{1}{r^2}\right) \exp\left[-\left(-\frac{8\mu_{\text{TFDW}}}{\lambda}\right)^{1/2} r\right], \quad \text{for } r \to \infty \qquad (6.8.3)$$

in contrast with the improper inverse power decay of (6.2.19). Also, inclusion of the gradient term in the energy functional invalidates Teller's nonbinding theorem (see §6.3) and makes molecular binding possible in

Table 6.5 Atomic Energies from Various TFD-λW Models (a.u.)[a,b,c]

	TF	TFD	TFD$\frac{1}{9}$W	TFD$\frac{1}{6}$W	TFD(0.186)W	TFD$\frac{1}{5}$W	TFD$\frac{1}{3}$W	TFDW	HF
Ne	−165.61	−176.3	−139.91	−132.53	−130.33	−128.83	−117.09	−86.43	−128.55
Ar	−652.72	−680.7	−561.98	−537.34	−529.94	−524.91	−485.00	−378.51	−526.82
Kr	−3289.50	−3377.9	−2898.54	−2796.95	−2766.29	−2745.60	−2578.10	−2132.19	−2752.05
Xe	−8472.46	−8646.1	−7563.13	−7330.95	−7260.72	−7213.92	−6827.59	−5828.96	−7232.13

[a] Values for TFD, TFD$\frac{1}{5}$W, and TFDW from Tomishima and Yonei (1966); HF results from Clementi and Roetti (1974).
[b] TF results from (6.2.23) of the text.
[c] TFD$\frac{1}{9}$W, TFD$\frac{1}{6}$W, TFD(0.186)W, TFD$\frac{1}{3}$W results. From Yang (1986).

the TFD-λW model. However, because of numerical difficulties, even for diatomic molecules, the strength of binding in the TFD-λW model is still uncertain (Yonei 1971, Gross and Dreizler 1979, Berk 1983).

There are two main ways to solve (6.8.2): the finite-difference method (Tomishima and Yonei 1966, Yang 1986) and the spline-representation method (Stich, Gross, Malzacher, and Dreizler 1982). The procedure in the first case is the following. (1) Make an initial guess of the density $\rho_0(\mathbf{r})$. (2) Compute the electrostatic potential $\phi_0(\mathbf{r})$ from this guessed $\rho_0(\mathbf{r})$. (3) Use this $\rho_0(\mathbf{r})$ in (6.8.2) to solve for a new density $\rho(\mathbf{r})$ whose normalization determines the value of $\mu_{\text{TFD-}\lambda\text{W}}$. (4) Insert $\rho(\mathbf{r})$ back in step (1) and repeat until self-consistency is reached. Total atomic energies for the inert gas atoms Ne, Ar, Kr, and Xe in the TFD-λW model are listed in Table 6.5, for various λ values. The behavior of the resulting electron density is shown in Figure 6.3. The paper of Stich, Gross, Malzacher, and Dreizler (1982) contains more atomic calculations with $\lambda = 1/5$ and also some results for positive ions.

From Table 6.5, it is evident that the TFD-λW models (except for $\lambda = 1$) all considerably improve the total energies of the TF or TFD model, and that indeed $\lambda = 1/5$ is an optimal choice for approximation of total energy, in agreement with the argument of Yonei and Tomishima (1965). However, analysis of the various energy components shows no strong preference for the value $\lambda = 1/5$ (Yang 1986). As shown in Figure 6.3, the TFD-λW atomic density is a beautiful average of the HF density, smoothing out the atomic shell structure.

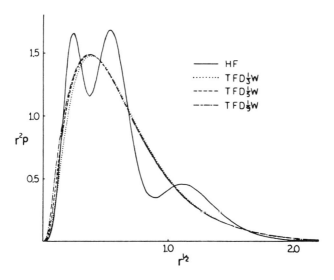

Figure 6.3 Electron density for argon from various models (after Yang 1986.)

The TFD-λW model, very recently extended by Engel and Dreizler (1989) to include fourth-order gradient correction, is the most sophisticated of the TF-type models for many-electron systems that has been solved self-consistently. Note that inclusion of higher-order gradient correction in the energy functional (6.8.1) tremendously complicates the resulting Euler equation. The TFD-λW model is still very unsatisfactory and cannot be regarded as a rigorous approach. Atomic shell structure is absent and molecular binding is in doubt. Also, negative ions are not stable (Stich, Gross, Malzacher, and Dreizler 1982).

6.9 Various related considerations

The gradient expansion to fourth order of §6.7 reproduces well the electronic kinetic energies of atoms and molecules if one inputs into it Hartree–Fock electron densities (Wang, Parr, Murphy, and Henderson 1976, Shih 1976, Murphy and Parr 1979, Murphy and Wang 1980, Wang and Rasolt 1976, Allan, West, Cooper, Grout, and March 1985, Lee and Ghosh 1986). From (6.7.23) the gradient expansion of $T[\rho]$ to fourth order may be written

$$T^{(4)}[\rho] = T_0[\rho] + T_2[\rho] + T_4[\rho] \tag{6.9.1}$$

where

$$T_0[\rho] = T_{\text{TF}}[\rho], \quad T_2[\rho] = \tfrac{1}{9} T_W[\rho], \quad T_4[\rho] = T_H[\rho] \tag{6.9.2}$$

Tables 6.6 and 6.7 present some numerical results for atoms (Murphy and Wang 1980) and molecules (Lee and Ghosh 1986). The results are remarkably good, although hardly quantitative. Extending this approach is neither feasible nor promising, however. $T_6[\rho]$ diverges for atoms.

Closer to what one wants, of course, is accuracy in the *local kinetic energy density* $t[\rho(\mathbf{r})] = t(\mathbf{r})$. This is defined up to an additive contribution $C\nabla^2\rho$, where C is an arbitrary constant, by the formula $T[\rho] = \int t(\mathbf{r}) \, d\mathbf{r}$. To complete the definition, one may take [cf. (6.5.6), where for another purpose a different amount of $\nabla^2\rho$ was included]

$$\begin{aligned} t(\mathbf{r}) &= \tfrac{1}{2}[\nabla_1 \nabla_2 \rho_1(\mathbf{r}_1, \mathbf{r}_2)]_{\mathbf{r}_1=\mathbf{r}_2=\mathbf{r}} \\ &= \tfrac{1}{8} \sum_i n_i \left[\frac{\nabla \rho_i(\mathbf{r}) \cdot \nabla \rho_i(\mathbf{r})}{\rho_i(\mathbf{r})} \right] \end{aligned} \tag{6.9.3}$$

Here the sum is over natural spin-orbital components; in Hartree–Fock approximation it is over the N Hartree–Fock orbital components with $n_i = 1$. With this definition,

$$t(\mathbf{r}) \geq 0 \tag{6.9.4}$$

which is a strong and useful requirement on any approximation. Equation (6.9.3) has been shown to nicely exhibit shell structure for atoms, and

Table 6.6 Kinetic Energy Gradient-Expansion Components for Hartree–Fock Noble-gas Atoms. Percentage Errors in Parentheses

Atom (Z)	$T_{HF}{}^a$	T_0		T_2	T_4	$T_0 + T_2$		$T_0 + T_2 + T_4$	
He (2)	2.8617	2.5605	(−10.5)	0.3180	0.0846	2.8785	(0.59)	2.9631	(3.54)
Ne (10)	128.55	117.77	(−8.39)	10.07	1.94	127.84	(−0.55)	129.78	(0.96)
Ar (18)	526.82	489.95	(−7.00)	34.27	6.21	524.22	(−0.49)	530.43	(0.69)
Kr (36)	2752.1	2591.2	(−5.85)	141.9	24.1	2733.0	(−0.69)	2575.1	(0.18)
Xe (54)	7232.1	6857.7	(−5.18)	325.8	53.8	7183.5	(−0.67)	7237.3	(0.07)

$^a T_{HF} = -E_{HF}$, E_{HF} from Clementi and Roetti (1974).

Table 6.7 Kinetic Energies for Molecules. Quantities in Parentheses Indicate Relative Errors, in Percent, With Respect to Hartree–Fock Values

Molecule	T_{HF}	T_0	$T_0 + T_2$	$T_0 + T_2 + T_4$
H_2	1.128	0.981 (−13.0)	1.107 (−1.88)	1.178 (4.44)
BH_3	26.34	23.83 (−9.5)	26.38 (0.17)	26.94 (2.29)
CH_4	40.17	36.45 (−9.3)	40.10 (−0.19)	40.91 (1.85)
HF	100.0	91.30 (−8.7)	99.42 (−0.59)	101.0 (1.00)
NH_3	56.16	50.97 (−9.2)	55.88 (−0.49)	56.90 (1.32)
H_2O	76.08	69.19 (−9.1)	75.61 (−0.62)	76.92 (1.11)
BF	124.2	113.2 (−8.9)	123.7 (−0.41)	125.8 (1.30)
CO	112.7	102.2 (−9.3)	112.1 (−0.52)	114.1 (1.28)
C_2H_2	76.70	69.52 (−9.4)	76.52 (−0.25)	77.93 (1.60)
N_2	108.7	98.53 (−9.4)	108.1 (−0.54)	110.1 (1.25)
H_2CO	113.8	103.3 (−9.2)	113.2 (−0.54)	115.3 (1.30)
F_2	198.5	180.6 (−9.0)	196.8 (−0.85)	199.9 (0.66)
CO_2	187.5	170.3 (−9.2)	186.42 (−0.60)	189.5 (1.05)
N_2O	183.6	166.7 (−9.2)	182.5 (−0.59)	186.2 (1.38)

useful comparisons between (6.9.3) and second-order gradient-expansion approximations to it have been made (Yang, Parr, and Lee 1986).

There are many other studies on kinetic energy involving the local behavior of kinetic energy density (for example, the important early paper by Goodisman 1970, Baltin 1972, Alonso and Girifalco 1978a, Tal and Bader 1978, Murphy and Parr 1979, Zorita, Alonso, and Balbás 1985, Pearson and Gordon 1985). Various forms of kinetic energy functional have been proposed (Haq, Chattaraj, and Deb 1984, Ghosh and Balbás 1985, Herring 1986, Yang, Parr, and Lee 1986, DePristo and Kress 1986, Herring and Chopra 1988, Plindov and Pogrebnya 1988).

If one forgets the Thomas–Fermi reference point for a moment and just asks what amount of $T_W[\rho]$ one would expect in a *good* kinetic energy functional, the answer at first glance would appear to be "all of it," for the correct long-range behavior and the correct cusps at nuclei both follow from such a term [if other terms are not present that swamp the effect of (6.7.1)]. Also, there is an argument from information theory that this term should be a leading component of $T[\rho]$ (Sears, Parr, and Dinur 1980; see also Gázquez and Ludena 1981). So one might conjecture that $T[\rho] = T_W[\rho] +$ "rest." What would be the rest? Hopefully, or at least perhaps, a "statistical" estimate will suffice for it. $T_0[\rho]$ itself is too much. But a formula for atoms and ions,

$$T[\rho] = T_W[\rho] + \gamma(N, Z)T_0[\rho] \qquad (6.9.5)$$

does well (Acharya, Bartolotti, Sears, and Parr 1980). Figure 6.4 shows the $\gamma(N, N)$ that exactly fit the Hartree–Fock kinetic energies of neutral atoms. To high accuracy, as the figure shows, $\gamma(N, N)$ has a nice functional form,

$$\gamma(N, N) = 1 - \frac{1.412}{N^{1/3}} \qquad (6.9.6)$$

This was discovered empirically, but Gázquez and Robles (1982) later provided a derivation of much the same result (see also Acharya 1983). Variational calculations with (6.9.5) have been done (Bartolotti and Acharya 1982), giving modest but not impressive accuracy: no shell structure is obtained. The good fit in Figure 6.4 provides no assurance, of course, that variational calculations using (6.9.5) will give good results.

One gets considerably improved results, including shell structure, if one replaces (6.9.5) with a formula of the form

$$T[\rho] = T_W[\rho] + C_F \int f(\mathbf{r})\rho^{5/3}(\mathbf{r})\, d\mathbf{r} \qquad (6.9.7)$$

where $f(\mathbf{r})$ is suitably determined (Deb and Ghosh 1983). An instructive rationale for (6.9.7) [or the less accurate (6.9.5)] (Acharya, Bartolotti, Sears, and Parr 1980) is that $T_W[\rho]$ is a correct estimate of the kinetic energy of the atomic $1s^2$ shell and $C_F \int [1 - f(\mathbf{r})]\rho^{5/3}(\mathbf{r})\, d\mathbf{r}$ is an incorrect statistical estimate of that energy. If we rewrite (6.9.7) in the form

$$T[\rho] = T_0[\rho] - C_F \int [1 - f(\mathbf{r})]\rho^{5/3}(\mathbf{r})\, d\mathbf{r} + T_W[\rho] \qquad (6.9.8)$$

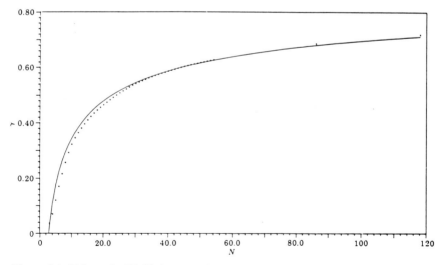

Figure 6.4. Values of $\gamma(N, N)$ for neutral atoms, calculated from Hartree–Fock electron densities and (6.9.5). The curve is the one-parameter fit of (6.9.6).

we see that this amounts to replacing the incorrect statistical estimate for the innermost shell with an improved value, accepting the statistical picture for the rest of the electrons. This idea is present in recent work by Englert and Schwinger (1982, 1984a,b,c), and was anticipated in a study by Tal and Bader (1978), who used $T_W[\rho]$ close to the nucleus and $T_0[\rho]$ outside. Nyden and Acharya (1981) did away with the "core" entirely. Many other works also favor the use of $T_W[\rho]$ as a kinetic energy component (Ludena 1983, Levy, Perdew, and Sahni 1984, Plumer and Stott 1985, March 1986a,b, Baltin 1987).

Deb and Chattaraj (1988) have proposed a first-order gradient correction to the kinetic energy of the form $-(1/40)\int \mathbf{r}\cdot(\nabla\rho/r^2)\,d\mathbf{r}$. This amounts to a generalization of a proposal by Langer (1937) to include a term $\int(\rho/r^2)\,d\mathbf{r}$ for atoms, studied also by Cummins and Nordholm (1984). Chattaraj and Deb (1984) have written an excellent review of the $T[\rho]$ problem.

The exchange energy $K[\rho]$ also possesses gradient expansions. To second order, the standard form is

$$K^{(2)}[\rho] = K_0[\rho] + K_2[\rho] \tag{6.9.9}$$

where

$$K_0[\rho] = K_{\text{TFD}}[\rho] = C_x \int \rho^{4/3}(\mathbf{r})\,d\mathbf{r} \tag{6.9.10}$$

and

$$K_2[\rho] = \beta \int \frac{\nabla\rho(\mathbf{r})\cdot\nabla\rho(\mathbf{r})}{\rho^{4/3}(\mathbf{r})}\,d\mathbf{r} \tag{6.9.11}$$

with β a constant. A purely theoretical value of β (Sham 1971; Gross and Dreizler 1981) is

$$\beta_s = \tfrac{7}{144}(81\pi^5)^{-1/3} = 1.667 \times 10^{-3} \tag{6.9.12}$$

Kleinman (1984) gave a value of $\tfrac{8}{7}\beta_s$ for infinite systems. Empirically, a value for β of about 5.5×10^{-3} is better (Herman, Van Dyke, and Ortenburger 1969, Shih, Murphy, and Wang 1980). The derivation given by Gross and Dreizler (1981) is very similar to the derivation in §6.7 for the second-order gradient correction to Thomas–Fermi kinetic energy. However, to do away with an apparent divergence. Gross and Dreizler replace the Coulomb interaction $1/r$ by a screened interaction $\exp(-\alpha r)/r$ and take the limit $\alpha \to 0$ at the end of the calculation.

There exist various other approximate exchange-energy functionals. Indeed, for atoms, a generally better form for $K[\rho]$ than (6.9.9) is (Bartolotti 1982a).

$$K[\rho] = \int \rho^{4/3}(\mathbf{r})[C(N) + D(N)y^2]\,d\mathbf{r}, \qquad y = \frac{d\ln\rho}{d\ln r} \tag{6.9.13}$$

with $C(N) = C_F - 0.3599 N^{-2/3}$, $D(N) = 2.009 \times 10^{-3} N^{-2/3}$. From a real-space analysis following Gross and Dreizler (1981), Perdew (1985a) provided a model that approximates the Hartree–Fock exchange energy to within 1%. Ghosh and Parr (1986) developed a phase-space approach based on a maximum entropy principle and obtained the nonlocal form (6.5.13). And both DePristo and Kress (1987) and Cedillo, Robles, and Gázquez (1988) construct rational-function approximations for the exchange energy functional. In §8.7, the effects of electron exchange will be readdressed in the context of a full exchange-correlation functional.

Efforts to find an accurate functional $T[\rho]$ by extensions of the Thomas–Fermi–Dirac–Weizsacker model have continued for many years—the problem is a very difficult one at best. An alternative approach is developed in the following chapter, the Kohn–Sham (1965) formulation of the density-functional theory, in which there is introduced the concept of corresponding non-interacting system, with kinetic energy $T_s[\rho]$ that can be accurately computed and is close to the exact $T[\rho]$. The Kohn–Sham idea dominates present-day density-functional calculations.

7
THE KOHN-SHAM METHOD: BASIC PRINCIPLES

7.1 Introduction of orbitals and the Kohn–Sham equations

It is indeed appealing that the ground-state energy of a many-electron system can be obtained as the minimum of the energy functional

$$E[\rho] = \int \rho(\mathbf{r})v(\mathbf{r}) \, d\mathbf{r} + F[\rho] \qquad (7.1.1)$$

where

$$F[\rho] = T[\rho] + V_{ee}[\rho] \qquad (7.1.2)$$

all terms having been defined in Chapter 3. The ground-state electron density is the density that minimizes $E[\rho]$ and hence satisfies the Euler equation

$$\mu = v(\mathbf{r}) + \frac{\delta F[\rho]}{\delta \rho(\mathbf{r})} \qquad (7.1.3)$$

where μ is the Lagrange multiplier associated with the constraint

$$\int \rho(\mathbf{r}) \, d\mathbf{r} = N \qquad (7.1.4)$$

Among all possible solutions of (7.1.3), one takes that which minimizes $E[\rho]$ (as opposed to those associated with other extrema).

We have seen in Chapter 6 how one can proceed to an approximate implementation by making drastic assumptions. How can we do better? How can we avoid the great loss in accuracy associated with the Thomas–Fermi model and its derivative models?

Thomas–Fermi and related models constitute a direct approach, whereby one constructs explicit approximate forms for $T[\rho]$ and $V_{ee}[\rho]$. This produces a nice simplicity—the equations involve electron density alone. Unfortunately, however, as has been said in Chapter 6, there are seemingly insurmountable difficulties in going beyond the crude level of approximation. In a trade of simplicity for accuracy, Kohn and Sham (1965) invented an ingenious indirect approach to the kinetic-energy functional $T[\rho]$, the Kohn–Sham (KS) method. They thereby turned density functional theory into a practical tool for rigorous calculations. The Kohn–Sham method is developed in this and the next chapter.

Kohn and Sham proposed introducing orbitals into the problem in such a way that the kinetic energy can be computed simply to good accuracy, leaving a small residual correction that is handled separately. To understand what is involved and what Kohn and Sham did, it is convenient to begin with the exact formula for the ground-state kinetic energy,

$$T = \sum_i^N n_i \langle \psi_i | -\tfrac{1}{2}\nabla^2 | \psi_i \rangle \tag{7.1.5}$$

where the ψ_i and n_i are, respectively, natural spin orbitals and their occupation numbers. As was described in §2.3, the Pauli principle requires that $0 \leq n_i \leq 1$; we are assured from the Hohenberg–Kohn theory that this T is a functional of the total electron density

$$\rho(\mathbf{r}) = \sum_i^N n_i \sum_s |\psi_i(\mathbf{r}, s)|^2 \tag{7.1.6}$$

For any *interacting* system of interest, there are an infinite number of terms in (7.1.5) or (7.1.6), which is ponderous at best. Kohn and Sham (1965) showed that one can build a theory using simpler formulas, namely

$$T_s[\rho] = \sum_i^N \langle \psi_i | -\tfrac{1}{2}\nabla^2 | \psi_i \rangle \tag{7.1.7}$$

and

$$\rho(\mathbf{r}) = \sum_i^N \sum_s |\psi_i(\mathbf{r}, s)|^2 \tag{7.1.8}$$

Equations (7.1.7) and (7.1.8) are the special case of (7.1.5) and (7.1.6) having $n_i = 1$ for N orbitals and $n_i = 0$ for the rest; this representation of kinetic energy and density holds true for the determinantal wave function that exactly describes N *noninteracting* electrons (see §1.3). A more general choice in the forms of (7.1.5) and (7.1.6) will be addressed in §7.6.

We know from §3.3 that any nonnegative, continuous, and normalized density ρ is N-representable and always can be decomposed according to (7.1.8). But given a $\rho(\mathbf{r})$, how can we have a *unique decomposition* in terms of orbitals so as to give a unique value to $T_s[\rho]$ through (7.1.7)?

In analogy with the Hohenberg–Kohn definition of the universal functional $F_{HK}[\rho]$ (see §3.2 and §3.3), Kohn and Sham invoked a corresponding *noninteracting reference system*, with the Hamiltonian

$$\hat{H}_s = \sum_i^N (-\tfrac{1}{2}\nabla_i^2) + \sum_i^N v_s(\mathbf{r}_i) \tag{7.1.9}$$

in which there are no electron–electron repulsion terms, and for which *the ground-state electron density is exactly* ρ. For this system there will be an exact determinantal ground-state wave function

$$\Psi_s = \frac{1}{\sqrt{N!}} \det [\psi_1 \psi_2 \cdots \psi_N] \qquad (7.1.10)$$

where the ψ_i are the N lowest eigenstates of the one-electron Hamiltonian $\hat{h}_s$:

$$\hat{h}_s \psi_i = [-\tfrac{1}{2}\nabla^2 + v_s(\mathbf{r})]\psi_i = \varepsilon_i \psi_i \qquad (7.1.11)$$

The kinetic energy is $T_s[\rho]$, given by (7.1.7),

$$T_s[\rho] = \langle \Psi_s | \sum_i^N (-\tfrac{1}{2}\nabla_i^2) | \Psi_s \rangle$$

$$= \sum_{i=1}^N \langle \psi_i | -\tfrac{1}{2}\nabla^2 | \psi_i \rangle \qquad (7.1.12)$$

and the density is decomposed as in (7.1.8).

The foregoing definition of $T_s[\rho]$ leaves an undesirable restriction on the density—it needs to be *noninteracting v-representable*; that is, there must exist a noninteracting ground state with the given $\rho(\mathbf{r})$. This restriction on the domain of definition of $T_s[\rho]$ can be lifted, and $T_s[\rho]$ of the form (7.1.7) can be defined for any density derived from an antisymmetric wave function. We shall return to this point in §7.3.

The quantity $T_s[\rho]$, although uniquely defined for any density, is still not the exact kinetic-energy functional $T[\rho]$ defined in §3.2. The very clever idea of Kohn and Sham (1965) is to set up a problem of interest in such a way that $T_s[\rho]$ is its kinetic-energy component, exactly. The resultant theory turns out, as we shall immediately see, to be of independent-particle form. Nevertheless, it is exact.

To produce the desired separation out of $T_s[\rho]$ as the kinetic energy component, rewrite (7.1.2) as

$$F[\rho] = T_s[\rho] + J[\rho] + E_{xc}[\rho] \qquad (7.1.13)$$

where

$$E_{xc}[\rho] \equiv T[\rho] - T_s[\rho] + V_{ee}[\rho] - J[\rho] \qquad (7.1.14)$$

The defined quantity $E_{xc}[\rho]$ is called the *exchange-correlation energy*; it contains the difference between T and T_s, presumably fairly small (see §7.3 below), and the nonclassical part of $V_{ee}[\rho]$.

The Euler equation (7.1.3) now becomes

$$\mu = v_{\text{eff}}(\mathbf{r}) + \frac{\delta T_s[\rho]}{\delta \rho(\mathbf{r})} \qquad (7.1.15)$$

where the *KS effective potential* is defined by

$$v_{\text{eff}}(\mathbf{r}) = v(\mathbf{r}) + \frac{\delta J[\rho]}{\delta \rho(\mathbf{r})} + \frac{\delta E_{xc}[\rho]}{\delta \rho(\mathbf{r})}$$

$$= v(\mathbf{r}) + \int \frac{\rho(\mathbf{r}')}{|\mathbf{r} - \mathbf{r}'|} d\mathbf{r}' + v_{xc}(\mathbf{r}) \quad (7.1.16)$$

with the *exchange-correlation potential*

$$v_{xc}(\mathbf{r}) = \frac{\delta E_{xc}[\rho]}{\delta \rho(\mathbf{r})} \quad (7.1.17)$$

We do not at first attempt the direct solution of (7.1.15), for (7.1.15) is merely a rearrangement of (7.1.3) and the explicit form of $T_s[\rho]$ in terms of density is as yet unknown. Rather, we follow the indirect approach designed by Kohn and Sham; see §7.5 for a direct approach.

The Kohn–Sham treatment runs as follows. Equation (7.1.15) with the constraint (7.1.4) is precisely the same equation as one obtains from conventional density-functional theory when one applies it to a system of *noninteracting* electrons moving in the external potential $v_s(\mathbf{r}) = v_{\text{eff}}(\mathbf{r})$. Therefore, for a given $v_{\text{eff}}(\mathbf{r})$, one obtains the $\rho(\mathbf{r})$ that satisfies (7.1.15) simply by solving the N one-electron equations

$$[-\tfrac{1}{2}\nabla^2 + v_{\text{eff}}(\mathbf{r})]\psi_i = \varepsilon_i \psi_i \quad (7.1.18)$$

and setting

$$\rho(\mathbf{r}) = \sum_i^N \sum_s |\psi_i(\mathbf{r}, s)|^2 \quad (7.1.19)$$

Here, v_{eff} depends on $\rho(\mathbf{r})$ through (7.1.17); hence, (7.1.16), (7.1.18) and (7.1.19) must be solved self-consistently. One begins with a guessed $\rho(\mathbf{r})$, constructs $v_{\text{eff}}(\mathbf{r})$ from (7.1.16), and then finds a new $\rho(\mathbf{r})$ from (7.1.18) and (7.1.19). The total energy can be computed directly from (7.1.1) with (7.1.13), or indirectly from (7.2.10) below.

Equations (7.1.16)–(7.1.19) are the celebrated Kohn–Sham equations. They deserve our careful derivation and analysis, to which we proceed.

7.2 Derivation of the Kohn–Sham equations

We here cast the Hohenberg–Kohn variational problem of (7.1.3) in terms of the Kohn–Sham orbitals appearing in (7.1.10). The energy functional (7.1.1) can be rewritten as

$$E[\rho] = T_s[\rho] + J[\rho] + E_{xc}[\rho] + \int v(\mathbf{r})\rho(\mathbf{r}) d\mathbf{r}$$

$$= \sum_i^N \sum_s \int \psi_i^*(\mathbf{x})(-\tfrac{1}{2}\nabla^2)\psi_i(\mathbf{x}) d\mathbf{x} + J[\rho] + E_{xc}[\rho] + \int v(\mathbf{r})\rho(\mathbf{r}) d\mathbf{r} \quad (7.2.1)$$

and the electron density as

$$\rho(\mathbf{r}) = \sum_{i}^{N} \sum_{s} |\psi_i(\mathbf{r}, s)|^2 \qquad (7.2.2)$$

We thus have the energy expressed in terms of N orbitals.

If the N orbitals are allowed to vary over the space of functions that are continuous—to have finite kinetic energy, and be square integrable—to guarantee normalization, then the density ρ of (7.2.2) covers all N-representable densities, the appropriate domain of definition of $E[\rho]$ in (7.2.1). That is to say, variational search for the minimum of $E[\rho]$ can be equivalently effected in the space of orbitals $\{\psi_i\}$. In doing this one must actually constrain the orbitals to be orthonormal, namely,

$$\int \psi_i^*(\mathbf{x}) \psi_j(\mathbf{x}) \, d\mathbf{x} = \delta_{ij} \qquad (7.2.3)$$

because otherwise the kinetic energy formula of (7.1.7) would not be valid. Note that (7.2.3) implies that $\rho(\mathbf{r})$ of (7.2.2) remains normalized as required by (7.1.4).

Define the functional of the N orbitals

$$\Omega[\{\psi_i\}] = E[\rho] - \sum_{i}^{N} \sum_{j}^{N} \varepsilon_{ij} \int \psi_i^*(\mathbf{x}) \psi_j(\mathbf{x}) \, d\mathbf{x} \qquad (7.2.4)$$

where $E[\rho]$ is the functional of the ψ_i expressed in (7.2.1) and (7.2.2), and the ε_{ij} are Lagrange multipliers for the constraints (7.2.3). For $E[\rho]$ to be a minimum, it is necessary that

$$\delta \Omega[\{\psi_i\}] = 0 \qquad (7.2.5)$$

which leads to the equations

$$\hat{h}_{\text{eff}} \psi_i = [-\tfrac{1}{2} \nabla^2 + v_{\text{eff}}] \psi_i = \sum_{j}^{N} \varepsilon_{ij} \psi_j \qquad (7.2.6)$$

with the effective potential $v_{\text{eff}}(\mathbf{r})$ determined from the density through (7.1.16); $v_{\text{eff}}(\mathbf{r})$ is a local operator in the sense of (2.1.24), and the one-electron Hamiltonian $\hat{h}_{\text{eff}}$ is a Hermitian operator. Hence (ε_{ij}) is a Hermitian matrix, and can be diagonalized by a unitary transformation of the orbitals. Such a transformation leaves invariant the determinant of (7.1.10), the density of (7.2.2), and hence the Hamiltonian of (7.2.6). The Kohn–Sham (KS) orbital equations are thus obtained in their canonical form:

$$[-\tfrac{1}{2} \nabla^2 + v_{\text{eff}}] \psi_i = \varepsilon_i \psi_i \qquad (7.2.7)$$

$$v_{\text{eff}}(\mathbf{r}) = v(\mathbf{r}) + \int \frac{\rho(\mathbf{r}')}{|\mathbf{r} - \mathbf{r}'|} d\mathbf{r}' + v_{xc}(\mathbf{r}) \qquad (7.2.8)$$

$$\rho(\mathbf{r}) = \sum_i^N \sum_s |\psi_i(\mathbf{r},s)|^2 \qquad (7.2.9)$$

In (7.2.6) and (7.2.7), the solutions ψ_i can be different. These equations are nonlinear and must be solved iteratively. The total energy can be determined from the resultant density via (7.2.1) or from the formula

$$E = \sum_i^N \varepsilon_i - \frac{1}{2}\int \frac{\rho(\mathbf{r})\rho(\mathbf{r}')}{|\mathbf{r}-\mathbf{r}'|}\,d\mathbf{r}\,d\mathbf{r}' + E_{xc}[\rho] - \int v_{xc}(\mathbf{r})\rho(\mathbf{r})\,d\mathbf{r} \qquad (7.2.10)$$

Here

$$\sum_i^N \varepsilon_i = \sum_i^N \langle \psi_i | -\tfrac{1}{2}\nabla^2 + v_{\text{eff}}(\mathbf{r}) | \psi_i \rangle$$

$$= T_s[\rho] + \int v_{\text{eff}}(\mathbf{r})\rho(\mathbf{r})\,d\mathbf{r} \qquad (7.2.11)$$

Just as in Hartree–Fock theory, the total electronic energy is not the sum of the orbital energies. Compare (1.3.15).

We now come to the interpretation of the KS equations (7.2.7) to (7.2.9). First, comparing with the single Euler equation (7.1.15), we see a major advance: Through the introduction of the N orbitals, the equations handle $T_s[\rho]$, the dominant part of the true kinetic energy $T[\rho]$, *indirectly* but *exactly*. The price for this gain in accuracy is that there are now N equations to solve as opposed to only one equation for the total density derived from direct approximation on $T_s[\rho]$ of the Thomas–Fermi type. Note that at this stage the other unknown part $E_{xc}[\rho]$ remains intact, entering (7.1.15) and (7.2.7) in the same way.

Second, the KS equations have the same form as the Hartree equations (1.3.38), except that they contain a more general local potential $v_{\text{eff}}(\mathbf{r})$. The computational effort to solve KS equations is not much more than to solve the Hartree equations—less than for the Hartree–Fock equations. The Hartree–Fock equations (1.3.23) contain a nonlocal potential operator in the one-electron Hamiltonian and hence are not a special case of the KS equations. Nevertheless, all three theories—Hartree, Hartree–Fock, and Kohn–Sham—provide one-electron equations for describing many-electron systems. The KS theory, exact in principle, is distinguished from the Hartree–Fock theory in its capacity to fully incorporate the exchange-correlation effect of electrons. In Hartree–Fock theory, approximate by definition, electron correlation effects are lacking and their incorporation is by no means an easy task, demanding involved wave-function techniques (see §1.4) or the addition of a correlation-energy functional (see §8.4). The KS equations are open for improvement with each successive better approximation to $E_{xc}[\rho]$ and would give exact ρ and E if $E_{xc}[\rho]$ were known precisely. As we shall see in the next chapter, the form of $E_{xc}[\rho]$ continues to be upgraded.

Third, we address the electron-spin degrees of freedom. Since the effective potential $v_{\text{eff}}(\mathbf{r})$ does not contain electron spin, the solutions of (7.2.7) are doubly degenerate; namely, for each eigenvalue ε_i, there are two independent solutions sharing the same spatial part. They can be chosen as $\phi_i(\mathbf{r})\alpha(s)$ and $\phi_i(\mathbf{r})\beta(s)$. For the case of an even number of electrons, the α-spin electron density usually is equal to the β-spin density; thus

$$\rho(\mathbf{r}) = 2\rho^\alpha(\mathbf{r}) = 2\rho^\beta(\mathbf{r}) = 2\sum_i^{N/2} |\phi_i(\mathbf{r})|^2 \qquad (7.2.12)$$

For the case of an odd number of electrons,

$$\rho(\mathbf{r}) = \rho^\alpha(\mathbf{r}) + \rho^\beta(\mathbf{r}) \qquad (7.2.13)$$

where $\rho^\alpha(\mathbf{r})$ usually differs from $\rho^\beta(\mathbf{r})$ only by one excess orbital. This prescription for the spatial parts of orbitals coincides with that of the restricted Hartree–Fock method. However, the Kohn–Sham and Hartree–Fock cases differ in nature. The restriction on spatial orbitals in the Kohn–Sham equations is the natural consequence of the theory, whereas the restriction in the restricted Hartree–Fock method is a further qualification on the determinantal-wave-function approximation of the Hartree–Fock method. In the next chapter, we shall study the spin-density-functional theory, which is analogous to the unrestricted Hartree–Fock method and allows spin dependence in the effective potential.

Finally, some remarks are appropriate concerning the KS noninteracting reference system. The KS Hamiltonian $\hat{H}_s$ is the Hamiltonian of (7.1.9), with $v_s(\mathbf{r}) = v_{\text{eff}}(\mathbf{r})$, and the Slater determinant (7.1.10) composed from the N KS orbitals $\{\psi_i\}$ in the KS equations is clearly an eigenstate of $\hat{H}_s$ with total energy the sum of KS orbital energies $\{\varepsilon_i\}$. The Hamiltonian $\hat{H}_s$ defines the so-called KS *noninteracting reference system*. A question arises: Which N orbitals should one choose to form the density in solving the KS equations? The rigorous answer is the set of N orbitals determined self-consistently from (7.2.7)–(7.2.9) that minimize the total energy (7.2.1). In practical calculations, however, the set of N lowest eigenstates ε_i is chosen. This is the original proposal of Kohn and Sham (1965) (see the KS definition of $T_s[\rho]$ in §7.1), as reflected in the original derivation of the KS equations given at the end of §7.1.

As has been discussed by Levy and Perdew (1985b), however, there could very well be some many-electron system, the ground-state density of which is not the same as the ground-state density of any noninteracting system; that is to say, one that would not be noninteracting v-

representable. For such a rare case, one must abandon the set of N lowest eigenstates of $\hat{h}_{\text{eff}}$ in favor of choosing a set that minimizes $E[\rho]$ of (7.2.1), which implies a correspondence to some excited state of $\hat{H}_s$.

The KS noninteracting reference system should present no mystery. In Hartree–Fock theory, one may define a Hartree–Fock Hamiltonian $\hat{H}_{\text{HF}}$ that is the sum of N Fock operators (Szabo and Ostlund, 1982, p. 130). $\hat{H}_{\text{HF}}$ may be regarded as defining the HF noninteracting reference system. A similar question then exists to that already posed, as to whether the ground state of $\hat{H}_{\text{HF}}$ always corresponds to the ground-state many-body wave function of HF theory—the minimum of the HF energy functional (1.3.2). This question never has been answered, but in practice the ground state of $\hat{H}_{\text{HF}}$ is always taken. The only difference between the two situations is that in the KS case only the total electron density is shared, while in the HF case, the determinantal wave function is shared, between interacting and noninteracting systems.

Given the auxiliary nature of the KS orbitals—just N orbitals the sum of squares of which add up to the true total electron density—one should expect no simple physical meaning for the Kohn–Sham orbital energies. There is none. However, from the exact long-range behavior of the exact electron density, (1.5.4), and (7.1.2), or from (7.1.15), one can infer that

$$\varepsilon_{\max} = \mu = -I. \tag{7.2.14}$$

Accordingly, $\varepsilon_{\max}$ may be expected to manifest the discontinuities at integral N values described by (4.3.15). Also, a proof that ε_i is equal to the derivative of total energy with respect to orbital occupation number is given by Janak (1978) (see §7.6). Equation (7.2.14) requires that $\lim_{|\mathbf{r}| \to \infty} v_{\text{eff}}(\mathbf{r}) = 0$.

7.3 More on the kinetic-energy functional

The Kohn–Sham definition of $T_s[\rho]$ given in the preceding section is restrictive in that it holds only for noninteracting v-representable densities; that is, for densities that are ground-state densities for some well-defined noninteracting system. This is reminiscent of the similar limitation on the Hohenberg–Kohn definition of the universal functional $F[\rho]$, which was described in §3.2. To carry out the variational search for the true ground state we need functionals defined on the larger domain of all N-representable densities (cf. §3.3). This is made possible through the constrained-search formulation, given for $F[\rho]$ in §3.4. There are many discussions of how to accomplish the same thing for $T_s[\rho]$: Percus (1978), Levy (1979, 1982), Perdew and Zunger (1981), Bartolotti (1982b), Lieb (1982), Hadjisavvas and Theophilou (1984), Levy and Perdew (1985b), Harriman (1985), Yang and Harriman (1986).

In analogy to (3.4.5), $T_s[\rho]$ can be defined by the constrained-search formula

$$T_s[\rho] = \operatorname*{Min}_{\Psi_D \to \rho} \langle \Psi_D | \hat{T} | \Psi_D \rangle$$

$$= \operatorname*{Min}_{\Sigma |\psi_i|^2 = \rho} \left[\sum_{i=1}^{N} \int \psi_i^*(\mathbf{x})(-\tfrac{1}{2}\nabla^2)\psi_i(\mathbf{x}) \, d\mathbf{x} \right] \quad (7.3.1)$$

where the search is over all single-determinantal wave functions Ψ_D that yield the given density ρ. $T_s[\rho]$ delivers the minimum of $\langle \Psi_D | \hat{T} | \Psi_D \rangle$, where $\hat{T}$ is the N-electron kinetic-energy operator. In (7.3.1), the second equality follows from the first simply by expressing the determinantal wave function Ψ_D in terms of its N orbitals. Note that this definition is applicable for all N-representable densities, since every such density can always be decomposed into N orthonormal orbitals (see §3.3).

The existence of the minimum in (7.3.1) has been proved by Lieb (1982). Therefore, in (7.3.1) is achieved a *unique* decomposition of a given N-representable density in terms of orbitals, in the form (7.1.8). There is thereby also achieved evaluation of the noninteracting kinetic energy as the sum of orbital kinetic energies, in (7.1.7).

A slightly different $T_s[\rho]$ may be defined by letting the search in (7.3.1) extend over all antisymmetric wave functions Ψ rather than over determinants only. That is,

$$\tilde{T}_s[\rho] = \operatorname*{Min}_{\Psi \to \rho} \langle \Psi | \hat{T} | \Psi \rangle \quad (7.3.2)$$

We certainly always have

$$\tilde{T}_s[\rho] \leq T_s[\rho] \quad (7.3.3)$$

since the search in (7.3.2) is over a larger set than is the search in (7.3.1). However, for any density ρ that is noninteracting v-representable and goes with a nondegenerate ground state, Lieb (1982) proved that

$$\tilde{T}_s[\rho] = T_s[\rho] \quad (7.3.4)$$

This can be shown as follows:

$$\operatorname*{Min}_{\Psi \to \rho} \langle \Psi | \hat{T} | \Psi \rangle = \operatorname*{Min}_{\Psi \to \rho} \langle \Psi | \hat{T} + \sum_{i=1}^{N} v(\mathbf{r}_i) | \Psi \rangle - \int v(\mathbf{r}) \rho(\mathbf{r}) \, d\mathbf{r} \quad (7.3.5)$$

Since ρ is noninteracting v-representable, we can choose $v(\mathbf{r})$ in this equation to be the potential for which $\rho(\mathbf{r})$ is the noninteracting N-electron ground-state density. Consequently, the minimum is achieved at the corresponding *determinantal* wave function. (The nondegenerate requirement is important; otherwise, Ψ might be a linear combination of determinants.)

7.3 THE KOHN–SHAM METHOD: BASIC PRINCIPLES

One gains further understanding by deriving the necessary conditions that the N orbitals must satisfy to be the minimizing orbitals for (7.3.1). Following Levy and Perdew (1985b), we introduce a Lagrange multiplier (function) $\lambda(\mathbf{r})$ for the condition that the sum of orbital densities is equal to the given density $\rho(\mathbf{r})$, and Lagrange multipliers ε_{ij} for the orthonormalization conditions on ψ_i and ψ_j. This gives the functional

$$\Omega[\psi_1, \psi_2, \ldots, \psi_N] = \sum_i^N \int \psi_i^*(\mathbf{x})(-\tfrac{1}{2}\nabla^2)\psi_i(\mathbf{x})\,d\mathbf{x}$$

$$+ \int \lambda(\mathbf{r})\left\{\sum_{i,s}^N |\psi_i(\mathbf{x})|^2 - \rho(\mathbf{r})\right\} d\mathbf{r} - \sum_{ij} \varepsilon_{ij} \int \psi_i^*(\mathbf{x})\psi_j(\mathbf{x})\,d\mathbf{x} \quad (7.3.6)$$

and the minimization condition

$$\frac{\delta\Omega}{\delta\psi_k^*(\mathbf{x})} = 0 \quad (7.3.7)$$

There result the equations

$$\hat{h}_s \psi_k = \sum_l^N \varepsilon_{kl}\psi_l \quad (7.3.8)$$

where

$$\hat{h}_s = -\tfrac{1}{2}\nabla^2 + \lambda(\mathbf{r}) \quad (7.3.9)$$

Since $\hat{h}_s$ is Hermitian, it is possible, via a unitary transformation, to convert (7.3.8) to the canonical form

$$\hat{h}_s \psi_k = \varepsilon_k \psi_k \quad (7.3.10)$$

This shows that for the orbitals to minimize $\langle \hat{T} \rangle$ they must be the eigenstates of a one-electron Hamiltonian having a local potential precisely equal to the local Lagrangian multiplier $\lambda(\mathbf{r})$. Not surprisingly, the equations (7.3.10) have just the form of the Kohn–Sham equations (7.2.7).

If a given density ρ is noninteracting v-representable, then the ψ_k are the N lowest eigenstates of $\hat{h}_s$ with $\lambda(\mathbf{r})$ the potential that produces ρ. That is, for any noninteracting v-representable density, the definition (7.3.1) reduces to the original KS definition of §7.1. If, on the other hand, a given density ρ is not noninteracting v-representable, then the ψ_k are not the N lowest eigenstates of any $\hat{h}_s$, but instead some N eigenstates of a certain $\hat{h}_s$. In other words, the resulting determinant formed from these N orbitals is some excited state of $\hat{H}_s$, with $\hat{H}_s = \sum_i^N \hat{h}_s(\mathbf{r}_i)$ (see Levy and Perdew 1985b). Finally, if a given density ρ is the true ground state for an N-electron interacting system with potential $v(\mathbf{r})$, then the equations (7.3.10) are precisely the Kohn–Sham equations (7.2.7) with $\lambda(\mathbf{r})$ equal to the local KS effective potential $v_{\text{eff}}(\mathbf{r})$.

The procedure from (7.3.6) to (7.3.10) should not be regarded as a practical proposal for obtaining the kinetic energy associated with a given ρ, for this would require solving (7.3.10) for all local potentials $\lambda(\mathbf{r})$ until the correct $\lambda(\mathbf{r})$ is found. But see Aryasetiawan and Stott (1988).

As was remarked in §7.2, the essence of the Kohn–Sham scheme is to design $T_s[\rho]$, a major portion of the true kinetic energy $T[\rho]$, such that $T_s[\rho]$ is explicitly known in terms of the auxiliary Kohn–Sham orbitals. But how close is $T_s[\rho]$ to $T[\rho]$? We now consider that question.

We have

$$\tilde{T}_s[\rho] \leq T[\rho] \tag{7.3.11}$$

since $\tilde{T}_s[\rho]$ is the expectation value of $\hat{T}$ for the wave function that minimizes $\langle \Psi | \hat{T} | \Psi \rangle$, while $T[\rho]$ is the expectation value of $\hat{T}$ for the wave function that minimizes $\langle \Psi | \hat{T} + \hat{V}_{ee} | \Psi \rangle$ (Theophilou 1979, Levy and Perdew 1985b). The intuitive argument that allows one to remember (7.3.11) is that an interacting problem involves partial occupancy of some orbitals with high eigenvalues. Thus, for a density that is noninteracting v-representable by a nondegenerate ground state, (7.3.4) and (7.3.11) together give

$$T_s[\rho] \leq T[\rho] \tag{7.3.12}$$

The inequality (7.3.12) has a significant consequence; namely, that the virial theorem for an atom or molecule [see (1.6.10)–(1.6.12)] will not hold if the "kinetic energy" is taken as T_s and the "potential energy" is taken as the rest of the energy. The exchange-correlation energy defined in (7.1.14) contains a positive kinetic-energy component. Scaling properties of density functionals will be discussed again in Chapter 11.

As to $T - T_s$ in actual cases, there have been too few studies of this important question. Some values are given in Table 7.1, taken from Ambladh and Pedroza (1984). $T - T_s$ is just a few electron volts for most atomic and molecular systems, apparently—on the order of magnitude of the correlation energy itself.

Finally, we remark that the direct approach to the kinetic energy functional taken in Chapter 6, in the Thomas–Fermi and related models, is best thought of as an approach to $T_s[\rho]$, not to $T[\rho]$. This may be seen from the derivation in §6.7 of the Thomas–Fermi–Weizsacker approximation through conventional gradient expansion.

7.4 Local-density and Xα approximations

The KS equations (7.2.7)–(7.2.9), while exactly incorporating the kinetic energy $T_s[\rho]$, still leave the exchange-correlation functional $E_{xc}[\rho]$ of (7.1.13) unsettled. An explicit form for $E_{xc}[\rho]$ is needed to specify the KS

Table 7.1 $T[\rho] - T_s[\rho]$ for Some Atoms[a]

Atom	$T[\rho] - T_s[\rho]$ (eV)
H$^-$	0.8
He	1.0
Li$^+$	1.1
Be^{2+}	1.1
Li	1.7
Be	2.0

[a] Values from Almbladh and Pedroza (1984).

equations. The search for an accurate $E_{xc}[\rho]$ has encountered tremendous difficulty and continues to be the greatest challenge in the density-functional theory. We describe in this section the simplest approximation, the local approximation proposed by Kohn and Sham (1965).

Recall how in §6.1 the uniform-electron-gas formula was used locally to obtain the Thomas–Fermi functional (6.1.19) for kinetic energy and the Dirac functional (6.1.20) for exchange energy. Now that the kinetic energy $T_s[\rho]$ is rigorously treated in the KS scheme, we can use the uniform-electron-gas formula solely for the unknown part of the rest of the energy functional. Thus we introduce the *local-density approximation* (LDA) for exchange and correlation energy,

$$E_{xc}^{\text{LDA}}[\rho] = \int \rho(\mathbf{r}) \varepsilon_{xc}(\rho) \, d\mathbf{r} \tag{7.4.1}$$

where $\varepsilon_{xc}(\rho)$ indicates the exchange and correlation energy per particle of a uniform electron gas of density ρ. The corresponding exchange-correlation potential of (7.1.16) then becomes

$$v_{xc}^{\text{LDA}}(\mathbf{r}) = \frac{\delta E_{xc}^{\text{LDA}}}{\delta \rho(\mathbf{r})} = \varepsilon_{xc}(\rho(\mathbf{r})) + \rho(\mathbf{r}) \frac{\partial \varepsilon_{xc}(\rho)}{\partial \rho} \tag{7.4.2}$$

and the KS orbital equations read

$$\left[-\frac{1}{2} \nabla^2 + v(\mathbf{r}) + \int \frac{\rho(\mathbf{r}')}{|\mathbf{r} - \mathbf{r}'|} d\mathbf{r}' + v_{xc}^{\text{LDA}}(\mathbf{r}) \right] \psi_i = \varepsilon_i \psi_i. \tag{7.4.3}$$

The self-consistent solution of (7.4.3) defines the Kohn–Sham local-density approximation (KS-LDA), which in the literature is usually simply called the *LDA method*.

The function $\varepsilon_{xc}(\rho)$ can be divided into exchange and correlation contributions,

$$\varepsilon_{xc}(\rho) = \varepsilon_x(\rho) + \varepsilon_c(\rho) \tag{7.4.4}$$

The exchange part is already known, given by the Dirac exchange-energy functional of (6.1.22),

$$\varepsilon_x(\rho) = -C_x \rho(\mathbf{r})^{1/3}, \qquad C_x = \frac{3}{4}\left(\frac{3}{\pi}\right)^{1/3} \tag{7.4.5}$$

As discussed in Appendix E, accurate values of $\varepsilon_c(\rho)$ are available, thanks to the quantum Monte Carlo calculations of Ceperley and Alder (1980). These values have been interpolated to provide an analytic form for $\varepsilon_c(\rho)$ (Vosko, Wilk, and Nusair 1980); see Appendix E for more details.

Application to an atom or molecule or solid of the local-density approximation of (7.4.1) amounts to assuming that the exchange-correlation energy for a nonuniform system can be obtained by applying uniform-electron-gas results to infinitesimal portions of the nonuniform-electron distribution, each having $\rho(\mathbf{r})\,d\mathbf{r}$ electrons, and then summing over all space the individual contributions $\varepsilon_c(\rho)\rho(\mathbf{r})\,d\mathbf{r}$. Such a procedure was explicitly carried out in §3.1 for the kinetic energy (see also §6.1).

The LDA is applicable to systems with slowly-varying densities but cannot be formally justified for highly inhomogeneous systems such as atoms and molecules. The essential justification for its use in atoms and molecules comes from successful numerical applications, as we shall illustrate later in this chapter. Long before the KS-LDA method was invented, there had existed the *Xα method* proposed by Slater (1951) as a simplification of the Hartree–Fock method. The idea of Slater was to approximate the complicated nonlocal Fock operator (1.3.9) by a simple local operator. He invoked the uniform-electron-gas model to produce the simplification, resulting in the *Xα equation* (sometimes called the Hartree–Fock–Slater equation)

$$\left[-\frac{1}{2}\nabla^2 + v(\mathbf{r}) + \int \frac{\rho(\mathbf{r}')}{|\mathbf{r}-\mathbf{r}'|}d\mathbf{r}' + v_{x\alpha}(\mathbf{r})\right]\psi_i = \varepsilon_i \psi_i \tag{7.4.6}$$

with the *Xα local potential*

$$v_{x\alpha}(\mathbf{r}) = -\frac{3}{2}\alpha\left\{\frac{3}{\pi}\rho(\mathbf{r})\right\}^{1/3} \tag{7.4.7}$$

where the parameter α was set equal to 1 in the original prescription.

Kohn and Sham (1965) realized that the *Xα* equation is equivalent to their local-density approximation (7.4.3) if correlation is ignored in

(7.4.3). Namely, if one uses the Dirac exchange formula (6.1.20),

$$E_x^{LDA}[\rho] = -K_D[\rho] = -\frac{3}{4}\left(\frac{3}{\pi}\right)^{1/3} \int \rho(\mathbf{r})^{4/3}\, d\mathbf{r} \qquad (7.4.8)$$

and

$$v_x^{LDA}(\mathbf{r}) = -\left(\frac{3}{\pi}\rho(\mathbf{r})\right)^{1/3} \qquad (7.4.9)$$

then the resulting KS equation is precisely the $X\alpha$ equation (7.4.6) with $\alpha = 2/3$. This value of α also agrees with an argument of Gáspár (1954) from the variational principle.

The two different values of α stem from applying the uniform-electron-gas approximation in different places: Slater in the exchange potential, giving $\alpha = 1$; Gáspár and Kohn and Sham in the exchange energy, giving $\alpha = 2/3$. The former concerns the one-electron equation; the latter emphasizes the total energy. This ambiguity has led to the use of α as an adjustable parameter in many $X\alpha$ calculations. It is clear now that, for atoms and molecules at least, a value of $\alpha \sim 0.75$ is better than $\alpha = 1$ or $2/3$ (Slater 1972, 1974); see Table 7.2.

The $X\alpha$ method may be regarded as a density-functional scheme with neglect of correlation and with approximation to the exchange-energy

Table 7.2 Recommended α Values for Neutral Atoms[a,b]

Atom (Z)	α
H (1)	0.978
He (2)	0.773
Li (3)	0.781
Be (4)	0.768
B (5)	0.765
C (6)	0.759
N (7)	0.752
O (8)	0.744
F (9)	0.737
Ne (10)	0.731
Ar (18)	0.722
Kr (36)	0.706

[a] Values taken from a more extensive table in Schwarz (1972).

[b] Values determined self-consistently by Schwarz (1972), so that the optimized total $X\alpha$ energy equals the Hartree–Fock energy.

functional

$$E_x^{\text{LDA}}[\rho] = -\frac{3}{2}\alpha K_D[\rho] = -\frac{9}{8}\alpha\left(\frac{3}{\pi}\right)^{1/3}\int \rho(\mathbf{r})^{4/3}\,d\mathbf{r} \quad (7.4.10)$$

which gives the $X\alpha$ potential (7.4.7) upon functional differentiation (Connolly 1977). The $X\alpha$ method, particularly in light of the Gaspar–Kohn–Sham derivation, differs from the Thomas–Fermi–Dirac model of §6.1 only in the kinetic-energy functional. The much more accurate $T_s[\rho]$ is used in the $X\alpha$ method in place of $T_{\text{TF}}[\rho]$ in the TFD model (Slater 1975). This makes a great difference. The $X\alpha$ method constitutes a quite efficient technique for electronic-structure calculations for molecules and solids. It is easier to implement than the Hartree–Fock method, particularly for larger molecules and solids, and it does not entail much loss in accuracy. Techniques for solving the $X\alpha$ equation, choosing α values, and many calculated results, are described in detail in the reviews of Johnson (1973) and Connolly (1977), and in the book by Slater (1974). In spite of suggestions to the contrary, the $X\alpha$ formula (7.4.10) probably should not be interpreted as including effects of correlation (Cook and Karplus 1987).

The first KS-LDA calculation was performed by Tong and Sham (1966). They numerically solved the KS-LDA equations for several atoms and compared the calculated values with Hartree–Fock and experimental ones. Their results are summarized in Table 7.3.

Electron densities from the Tong–Sham calculation are close to the Hartree–Fock densities, displaying the proper shell structure that is missing in Thomas–Fermi and related models. Table 7.3 shows that the value of $\alpha = 2/3$ [corresponding to (7.4.8) and (7.4.9)] gives slightly better total energies than the original Slater value, $\alpha = 1$. However, the use of the LDA exchange energy (7.4.8) still leaves about 10% error in the Hartree–Fock exchange energy—see Table 7.4. This is the major source of error in LDA, since correlation energy is a magnitude smaller than exchange energy. The inclusion of ε_c in the Tong–Sham calculations

Table 7.3 Total Energies of Atoms, Computed by Various Methods (atomic units)[a]

Atom	$X\alpha\ (\alpha = \tfrac{2}{3})$[b]	$X\alpha\ (\alpha = 1)$[b]	LDA	Hartree–Fock	Experiment
He	−2.72	−2.70	−2.83	−2.86	−2.90
Li	−7.17	−7.15	−7.33	−7.43	−7.48
Ne	−127.49	−127.38	−128.12	−128.55	−128.94
Ar	−524.51	−524.35	−525.85	−526.82	−527.60

[a] Atomic $X\alpha$ and LDA calculations by Tong and Sham (1966).
[b] Value of α in (7.4.7). In both cases $\alpha = 1$ in (7.4.10).

Table 7.4 LDA Calculations for N_2 and CO[a,b]

Molecule	Binding energy (eV)			Bond Length (a_0)			Vibrational Frequency (cm^{-1})		
	LDA	HF	Exptl.	LDA	HF	Exptl.	LDA	HF	Exptl.
N_2	7.8	5.3	9.8	2.16	2.01	2.07	2070	2730	2358
CO	9.6	7.9	11.2	2.22	2.08	2.13	2090	2131	2170

[a] Gunnarsson, Harris, and Jones (1977).
[b] HF designates Hartree–Fock method.

did not prove to be particularly beneficial; the correlation energies obtained are a factor of 2 too large. Tong and Sham used a relatively inaccurate expansion for $\varepsilon_c(\mathbf{r})$ (Pines 1963). Tong (1972) has also made a similar KS-LDA calculation for sodium metal.

Laufer and Krieger (1986) have given an interesting comparison between the LDA and the exact solution in a soluble model of two interacting electrons in a quadratic external potential. In this case, the LDA gives accurate total energies but poor Kohn–Sham eigenvalues, due to the errors in the exchange-energy functional.

Examples of KS-LDA calculations for molecular systems are given by Gunnarsson, Harris, and Jones (1977). Their results show that the KS-LDA method is capable of describing molecular bonding reasonably well, which contrasts with the nonbonding defect in Thomas–Fermi theory. Results for N_2 and CO are given in Table 7.4. LDA may exceed the Hartree–Fock method in accuracy.

An important early application of the LDA idea to molecular systems was made by Gordon and Kim (1972). The Thomas–Fermi–Dirac energy functional of (6.1.22), plus a correlation correction in local-density approximation, was used to calculate the interaction energy between closed-shell atoms. The Gordon–Kim and subsequent density-functional calculations on such systems will be described in §10.2.

By the middle of the 1970s, both the successes and the limitations of the KS-LDA method were realized. Meanwhile, the spin-polarized extension of density-functional theory was well established and the corresponding Kohn–Sham local-spin-density approximation (KS-LSDA) turned out to be superior to KS-LDA. The KS-LSDA method will be a principal subject of the next chapter, where also we shall discuss approximations beyond LDA.

7.5 The integral formulation

The Kohn–Sham development of Hohenberg–Kohn density-functional theory solves the problem of unknown kinetic-energy functionals by resolving the electron density into N orbital components. It is an immense advantage that the N Kohn–Sham orbital equations can in principle include electron exchange-correlation effects and deliver the exact ground-state energy and electron density. However, it is this very dependence on the set of N orbitals that diminishes the simplicity of density-functional theory as expressed in the single Euler equation (7.1.15),

$$\mu = v_{\text{eff}}(\mathbf{r}) + \frac{\delta T_s[\rho]}{\delta \rho(\mathbf{r})} \qquad (7.5.1)$$

where the potential $v_{\text{eff}}(\mathbf{r})$ is in turn dependent on $\rho(\mathbf{r})$.

In the present section we describe how one can solve (7.5.1) itself directly without evaluating $\delta T_s / \delta \rho$. The way to bypass orbitals has been known for a long time (Golden 1960); namely, one equates the electron density to the inverse Laplace transform of the Green's function and then relates the Green's function directly to the potential. The gradient correction to the Thomas–Fermi theory of §6.7 is one way to this, based on the $\hbar$ semiclassical expansion of the Green's function. As we have seen, however, this is not a convergent procedure. There is another procedure based on a convergent sequence of relations between the Green's function and the local potential. Golden (1957a,b) first suggested the use of the product approximation to improve the Thomas–Fermi theory, and Handler (1965) carried out a numerical calculation in second-order approximation for the problem of noninteracting electrons in a quadratic external potential. The product approximation for the Green's function was shown to converge to the correct result for noninteracting electrons in a linear potential by Handler and Wang (1972) and in a quadratic potential by Handler (1973). Handler (1973) also gave the expression for the density matrix in terms of the local potential in the product representation via the inverse Laplace transform. Starting from the path-integral representation of the Green's function, which is essentially equivalent to the product representation, Harris and Pratt (1985) carried out the inverse Laplace transform analytically and expressed the electron density as a multidimensional integral within the framework of the many-electron Hartree approximation. They also proposed use of such a relation in an integral equation for the local potential. Proceeding from these works, Yang (1987) constructed the relation between electron density and local potential in Kohn–Sham theory, thereby generalizing the work of Harris and Pratt to include electron exchange and correlation. The possibility of this generalization was also mentioned by Pratt, Hoffman and Harris (1988). Yang also gave the spin-polarized version of the theory and called attention to the possible application of the theory to large molecules.

7.5 THE KOHN–SHAM METHOD: BASIC PRINCIPLES

We now describe the theory. What (7.5.1) contains is that $\rho(\mathbf{r})$ is a unique functional of $\mu - v_{\text{eff}}(\mathbf{r})$:

$$\rho(\mathbf{r}) = \rho[\mu - v_{\text{eff}}(\mathbf{r}); \mathbf{r}] \tag{7.5.2}$$

The use of the set of N Kohn–Sham orbitals in KS theory is only a device for obtaining the relation (7.5.2). An explicit but crude approximation to (7.5.2) is the Thomas–Fermi equation (3.1.13), which can be recast in the form

$$\rho_{\text{TF}}(\mathbf{r}) = \left[\frac{3}{5C_F}(\mu - v_{\text{eff}}(\mathbf{r}))\right]^{3/2} \tag{7.5.3}$$

Note that the exchange-correlation effect missing in (3.1.13) is now included via $v_{\text{eff}}(\mathbf{r})$.

The Kohn–Sham method and Thomas–Fermi method are just two versions of the relation (7.5.2): one exact but using orbitals, the other crude but direct. We seek something intermediate, combining the advantages of the two approaches. This will be the integral formulation of KS theory.

Consider, for simplicity, spin-compensated situations, that is, a closed-shell atom or molecule. In the KS scheme, (7.2.7)–(7.2.9), the ground-state electron density is the sum of spatial orbital densities and can be expressed also as the diagonal element of the KS one-electron reduced density matrix

$$\rho(\mathbf{r},\mathbf{r}') = 2\sum_i^{N/2} \phi_i(\mathbf{r})\phi_i^*(\mathbf{r}') = 2\sum_i^{\infty} \phi_i(\mathbf{r})\phi_i^*(\mathbf{r}')\eta(\varepsilon_F - \varepsilon_i)$$
$$= 2\langle\mathbf{r}|\eta(\varepsilon_F - \hat{h}_{\text{eff}})|\mathbf{r}'\rangle \tag{7.5.4}$$

where the prefactor 2 accounts for the double degeneracy due to electron spin, $\eta(x)$ is the Heaviside step function (6.7.4), the ε_i and the ψ_i are the eigenvalues and eigenfunctions of the KS Hamiltonian $\hat{h}_{\text{eff}}$ of (7.2.6), and ε_F can be any number between the highest occupied eigenvalue and the lowest unoccupied eigenvalue.

We follow the development of the conventional gradient correction in §6.7, and obtain

$$\rho(\mathbf{r},\mathbf{r}') = \frac{2}{2\pi i}\int_{\gamma-i\infty}^{\gamma+i\infty}\frac{d\beta}{\beta}e^{\beta\varepsilon_F}G(\mathbf{r},\mathbf{r}';\beta); \quad \gamma>0 \tag{7.5.5}$$

where the Green's function is defined as

$$G(\mathbf{r},\mathbf{r}';\beta) = \langle\mathbf{r}|\exp(-\beta\hat{h}_{\text{eff}})|\mathbf{r}'\rangle$$
$$= \sum_i^{\infty}\phi_i(\mathbf{r})\phi_i^*(\mathbf{r}')e^{-\beta\varepsilon_i} \tag{7.5.6}$$

The gradient expansion here now uses Wigner-semiclassical expansion of G. Instead, we use the convergent path-integral representation of G. We write

$$G(\mathbf{r}, \mathbf{r}'; \beta) = \langle \mathbf{r}| \exp(-\beta \hat{h}_{\text{eff}}) |\mathbf{r}'\rangle = \langle \mathbf{r}| [\exp(-\beta \hat{h}_{\text{eff}}/q)]^q |\mathbf{r}'\rangle$$
$$= \int d\mathbf{r}_1 \cdots d\mathbf{r}_{q-1} \prod_{j=0}^{q-1} \langle \mathbf{r}_{j+1}| \exp(-\beta \hat{h}_{\text{eff}}/q) |\mathbf{r}_j\rangle \quad (7.5.7)$$

where q is any positive integer greater than 1, $\mathbf{r}_0 = \mathbf{r}'$, $\mathbf{r}_q = \mathbf{r}$, and the second line follows from inserting the identity operator $\int d\mathbf{r}_j\, |\mathbf{r}_j\rangle\langle\mathbf{r}_j|$ between the products of operators $\exp(-\beta \hat{h}_{\text{eff}}/q)$ on the first line. No approximation has been made yet, but we have achieved expressing the function $G(\mathbf{r}, \mathbf{r}'; \beta)$ in terms of the function $G(\mathbf{r}, \mathbf{r}'; \beta/q)$ through $(q-1)$-dimensional integration. The latter has a time dependence of β/q, shorter than the time dependence β in the former. Therefore, a short-time approximation (lowest-order) can be employed

$$\langle \mathbf{r}_{j+1}| \exp(-\beta \hat{h}_{\text{eff}}/q) |\mathbf{r}_j\rangle = G_1(\mathbf{r}_{j+1}, \mathbf{r}_j; \beta/q) + O(\beta/q)^2 \quad (7.5.8)$$

Here and henceforth $O(x)^n$ represents terms of order equal to or higher than x^n; the subscript 1 on G_1 connotes first-order approximation.

A possible candidate for G_1 can be obtained by using the zeroth-order expansion of (6.7.13) in the transformation (6.7.10). The result is

$$G_{\text{TF}}(\mathbf{r}_{j+1}, \mathbf{r}_j; \beta/q) = \left(\frac{mq}{2\pi\beta\hbar^2}\right)^{3/2} \exp\left[-\frac{mq}{2\beta\hbar^2}(\mathbf{r}_{j+1} - \mathbf{r}_j)^2 - \frac{\beta}{q} v_{\text{eff}}\left(\frac{\mathbf{r}_{j+1} + \mathbf{r}_j}{2}\right)\right]$$
$$(7.5.9)$$

Here and in the rest of this section, atomic units are not used for the sake of explicitness. This Green's function is labeled TF because it gives the Thomas–Fermi approximation to the kinetic energy functional, as was discussed in §6.7. It has been recently shown, however, that G_{TF} is correct in the order of β only for its diagonal elements, and is less accurate for its off-diagonal elements (Yang 1986). The zeroth-order Green's function with consistent accuracy for both its diagonal and off-diagonal elements is (Fujiwara, Osborn, and Wilk 1982, Yang 1986)

$$G_1(\mathbf{r}_{j+1}, \mathbf{r}_j; \beta/q) = \left(\frac{mq}{2\pi\beta\hbar^2}\right)^{3/2} \exp\left[-\frac{mq}{2\beta\hbar^2}(\mathbf{r}_{j+1} - \mathbf{r}_j)^2 - \frac{\beta}{q} u(\mathbf{r}_{j+1}, \mathbf{r}_j)\right]$$
$$(7.5.10)$$

$$u(\mathbf{r}_{j+1}, \mathbf{r}_j) = \int_0^1 v_{\text{eff}}[\mathbf{r}_j + (\mathbf{r}_{j+1} - \mathbf{r}_j)\tau]\, d\tau \quad (7.5.11)$$

This approximation for G_1 reduces to G_{TF} if the τ integration in (7.5.11) is approximated by the midpoint rule. It has been shown also that G_1

gives the Thomas–Fermi-$\frac{1}{9}$-Weizsacker kinetic-energy functional (Yang 1986).

Now G_1 is used in (7.5.7), leading to the "qth-order" approximation

$$G_q(\mathbf{r}, \mathbf{r}'; \beta) = \left(\frac{mq}{2\pi\beta\hbar^2}\right)^{3q/2} \int d\mathbf{r}_1 \cdots d\mathbf{r}_{q-1}$$

$$\times \exp\left[-\frac{mq}{2\hbar^2\beta}\sum_{j=0}^{q-1}(\mathbf{r}_{j+1}-\mathbf{r}_j)^2 - \frac{\beta}{q}\sum_{j=0}^{q-1}u(\mathbf{r}_{j+1},\mathbf{r}_j)\right] \quad (7.5.12)$$

Because of the use of an approximate G_1, G_q is now approximate. As may be seen from (7.5.7) and (7.5.8), we obtain

$$G_q(\mathbf{r}, \mathbf{r}'; \beta) = G(\mathbf{r}, \mathbf{r}'; \beta) + O(\beta^2/q) \quad (7.5.13)$$

However, G_q is convergent to the exact G as $q \to \infty$, which provides the basis for the path-integral formulation of G.

Before proceeding further, we pause to mention the physical meaning of the Green's function G and its path-integral representation. G is also called the propagator because $|G(\mathbf{r}, \mathbf{r}'; iT/\hbar)|^2$ is the probability that an electron starting at $\mathbf{r}'$, time zero, traveling in the potential v_{eff}, arrives at $\mathbf{r}$ at time T (Feynmann and Hibbs 1965). This dynamic interpretation of G comes from the use of complex β, which is needed in (7.5.5). Equation (7.5.12) expresses G as the limit of the sequence G_q.

We now obtain the corresponding approximation for the density matrix (7.5.4) by inserting (7.5.12) into (7.5.5). The result is

$$\rho_q(\mathbf{r}, \mathbf{r}') = 2\int d\mathbf{r}_1 \cdots d\mathbf{r}_{q-1}\left(\frac{qk_q}{2\pi l_q}\right)^{3q/2} J_{3q/2}(k_q l_q)\eta(k_q^2) \quad (7.5.14)$$

where $J_{3q/2}(x)$ is the Bessel function of order $3q/2$, $\eta(x)$ is the step function, and

$$k_q^2 = \frac{2m}{\hbar^2}\left[\varepsilon_F - \frac{1}{q}\sum_{j=0}^{q-1}u(\mathbf{r}_{j+1}, \mathbf{r}_j)\right],$$

$$l_q^2 = q\sum_{j=0}^{q-1}(\mathbf{r}_{j+1}-\mathbf{r}_j)^2 \quad (7.5.15)$$

In deriving (7.5.14), an identity has been used,

$$\frac{1}{2\pi i}\int_{\gamma-i\infty}^{\gamma+i\infty}\frac{d\beta}{\beta^{\alpha+1}}e^{-\beta y^2+\beta x} = \left(\frac{x}{y}\right)^\alpha J_\alpha(2xy)\eta(x^2) > 0 \quad (7.5.16)$$

which can be proved by a transformation of variables from an integral representation of Bessel functions (Arfken 1980, p. 505). The step function appears in (7.5.16) because when x is negative the contour of integration in (7.5.16) can be closed on the right-hand side of the complex-β plane, and the integral vanishes because there is no singularity inside the closed contour. Harris and Pratt obtained (7.5.14) in the Hartree approximation.

The corresponding electron density may now be expressesd as an explicit functional of $v_{\text{eff}}(\mathbf{r})$; namely,

$$\rho_q(\mathbf{r}) = \rho_q(\mathbf{r}, \mathbf{r}) = \rho_q[\varepsilon_F - v_{\text{eff}}; \mathbf{r}]$$

$$= 2 \int d\mathbf{r}_1 \cdots d\mathbf{r}_{q-1} \left(\frac{q\bar{k}_q}{2\pi\bar{l}_q}\right)^{3q/2} J_{3q/2}(\bar{k}_q\bar{l}_q)\eta(\bar{k}_q^2) \quad (7.5.17)$$

where $\bar{k}_q = k_q|_{\mathbf{r}'=\mathbf{r}}$ and $\bar{l}_q = l_q|_{\mathbf{r}'=\mathbf{r}}$. This relation and the expression for v_{eff} [see (7.2.8)],

$$v_{\text{eff}}(\mathbf{r}) = v(\mathbf{r}) + \int \frac{\rho(\mathbf{r}')}{|\mathbf{r}-\mathbf{r}'|} d\mathbf{r}' + v_{xc}(\mathbf{r}) \quad (7.5.18)$$

comprise the integral formulation of Kohn–Sham density functional theory.

Self-consistent solution of (7.5.17) and (7.5.18), with ε_F determined by the normalization condition

$$\int \rho(\mathbf{r}) \, d\mathbf{r} = N \quad (7.5.19)$$

gives the electron density at the qth-order approximation to the Euler equation (7.5.1). From this $\rho_q(\mathbf{r})$, the total energy can be obtained from (7.2.1), with the kinetic energy given by

$$T_s[\rho] = \int d\mathbf{r} \, t_q(\mathbf{r})$$

where

$$t_q(\mathbf{r}) = \frac{\hbar^2}{2m}[-\nabla_{\mathbf{r}'}^2 \rho_q(\mathbf{r}, \mathbf{r}')]_{\mathbf{r}=\mathbf{r}'}$$

$$= -\frac{\hbar^2}{m} \int d\mathbf{r}_1 \cdots d\mathbf{r}_{q-1} \left(\frac{q\bar{k}_q}{2\pi\bar{l}_q}\right)^{3q/2} \eta(\bar{k}_q^2)$$

$$\times \left[\left(\frac{q\bar{k}_q}{\bar{l}_q}\right)^2 (\mathbf{r}-\mathbf{r}_1)^2 J_{(3q/2)+2}(\bar{k}_q\bar{l}_q) - \frac{3q\bar{k}_q}{\bar{l}_q} J_{(3q/2)+1}(\bar{k}_q\bar{l}_q)\right.$$

$$+ \frac{2m}{\hbar^2}(\mathbf{r}-\mathbf{r}_1) \cdot \nabla_r u(\mathbf{r}, \mathbf{r}_1) J_{3q/2}(\bar{k}_q\bar{l}_q) - \left(\frac{m\bar{l}_q}{\hbar^2 q \bar{k}_q}\right) \nabla_r^2 u(\mathbf{r}, \mathbf{r}_1) J_{(3q/2)-1}(\bar{k}_q\bar{l}_q)$$

$$\left. + \left(\frac{m\bar{l}_q}{\hbar^2 q \bar{k}_q}\right)^2 |\nabla_r u(\mathbf{r}, \mathbf{r}_1)|^2 J_{(3q/2)-2}(\bar{k}_q\bar{l}_q)\right] \quad (7.5.20)$$

There is also an indirect way to compute $T_s[\rho]$ (Yang 1988a) that is analogous to (7.2.10), namely,

$$T_s[\rho] = \int d\mathbf{r}\, \bar{t}_q(\mathbf{r})$$

where

$$\bar{t}_q(\mathbf{r}) = [\varepsilon_F - v_{\text{eff}}(\mathbf{r})]\rho_q(\mathbf{r})$$
$$- 2\int d\mathbf{r}_1 \cdots d\mathbf{r}_{q-1} \left(\frac{q\bar{k}_q}{2\pi l_q}\right)^{3q/2} \left(\frac{\hbar^2 \bar{k}_q}{m l_q}\right) J_{(3q/2)+1}(\bar{k}_q \bar{l}_q) \eta(\bar{k}_q^2) \quad (7.5.21)$$

In this formulation, the number of electrons N enters only as the normalization constant in (7.5.19). We thus have a full implementation of the fundamental idea of density-functional theory: use only $\rho(\mathbf{r})$ as the basic variable. This distinguishes the integral formulation from any other *ab initio* theory of electronic structure. One hopes that this theory will make possible ab initio calculations of electronic structure for systems of very many electrons, particularly macromolecules of biological interest.

A simplified picture results for the case $q = 1$. There is then no spatial integration in (7.5.17) and it reduces to the Thomas–Fermi relation (7.5.3). Results of Thomas–Fermi theory are known to become better and better as it is applied to systems with more and more electrons (Lieb 1981). This supports optimism as regards application of the integral theory to macromolecules.

The parameter q is an index representing the accuracy of the solution. On the one hand, it resembles K, the number of basis functions used in solving the Kohn–Sham equations, in that the limits $K \to \infty$ and $q \to \infty$ are the same—the exact solution. On the other hand, q can be any positive integer, whereas K has a minimum value, $N/2$.

In addition to (7.5.12), there are other path-integral representations, some having better convergence properties; for example, the potential-gradient-included polygonal path-integral and Fourier path-integral representations proposed by Yang (1988a). Recent successes of path-integral methods in quantum statistical mechanics (Kalos, 1984) provide additional support to the integral formulation. Cioslowski and Yang (1989) have calculated the density of fermions in a harmonic potential using (7.5.14) for $q = 1$, 2, and 3. The proper shell structure in the density is already evident at the level $q = 2$. Results for $q = 3$ provide further improvement. For Monte Carlo calculations on a similar system, see Hoffman, Pratt, and Harris (1988).

7.6 Extension to nonintegral occupation numbers and the transition-state concept

The Kohn–Sham theory as developed in the previous sections hinges on the introduction of a noninteracting kinetic energy functional $T_s[\rho]$ that can be explicitly expressed in terms of N one-electron orbitals. Recall that the Kohn–Sham choice of only N orbitals in (7.1.7) and (7.1.8) is merely a special case of the more general forms (7.1.5) and (7.1.6) involving an arbitrary number of orbitals and fractional occupation numbers. We here develop such a more general Kohn–Sham theory and show that it provides an appealing meaning for the eigenvalues of the Kohn–Sham equation.

The original idea is due to Janak (1978), deriving from a similar approach of Slater (1974) in the $X\alpha$ method. Refinements include those of Rajagopal (1980) and Perdew and Zunger (1981); see also J. Harris (1979, 1984) and Gopinathan and Whitehead (1980).

Based on the constrained-search definition for $T_s[\rho]$ of (7.3.1), we can introduce a generalization (J for Janak),

$$T_J[\rho] = \text{Min}\left[\sum_i n_i \int \psi_i^*(\mathbf{x})(-\tfrac{1}{2}\nabla^2)\psi_i(\mathbf{x})\,d\mathbf{x}\right] \qquad (7.6.1)$$

where the search is over all possible n_i ($0 \leq n_i \leq 1$) and the associated orthonormal orbitals ψ_i yielding the given density,

$$\rho(\mathbf{r}) = \sum_i n_i \sum_s |\psi_i(\mathbf{x})|^2 \qquad (7.6.2)$$

The number of orbitals allowed in (7.6.2) is arbitrary apart from being not less than N, the total number of electrons. N is given by

$$N = \sum_i n_i \qquad (7.6.3)$$

Supposing that for a given $\rho(\mathbf{r})$ the minimum in (7.6.1) is attained by a particular set $\{n_i, \psi_i\}$, then the generalized kinetic energy becomes

$$T_J[\rho] = \sum_i n_i \int \psi_i^*(\mathbf{x})(-\tfrac{1}{2}\nabla^2)\psi_i(\mathbf{x})\,d\mathbf{x} \qquad (7.6.4)$$

Recall from §2.6 that the first-order reduced density matrix

$$\Gamma_1(\mathbf{r},\mathbf{r}') = \sum_i n_i \sum_s \psi_i(\mathbf{x})\psi_i^*(\mathbf{x}'), \qquad 0 \leq n_i \leq 1 \qquad (7.6.5)$$

is always ensemble -N-representable. This implies that there always is an N-particle ensemble density matrix $\hat{\Gamma}_N$ that gives the $\Gamma_1(\mathbf{r},\mathbf{r}')$ of (7.6.5). This enables us to rewrite (7.6.1) as

$$T_J[\rho] = \underset{\hat{\Gamma}_N \to \rho(\mathbf{r})}{\text{Min}}\ \text{Tr}\,(\hat{\Gamma}_N \hat{T}) \qquad (7.6.6)$$

The equivalence of (7.6.1) and (7.6.6) was established by Perdew, Zunger, and Levy (Perdew and Zunger 1981, Appendix 4).

Now, replacing $T_s[\rho]$ with $T_J[\rho]$ in (7.2.1), we obtain the total-energy functional

$$E[\rho] = T_J[\rho] + E_{xc}[\rho] + \int v(\mathbf{r})\rho(\mathbf{r})\,d\mathbf{r}$$

$$= \sum_i -\tfrac{1}{2}n_i \int \psi_i^*(\mathbf{x})\nabla^2\psi_i(\mathbf{x})\,d\mathbf{x} + J[\rho] + E_{xc}[\rho] + \int v(\mathbf{r})\rho(\mathbf{r})\,d\mathbf{r}$$
(7.6.7)

where the definition of $E_{xc}[\rho]$ has been modified by replacing $T_s[\rho]$ with $T_J[\rho]$ in (7.1.13). Note that the density ρ is given in terms of the n_i and ψ_i via (7.6.2). Therefore, we now have the energy expressed in terms of the n_i and ψ_i, and we can obtain the ground-state energy by minimizing $E[\rho]$ with respect to the n_i and ψ_i. For a fixed set of n_i, the set of ψ_i must satisfy the Euler equation

$$\frac{\delta}{\delta\psi_i^*(\mathbf{x})}\left\{E[\rho] - \sum_i \varepsilon_i'\!\left(\int |\psi_i|^2\,d\mathbf{x} - 1\right)\right\} = 0 \quad (7.6.8)$$

which leads to

$$[-\tfrac{1}{2}n_i\nabla^2 + n_i v_{\text{eff}}(\mathbf{r})]\psi_i = \varepsilon_i'\psi_i \quad (7.6.9)$$

where v_{eff} is given in terms of $\rho(\mathbf{r})$ by (7.2.8). For $n_i \neq 0$, let $\varepsilon_i = \varepsilon_i'/n_i$. Then (7.6.9) becomes

$$[-\tfrac{1}{2}\nabla^2 + v_{\text{eff}}(\mathbf{r})]\psi_i = \varepsilon_i\psi_i \quad (7.6.10)$$

which is precisely the Kohn–Sham equation (7.2.7). Orthogonality of the ψ_i follows from the equation itself and need not be separately imposed.

To examine the dependence of E on the occupation numbers, differentiate E with respect to n_i. One obtains

$$\frac{\partial E}{\partial n_i} = -\frac{1}{2}\int \psi_i^*(\mathbf{x})\nabla^2\psi_i(\mathbf{x})\,d\mathbf{x} + \int \frac{\delta}{\delta\rho(\mathbf{r})}\left\{J[\rho] + E_{xc}[\rho] + \int v\rho\,d\mathbf{r}\right\}\frac{\partial\rho(\mathbf{r})}{\partial n_i}\,d\mathbf{r}$$

$$= -\frac{1}{2}\int \psi_i^*(\mathbf{x})\nabla^2\psi_i(\mathbf{x})\,d\mathbf{x} + \int v_{\text{eff}}(\mathbf{r})\psi_i^*(\mathbf{x})\psi_i(\mathbf{x})\,d\mathbf{x}$$

$$= \varepsilon_i \quad (7.6.11)$$

where the chain rule (A.33), and the Kohn–Sham equation (7.6.10), have been used. Equation (7.6.11) is sometimes called the *Janak theorem*. It is independent of the form of approximation to $E_{xc}[\rho]$, and it provides a meaning for the eigenvalues of the Kohn–Sham equation.

How can we infer from (7.6.11) the distribution of n_i that minimizes the total energy E? This is by no means a trivial problem, owing to the complex dependence of ε_i on all the occupation numbers. The choice of Kohn and Sham (1965) seems to work well in practice: set $n_i = 1$ for the

N lowest eigenstates and $n_i = 0$ for the rest. Nevertheless, there may be systems for which such a Kohn–Sham method does not work (these would be systems that are not noninteracting v-representable; see §7.3 and Perdew and Zunger 1981).

The above formalism also holds for the case where $N = \sum_i n_i$ is not an integer. This can be used to provide an interpolation scheme for systems with nonintegral number of electrons.

Integrating (7.6.11) between N and $N+1$, where n is an integer, we find in an obvious notation:

$$-A = E_{N+1} - E_N = \int_0^1 \varepsilon_{\text{LUMO}}(n)\, dn \qquad (7.6.12)$$

where A is the electron affinity and LUMO denotes the lowest unoccupied molecular orbital. The alternative to this exact formula is the numerical approximation,

$$-A = E_{N+1} - E_N \approx \varepsilon_{\text{LUMO}}\,(n = \tfrac{1}{2}) \qquad (7.6.13)$$

This is an example of a *transition state* formula (Slater 1972). To get $E_{N+1} - E_N$, determine the highest orbital energy for the intermediate state of $N + \tfrac{1}{2}$ electrons. We then obtain

$$-I = E_N - E_{N-1} = \int_0^1 \varepsilon_{\text{HOMO}}(n)\, dn \qquad (7.6.14)$$

and

$$-I = E_N - E_{N-1} \approx \varepsilon_{\text{HOMO}}\,(n = \tfrac{1}{2}) \qquad (7.6.15)$$

where I is the ionization potential and HOMO denotes highest occupied molecular orbital. These formulas are depicted in Figure 7.1 for a typical situation. Note that (7.6.12) and (7.6.14) are in principle exact. Note the focus on frontier orbitals.

The Mulliken electronegativity of (4.2.2) can be computed similarly. Of course, one can compute I and A separately and then average them. It is better, however, to compute the energy difference between positive and negative ions by the transition-state method (Bartolotti, Gadre, and Parr 1980, Manoli and Whitehead 1984), as shown in Figure 7.2. Indeed, the best route to A, since (7.6.13) often gives convergence difficulties, is to calculate I and χ_M by the transition state method and then employ

$$A = 2\chi_M - I \qquad (7.6.16)$$

One must be careful not necessarily to assign physical meaning to a nonintegral-N species as described by Janak's formulas; in general his formulas are not expected to conform to the interpolation analysis of

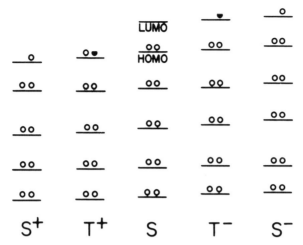

Figure 7.1 Kohn–Sham orbital energies for a system and its positive and negative ions, and the corresponding intermediate transition states. The ionization potential of S is the average of $-\varepsilon_{HOMO}$ between S^+ and S, approximately the value of $-\varepsilon_{HOMO}$ for the transition state T^+. The electron affinity of S is the average of $-\varepsilon_{LUMO}$ between S and S^-, approximately the value of $-\varepsilon_{LUMO}$ for the translation state T^- (after Bartolotti, Gadre, and Parr 1980).

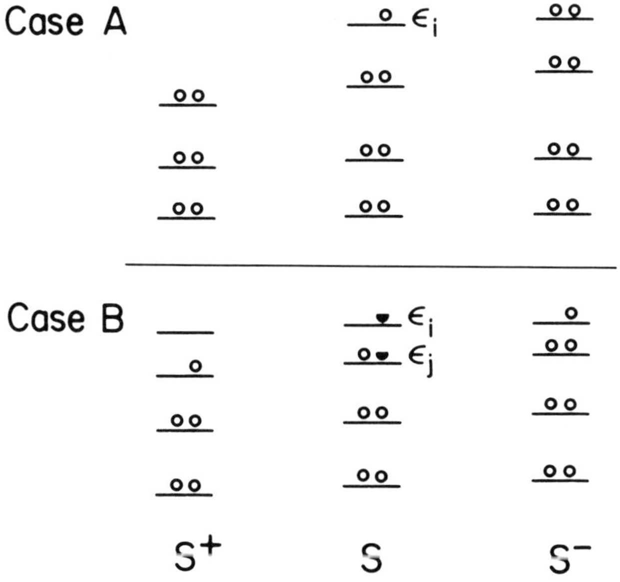

Figure 7.2 Transition states for determining electronegativities. The two main cases are displayed. In case A, the Mulliken electronegativity is (approximately) the negative of the Kohn–Sham orbital energy ε_i for the neutral-species configuration shown. In case B, it is (approximately) the average of the negative ε_i and ε_j shown (After Bartolotti, Gadre, and Parr 1980).

§4.3. There can be no quarrel with them for integral-N species, however, for they have been devised to describe those exactly.

The Janak extension of the Kohn–Sham theory also allows the evaluation of the frontier or Fukui function $f(\mathbf{r})$ of (5.4.1), which is the derivative of the electron density with respect to total number of electrons, since electron number can be continuous in the Janak formulation. This quantity is the reactivity index for a molecule and is related to the frontier orbitals of a molecule (see §5.4). The explicit formulas (Yang, Parr, and Pucci 1984) are as follows:

$$f^+(\mathbf{r}) = |\phi_{\text{LUMO}}(\mathbf{r})|^2 + \sum_{i=1}^{N} \frac{\partial}{\partial N} |\phi_i(\mathbf{r})|^2 \qquad (7.6.17)$$

$$f^-(\mathbf{r}) = |\phi_{\text{HOMO}}(\mathbf{r})|^2 + \sum_{i=1}^{N-1} \frac{\partial}{\partial N} |\phi_i(\mathbf{r})|^2 \qquad (7.6.18)$$

Here the ϕ are spatial orbitals for the neutral system. Note that the discontinuity $f^+ \neq f^-$: f^+ is for addition of orbitals, f^- for subtraction; HOMO refers to the highest occupied molecular orbital.

A theory related to Janak's is discussed in §9.5—density-matrix-functional theory.

8
THE KOHN–SHAM METHOD: ELABORATION

8.1 Spin-density-functional theory

The density-functional theory as it has been discussed up to this point is restricted to many-electron systems with only *scalar* external potential $v(\mathbf{r})$, usually the electrostatic potential due to the atomic nuclei. For such systems, we have seen that the electron density can be used as the basic variable, in place of the complicated N-electron wave function. There is an intimate association between electron density $\rho(\mathbf{r})$ and scalar potential $v(\mathbf{r})$.

We now take up systems with a more generalized potential, a magnetic field in addition to the usual scalar potential. To characterize the system completely we naturally need more information about the system than just the total electron density. As will be shown shortly what is now needed are the α-electron and β-electron densities $\rho^\alpha(\mathbf{r})$ and $\rho^\beta(\mathbf{r})$, or equivalently the electron spin density $Q(\mathbf{r}) = \rho^\alpha(\mathbf{r}) - \rho^\beta(\mathbf{r})$, plus the total electron density $\rho(\mathbf{r})$. As basic variables there now are two functions in 3-dimensional space, and the theory becomes a *spin-density-functional theory*.

The spin-density-functional theory is the necessary generalization for systems in the presence of an external magnetic field. It is also exceedingly important for systems in the absence of a magnetic field, because it allows one to build more physics into the approximate exchange-correlation functional through its spin dependence. As will be seen, this last aspect of the spin-density-functional theory greatly assists accurate calculations for atoms, molecules, and solids.

In the presence of a magnetic field $\mathbf{B}(\mathbf{r})$ that acts only on the spins of the electrons, the Hamiltonian of the system becomes

$$\hat{H} = -\frac{1}{2} \sum_i^N \nabla_i^2 + \sum_{i<j}^N \frac{1}{r_{ij}} + \sum_i^N v(\mathbf{r}_i) + 2\beta_e \sum_i^N \mathbf{B}(\mathbf{r}) \cdot \mathbf{s}_i \qquad (8.1.1)$$

where $\beta_e = e\hbar/2mc$ is the Bohr magneton and $\mathbf{s}_i$ is the electron spin angular momentum vector for the ith electron (Levine 1983, p. 285). Equation (8.1.1) is only a partial description of molecules with magnetic interactions, for the interaction of $\mathbf{B}(\mathbf{r})$ with the electronic current (orbital angular momentum) is neglected, as are magnetic interactions between electrons.

The density-functional theory for the ground state of the Hamiltonian of (8.1.1) was developed by von Barth and Hedin (1972), and Pant and Rajagopal (1972) (see also Stoddart and March 1971, Rajagopal and Callaway 1973). Note that the added magnetic interaction is still a one-electron operator, just like the nuclear potential $v(\mathbf{r})$. We can combine terms in the following convenient way:

$$\hat{V} = \sum_i^N v(\mathbf{r}_i) + 2\beta_e \sum_i^N \mathbf{B}(\mathbf{r}_i) \cdot \mathbf{s}_i$$

$$= \int v(\mathbf{r})\hat{\rho}(\mathbf{r}) \, d\mathbf{r} - \int \mathbf{B}(\mathbf{r}) \cdot \hat{\mathbf{m}}(\mathbf{r}) \, d\mathbf{r} \tag{8.1.2}$$

where $\hat{\rho}(\mathbf{r})$ is the operator for electron density,

$$\hat{\rho}(\mathbf{r}) = \sum_i^N \delta(\mathbf{r} - \mathbf{r}_i) \tag{8.1.3}$$

and $\hat{\mathbf{m}}(\mathbf{r})$ is the operator for the electron magnetization density,

$$\hat{\mathbf{m}}(\mathbf{r}) = -2\beta_e \sum_i^N \mathbf{s}_i \, \delta(\mathbf{r} - \mathbf{r}_i) \tag{8.1.4}$$

Note that both $\hat{\rho}(\mathbf{r})$ and $\hat{\mathbf{m}}(\mathbf{r})$ are local operators. The expectation value of $\hat{V}$ for the state $|\Psi\rangle$ is given by

$$\langle \Psi | \hat{V} | \Psi \rangle = \int v(\mathbf{r})\rho(\mathbf{r}) \, d\mathbf{r} - \int \mathbf{B}(\mathbf{r}) \cdot \mathbf{m}(\mathbf{r}) \, d\mathbf{r} \tag{8.1.5}$$

where the electron density is given by

$$\rho(\mathbf{r}) = \langle \Psi | \hat{\rho}(\mathbf{r}) | \Psi \rangle \tag{8.1.6}$$

and the magnetization density by

$$\mathbf{m}(\mathbf{r}) = \langle \Psi | \hat{\mathbf{m}}(\mathbf{r}) | \Psi \rangle \tag{8.1.7}$$

We shall discuss only the simple case of z-direction $b(\mathbf{r})$; the general $\mathbf{B}(\mathbf{r})$ needs straightforward extension involving the full matrix $\gamma_1(\mathbf{x}', \mathbf{x})$ in spin space. We then have

$$\langle \Psi | \hat{V} | \Psi \rangle = \int v(\mathbf{r})\rho(\mathbf{r}) \, d\mathbf{r} - \int b(\mathbf{r})m(\mathbf{r}) \, d\mathbf{r} \tag{8.1.8}$$

where

$$m(\mathbf{r}) = -2\beta_e \langle \Psi | \sum_i^N s_z(i) \delta(\mathbf{r} - \mathbf{r}_i) | \Psi \rangle$$

$$= -2\beta_e \int s_z \delta(\mathbf{r} - \mathbf{r}') \gamma_1(\mathbf{x}', \mathbf{x}') \, d\mathbf{x}'$$

$$= -2\beta_e \sum_{s=\alpha,\beta} s_z \gamma_1(\mathbf{r}s, \mathbf{r}s)$$

$$= -2\beta_e [\tfrac{1}{2}\gamma_1(\mathbf{r}\alpha, \mathbf{r}\alpha) + (-\tfrac{1}{2})\gamma_1(\mathbf{r}\beta, \mathbf{r}\beta)]$$

$$= \beta_e [\rho^\beta(\mathbf{r}) - \rho^\alpha(\mathbf{r})] \tag{8.1.9}$$

The second line follows from (2.3.24), the third line results from integration over the spatial coordinates, and the fourth line comes from the fact that $s_z = 1/2$ for the α(spin-up) state and $s_z = -1/2$ for the β (spin-down) state; ρ^α and ρ^β are the spin-up and spin-down electron densities, corresponding to the diagonal elements of γ_1 in spin space (see §2.4). It is now clear that $\rho(\mathbf{r})$ and $m(\mathbf{r})$, or $\rho^\alpha(\mathbf{r})$ and $\rho^\beta(\mathbf{r})$, play the same role as $\rho(\mathbf{r})$ does in the absence of a magnetic field. In other words, the basic variables now are $\rho(\mathbf{r})$ and $m(\mathbf{r})$, or $\rho^\alpha(\mathbf{r})$ and $\rho^\beta(\mathbf{r})$.

We obtain the spin-density-functional theory by breaking the minimum search for the ground-state energy into two steps, as an extension of (3.4.7). Namely,

$$E_0 = \min_\Psi \langle \Psi | \hat{T} + \hat{V}_{ee} + \sum_i^N v(\mathbf{r}_i) + 2\beta_e \sum_i^N \mathbf{b}(\mathbf{r}_i) \cdot \mathbf{s}_z(i) | \Psi \rangle$$

$$= \min_{\rho^\alpha, \rho^\beta} \left\{ \min_{\Psi \to \rho^\alpha, \rho^\beta} \langle \Psi | \hat{T} + \hat{V}_{ee} | \Psi \rangle + \int [v(\mathbf{r})\rho(\mathbf{r}) - b(\mathbf{r})m(\mathbf{r})] \, d\mathbf{r} \right\}$$

$$= \min_{\rho^\alpha, \rho^\beta} \left\{ F[\rho^\alpha, \rho^\beta] + \int d\mathbf{r} [(v(\mathbf{r}) + \beta_e b(\mathbf{r}))\rho^\alpha(\mathbf{r}) + (v(\mathbf{r}) - \beta_e b(\mathbf{r}))\rho^\beta(\mathbf{r})] \right\}$$

$$\tag{8.1.10}$$

where

$$F[\rho^\alpha, \rho^\beta] = \min_{\Psi \to \rho^\alpha, \rho^\beta} \langle \Psi | \hat{T} + \hat{V}_{ee} | \Psi \rangle \tag{8.1.11}$$

This provides constrained-search formulation of the universal functional $F[\rho^\alpha, \rho^\beta]$, an extension of (3.4.5). The functional $F[\rho^\alpha, \rho^\beta]$ searches all Ψ that yield the input $\rho^\alpha(\mathbf{r})$ and $\rho^\beta(\mathbf{r})$; then $F[\rho^\alpha, \rho^\beta]$ assumes the minimum of $\langle \hat{T} + \hat{V}_{ee} \rangle$. The last equality of (8.1.10) is the basis of the spin-density-functional theory: ρ^α and ρ^β are all that are needed to describe the ground state of the many-electron system in the presence of a magnetic field $b(\mathbf{r})$. However, $F[\rho^\alpha, \rho^\beta]$ is unknown, and approximation is necessary for the theory to be implemented.

The Kohn–Sham method can now be introduced to rigorously handle the kinetic-energy contribution to $F[\rho^\alpha, \rho^\beta]$. In the manner of (7.1.13), define

$$F[\rho^\alpha, \rho^\beta] = T_s[\rho^\alpha, \rho^\beta] + J[\rho^\alpha + \rho^\beta] + E_{xc}[\rho^\alpha, \rho^\beta] \quad (8.1.12)$$

where $T_s[\rho^\alpha, \rho^\beta]$ is the Kohn–Sham kinetic-energy functional corresponding to a system of noninteracting electrons with densities ρ^α and ρ^β, and $E_{xc}[\rho^\alpha, \rho^\beta]$ is the exchange-correlation-energy functional. A constrained-search definition of T_s can also be given (Perdew and Zunger 1981) as extensions of (7.3.1) and (7.6.1); namely,

$$T_s[\rho^\alpha, \rho^\beta] = \text{Min} \left[\sum_{i\sigma} n_{i\sigma} \int d\mathbf{r} \phi_{i\sigma}^*(\mathbf{r})(-\tfrac{1}{2}\nabla^2)\phi_{i\sigma}(\mathbf{r}) \right] \quad (8.1.13)$$

where the minimization is over the set of $n_{i\sigma}$ and $\phi_{i\sigma}$, with constraints

$$\sum_i n_{i\alpha} |\phi_{i\alpha}(\mathbf{r})|^2 = \rho^\alpha(\mathbf{r}), \quad \sum_i n_{i\beta} |\phi_{i\beta}(\mathbf{r})|^2 = \rho^\beta(\mathbf{r}) \quad (8.1.14)$$

Note that here we use the spatial part of the spin orbital $\psi_i(\mathbf{r}s) = \phi_{i\sigma}(\mathbf{r})\sigma(s)$. In practice, the occupation numbers $n_{i\sigma}$ are chosen so that the lowest eigenstates are occupied (with $n_{i\sigma} = 1$) and the rest are unoccupied (with $n_{i\sigma} = 0$). Suppose the set of $n_{i\sigma}$ and $\phi_{i\sigma}$ minimizes (8.1.13). Then

$$T_s[\rho^\alpha, \rho^\beta] = \sum_{i\sigma} n_{i\sigma} \int d\mathbf{r} \phi_{i\sigma}^*(\mathbf{r})(-\tfrac{1}{2}\nabla^2)\phi_{i\sigma}(\mathbf{r}) \quad (8.1.15)$$

We may express the energy (8.1.10) as a functional of the orbitals $\phi_{i\sigma}$:

$$E[\rho^\alpha, \rho^\beta] = \sum_{i\sigma} n_{i\sigma} \int d\mathbf{r} \phi_{i\sigma}^*(\mathbf{r})(-\tfrac{1}{2}\nabla^2)\phi_{i\sigma}(\mathbf{r}) + J[\rho^\alpha + \rho^\beta] + E_{xc}[\rho^\alpha, \rho^\beta]$$

$$+ \int d\mathbf{r}[(v(\mathbf{r}) + \beta_e b(\mathbf{r}))\rho^\alpha(\mathbf{r}) + (v(\mathbf{r}) - \beta_e b(\mathbf{r}))\rho^\beta(\mathbf{r})] \quad (8.1.16)$$

The variational search for the minimum of $E[\rho^\alpha, \rho^\beta]$ can then be carried out through orbitals, subject to normalization constraints:

$$\int \phi_{i\sigma}(\mathbf{r})\phi_{i\sigma}(\mathbf{r}) \, d\mathbf{r} = 1 \quad (8.1.17)$$

The resulting Kohn–Sham equations are

$$\hat{h}_{\text{eff}}^\alpha \phi_{i\alpha}(\mathbf{r}) = [-\tfrac{1}{2}\nabla^2 + v_{\text{eff}}^\alpha(\mathbf{r})]\phi_{i\alpha}(\mathbf{r})$$

$$= \frac{\varepsilon_{i\alpha}'}{n_{i\alpha}} \phi_{i\alpha}(\mathbf{r}) = \varepsilon_{i\alpha} \phi_{i\alpha}(\mathbf{r}), \quad i = 1, 2, \ldots, N^\alpha$$

$$\hat{h}_{\text{eff}}^\beta \phi_{j\beta}(\mathbf{r}) = [-\tfrac{1}{2}\nabla^2 + v_{\text{eff}}^\beta(\mathbf{r})]\phi_{j\beta}(\mathbf{r})$$

$$= \frac{\varepsilon_{j\beta}'}{n_{j\beta}} \phi_{j\beta}(\mathbf{r}) = \varepsilon_{j\beta} \phi_{j\beta}(\mathbf{r}), \quad j = 1, 2, \ldots, N^\beta \quad (8.1.18)$$

8.1 THE KOHN–SHAM METHOD: ELABORATION

where the spin-dependent effective potentials are

$$v_{\text{eff}}^{\alpha} = v(\mathbf{r}) + \beta_e b(\mathbf{r}) + \int \frac{\rho(\mathbf{r}')}{|\mathbf{r}-\mathbf{r}'|} d\mathbf{r}' + \frac{\delta E_{xc}[\rho^{\alpha}, \rho^{\beta}]}{\delta \rho^{\alpha}(\mathbf{r})}$$
$$v_{\text{eff}}^{\beta} = v(\mathbf{r}) - \beta_e b(\mathbf{r}) + \int \frac{\rho(\mathbf{r}')}{|\mathbf{r}-\mathbf{r}'|} d\mathbf{r}' + \frac{\delta E_{xc}[\rho^{\alpha}, \rho^{\beta}]}{\delta \rho^{\beta}(\mathbf{r})} \quad (8.1.19)$$

and the $\varepsilon'_{i\sigma}$ are Lagrangian multipliers for the constraints (8.1.17). Note that in (8.1.18), the numbers of electrons with α spin and β spin,

$$N^{\alpha} = \int d\mathbf{r}\, \rho^{\alpha}(\mathbf{r}), \qquad N^{\beta} = \int d\mathbf{r}\, \rho^{\beta}(\mathbf{r}) \quad (8.1.20)$$

need also to be varied to achieve minimum total energy under the constraint

$$N = N^{\alpha} + N^{\beta} \quad (8.1.21)$$

There is also a spin-polarized version of Janak's theorem (7.6.11),

$$\frac{\partial E}{\partial n_{i\sigma}} = \varepsilon_{i\sigma} \quad (8.1.22)$$

This can be proved in the same manner as (7.6.11).

The kinetic energy $T_s[\rho^{\alpha}, \rho^{\beta}]$ is exactly treated through the spin Kohn–Sham equations (8.1.18) and (8.1.19). The exact exchange-correlation-energy functional $E_{xc}[\rho^{\alpha}, \rho^{\beta}]$ exists, but it remains unknown. Studies of and approximation for $E_{xc}[\rho^{\alpha}, \rho^{\beta}]$ are the main subject of the rest of this chapter.

This spin-polarized Kohn–Sham theory possesses two main advantages over the spin-compensated Kohn–Sham theory. First, it is capable of describing many-electron systems in the presence of a magnetic field acting on the spins of electrons. Some of the magnetic properties of the system, for example the electron spin susceptibility, can be determined in such a theory (Kohn and Vashishta 1983). Also, spin-orbit-coupling and relativistic-correction effects can be included (Gunnarsson and Lundqvist 1976).

A second, major advantage of spin-density theory comes in its application in the absence of magnetic fields. Should not the spin-polarized Kohn–Sham results reduce to the spin-compensated results when $b(\mathbf{r}) = 0$? The answer is certainly yes under the condition that the exact exchange-correlation functionals $E_{xc}[\rho]$ and $E_{xc}[\rho^{\alpha}, \rho^{\beta}]$ are both known and used, for then both schemes would give the same total density ρ and energy E. But we do not know the exact functionals and in any actual calculation we have to use approximate forms. This produces a significant difference, because an approximate form of the spin-density functional $E_{xc}[\rho^{\alpha}, \rho^{\beta}]$ can be (and usually is) a better description of the real system than the corresponding approximate $E_{xc}[\rho]$. This is surely the case for spin-polarized systems such as open-shell atoms and molecules. In the local-density approximation $E_{xc}^{LDA}[\rho]$ of §7.4, for example, the exchange-correlation energy of the electrons is approximated locally by results for the *homogeneous spin-compensated electron gas*. Such a procedure is not appropriate for systems with unpaired electrons, like open-shell molecules. A better approximation for such systems will be obtained through the use of $E_{xc}^{LDA}[\rho^\alpha, \rho^\beta]$, the exchange-correlation energy of the *homogeneous spin-polarized electron gas*. This constitutes the *local spin-density approximation* (LSDA), which is the subject of the next section.

There is yet another situation in which the spin-polarized Kohn–Sham equation is superior. Even in the absence of an external magnetic field, the spin Kohn–Sham equations allow electrons of different spins to have different spatial densities, just as do the unrestricted Hartree–Fock equations. In (8.1.18) and (8.1.19) the effective potential $v_{\text{eff}}^\alpha(\mathbf{r})$ differs from $v_{\text{eff}}^\beta(\mathbf{r})$ if the input $\rho^\alpha(\mathbf{r})$ differs from $\rho^\beta(\mathbf{r})$, possibly leading to self-consistent solutions with $Q(\mathbf{r}) = \rho^\alpha(\mathbf{r}) - \rho^\beta(\mathbf{r}) \neq 0$. Thus, this theory can account for molecules at large bond length (near the dissociation limit) and also for spontaneous magnetization phenomena in magnetic materials, for which the usual restricted Hartree–Fock method fails and an unrestricted Hartree–Fock method is required.

The resolution of the total electron density into one-electron orbitals enables one to treat the kinetic energy of the electrons exactly (for the noninteracting reference system)—the key idea of Kohn–Sham theory. Resolution of the density into two parts of different spin immediately enhances the quality of even rudimentary approximations for the exchange-correlation effect.

8.2. Spin-density functionals and the local spin-density approximations

Before going into details of approximation, let us consider how spin-density functionals are related to the corresponding spin-compensated density functionals. The kinetic energy defined in (8.1.15) is separable into two different spin contributions,

$$T_s[\rho^\alpha, \rho^\beta] = T_s[\rho^\alpha, 0] + T_s[0, \rho^\beta] \tag{8.2.1}$$

where

$$T_s[\rho^\alpha, 0] = \sum_{i\alpha} n_{i\alpha} \int d\mathbf{r}\, \phi_{i\alpha}^*(\mathbf{r})(-\tfrac{1}{2}\nabla^2)\phi_{i\alpha}(\mathbf{r}) \tag{8.2.2}$$

and $T_s[0, \rho^\beta]$ has a similar formula. For the spin-compensated case,

$$\rho^\alpha(\mathbf{r}) = \rho^\beta(\mathbf{r}) = \tfrac{1}{2}\rho(\mathbf{r}) \tag{8.2.3}$$

$$T_s[\tfrac{1}{2}\rho, \tfrac{1}{2}\rho] = T_s[\tfrac{1}{2}\rho, 0] + T_s[0, \tfrac{1}{2}\rho]$$
$$= 2T_s[\tfrac{1}{2}\rho, 0] \tag{8.2.4}$$

Since the kinetic energy is spin-independent,
$$T_s[\tfrac{1}{2}\rho, 0] = T_s[0, \tfrac{1}{2}\rho] \tag{8.2.5}$$

For convenience, we write
$$T_s^0[\rho] = T_s[\tfrac{1}{2}\rho, \tfrac{1}{2}\rho] \tag{8.2.6}$$

where the superscript 0 indicates complete spin pairing (in the literature this is often referred to as the "paramagnetic" case). It then follows from (8.2.1), (8.2.4), and (8.2.6) that

$$T_s[\rho^\alpha, \rho^\beta] = \tfrac{1}{2} T_s^0[2\rho^\alpha] + \tfrac{1}{2} T_s^0[2\rho^\beta] \tag{8.2.7}$$

This argument was originally given by Oliver and Perdew (1979). Note that this derivation and the result (8.2.7) conform to the constrained-search definition of (8.1.13), since the minimization in (8.1.13) can be decomposed into separate minimizations for the different spins.

Although it is true for an even number of paired electrons, in general the exact Kohn–Sham kinetic energy of §7.3 is not equal to the spin-compensated kinetic energy of (8.2.7); that is,

$$T_s[\rho] \neq T_s[\tfrac{1}{2}\rho, \tfrac{1}{2}\rho] = T_s^0[\rho] \tag{8.2.8}$$

A counter-example is when ρ integrates to an odd number.

The Thomas–Fermi model and its gradient correction in Chapter 6 may be regarded as approximations to $T_s^0[\rho]$. Using (8.2.7), (6.1.19), and (6.7.1), we obtain the spin-polarized formulas,

$$T_{\text{TF}}[\rho^\alpha, \rho^\beta] = 2^{2/3} C_F \int [(\rho^\alpha)^{5/3} + (\rho^\beta)^{5/3}] \, d\mathbf{r} \tag{8.2.9}$$

and
$$T_W[\rho^\alpha, \rho^\beta] = \frac{1}{8} \int \frac{|\nabla \rho^\alpha|}{\rho^\alpha} d\mathbf{r} + \frac{1}{8} \int \frac{|\nabla \rho^\beta|^2}{\rho^\beta} d\mathbf{r} \tag{8.2.10}$$

Take the hydrogen atom as an example, which has exact $T = 0.5$. Since $\int \rho_{1s}^{5/3} \, d\mathbf{r} = 0.1007$, blind use of the original Thomas–Fermi formula gives

$$T_{\text{TF}}[\rho_{1s}] = C_F \cdot 0.1007 = 0.2891$$

while the spin-polarized result is

$$T_{\text{TF}}[\rho_{1s}, 0] = 2^{2/3} T_{\text{TF}}[\rho_{1s}] = 0.4590$$

If we include the gradient correction, noting $T_W[\rho_{1s}] = T = 0.5000$, we obtain the approximate kinetic energy

$$T_{\text{TF}}[\rho_{1s}, 0] + \tfrac{1}{9} T_W[\rho_{1s}] = 0.5146$$

This has about the same accuracy as the corresponding results for noble gas atoms in Table 6.7. The result well illustrates the point that for spin-polarized systems, the use of spin-density functionals is mandatory.

With the spin-polarized Kohn–Sham equations of §8.1, the kinetic energy is handled exactly and only the exchange-correlation energy remains to be determined. The exchange-correlation contribution can be separated into exchange and correlation pieces:

$$E_{xc}[\rho^\alpha, \rho^\beta] = E_x[\rho^\alpha, \rho^\beta] + E_c[\rho^\alpha, \rho^\beta] \qquad (8.2.11)$$

where the exchange part is defined from (2.5.24) as

$$E_x[\rho^\alpha, \rho^\beta] = -\frac{1}{2} \iint \frac{1}{r_{12}} [|\rho_1^{\alpha\alpha}(\mathbf{r}_1, \mathbf{r}_2)|^2 + |\rho_1^{\beta\beta}(\mathbf{r}_1, \mathbf{r}_2)|^2] \, d\mathbf{r}_1 \, d\mathbf{r}_2 \qquad (8.2.12)$$

with

$$\rho_1^{\alpha\alpha}(\mathbf{r}_1, \mathbf{r}_2) = \sum_i n_{i\alpha} \phi_{i\alpha}(\mathbf{r}_1) \phi_{i\alpha}^*(\mathbf{r}_2)$$

$$\rho_1^{\beta\beta}(\mathbf{r}_1, \mathbf{r}_2) = \sum_i n_{i\beta} \phi_{i\beta}(\mathbf{r}_1) \phi_{i\beta}^*(\mathbf{r}_2) \qquad (8.2.13)$$

The $n_{i\sigma}$ and $\phi_{i\sigma}$ are those giving the Kohn–Sham kinetic energy (8.1.13); they are determined by ρ^α and ρ^β. This definition of the exchange-energy functional is the natural extension of the definition in Hartree–Fock theory (Harris and Jones 1974, Levy and Perdew 1985b). For a given problem, note that the exchange energy as defined by (8.2.12) is *not* equal to the Hartree–Fock exchange energy, since Kohn–Sham orbitals are not identical with Hartree–Fock orbitals.

The argument leading from (8.2.1) to (8.2.7) can now be applied to (8.2.12), leading (Oliver and Perdew 1979) to

$$E_x[\rho^\alpha, \rho^\beta] = \tfrac{1}{2} E_x[\rho^\alpha, \rho^\alpha] + \tfrac{1}{2} E_x[\rho^\beta, \rho^\beta]$$
$$= \tfrac{1}{2} E_x^0[2\rho^\alpha] + \tfrac{1}{2} E_x^0[2\rho^\beta] \qquad (8.2.14)$$

where

$$E_x^0[\rho] = E_x[\tfrac{1}{2}\rho, \tfrac{1}{2}\rho] \qquad (8.2.15)$$

The Dirac local-density approximation (LDA) for exchange is for the spin-compensated case. Thus, using (8.2.14) and (6.1.20), we obtain the *local spin-density approximation* (LSD) for the exchange energy functional,

$$E_x^{\text{LSD}}[\rho^\alpha, \rho^\beta] = 2^{1/3} C_x \int [(\rho^\alpha)^{4/3} + (\rho^\beta)^{4/3}] \, d\mathbf{r} \qquad (8.2.16)$$

Define the spin polarization parameter ζ by

$$\zeta = \frac{\rho^\alpha - \rho^\beta}{\rho} = \frac{\rho^\alpha - \rho^\beta}{\rho^\alpha + \rho^\beta} \qquad (8.2.17)$$

Then $\rho^\alpha = \frac{1}{2}(1 + \zeta)\rho$, $\rho^\beta = \frac{1}{2}(1 - \zeta)\rho$, and the LSD exchange energy becomes

$$E_x^{\text{LSD}}[\rho^\alpha, \rho^\beta] = \tfrac{1}{2} C_x \int \rho^{4/3}[(1 + \zeta)^{4/3} + (1 - \zeta)^{4/3}] \, d\mathbf{r}$$

$$= \int \rho \varepsilon_x(\rho, \zeta) \, d\mathbf{r} \qquad (8.2.18)$$

where

$$\varepsilon_x(\rho, \zeta) = \varepsilon_x^0(\rho) + [\varepsilon_x^1(\rho) - \varepsilon_x^0(\rho)] f(\zeta) \qquad (8.2.19)$$

with the exchange density for the spin-compensated ("paramagnetic") homogeneous electron gas given by

$$\varepsilon_x^0(\rho) = \varepsilon_x(\rho, 0) = C_x \rho^{1/3} \qquad (8.2.20)$$

for spin-completely-polarized ("ferromagnetic") homogeneous electron gas

$$\varepsilon_x^1(\rho) = \varepsilon_x(\rho, 1) = 2^{1/3} C_x \rho^{1/3} \qquad (8.2.21)$$

and

$$f(\zeta) = \tfrac{1}{2}(2^{1/3} - 1)^{-1}[(1 + \zeta)^{4/3} + (1 - \zeta)^{4/3} - 2] \qquad (8.2.22)$$

The LSD exchange energy density was originally cast in the form (8.2.19) by von Barth and Hedin (1972); the quantity $f(\zeta)$ serves as a weight factor between the two extreme cases $\zeta = 0$ and $\zeta = 1$.

Now we turn to the correlation energy $E_c[\rho^\alpha, \rho^\beta]$ defined by (8.2.11). One cannot decompose $E_c[\rho^\alpha, \rho^\beta]$ into a sum of two different spin contributions, because correlation energy contains the effects of like-spin electron–electron interaction as well as the effects of unlike-spin electron–electron interaction. This is already clear from the exact expression for the electron–electron potential energy,

$$V_{ee} = \int \frac{1}{r_{12}} \rho_2(\mathbf{r}_1, \mathbf{r}_2) \, d\mathbf{r}_1 \, d\mathbf{r}_2 \qquad (8.2.23)$$

with the second-order density matrix given [see (2.4.15) and (2.4.19)] by

$$\rho_2(\mathbf{r}_1, \mathbf{r}_2) = \rho_2^{\alpha,\alpha}(\mathbf{r}_1, \mathbf{r}_2) + \rho_2^{\beta,\beta}(\mathbf{r}_1, \mathbf{r}_2) + \rho_2^{\alpha,\beta}(\mathbf{r}_1, \mathbf{r}_2) + \rho_2^{\beta,\alpha}(\mathbf{r}_1, \mathbf{r}_2) \qquad (8.2.24)$$

The correlation energy E_c contains the part of V_{ee} that is not included in the Coulomb potential energy $J[\rho]$ and the exchange energy $E_x[\rho^\alpha, \rho^\beta]$, plus a contribution from the difference between T and T_s.

In view of the foregoing, it is not surprising that there is no available closed form for the correlation energy for the homogeneous electron gas,

$$E_c^{\text{LSD}}[\rho^\alpha, \rho^\beta] = \int \rho \varepsilon_c(\rho, \zeta) \, d\mathbf{r} \qquad (8.2.25)$$

with $\varepsilon_c(\rho, \zeta)$ the correlation energy per electron. Many workers have provided approximate forms for ε_c (von Barth and Hedin 1972, Gunnarsson and Lundqvist 1976, Vosko, Wilk, and Nusair 1980). More details on E_c^{LSD} may be found in Vosko, Wilk, and Nusair (1980) and in Appendix E.

The use of

$$E_{xc}^{LSD}[\rho^\alpha, \rho^\beta] = E_x^{LSD}[\rho^\alpha, \rho^\beta] + E_c^{LSD}[\rho^\alpha, \rho^\beta] \qquad (8.2.26)$$

in the Kohn–Sham spin-density functional theory constitutes the so-called *local spin-density method* (LSD). This had been employed in many practical applications of the density-functional theory. The first LSD calculation was performed by Gunnarsson, Lundqvist, and Wilkens (1974). The efficiency of LSD has been most decisively demonstrated by Gunnarsson and Lundqvist (1976). The *spin-polarized Xα method*, defined as using only $E_x^{LSD}[\rho^\alpha, \rho^\beta]$ with a multiplicative factor and neglecting the correlation term, can be regarded as a special simple version of LSD; see the reviews of Johnson (1973) and Connolly (1977), and the book by Slater (1974).

To illustrate LSD calculations, first consider again the H atom, which has an exact energy -13.61 eV. LSD gives total energy -13.39 eV (Gunnarsson, Lundqvist, and Wilkens 1974), whereas LDA, using the spin-polarized $E_{xc}^{LSD}[\rho]$, gives -12 eV (Tong and Sham 1966). Gunnarsson and Lundqvist (1976) compared atomic ionization potentials

Table 8.1 Ionization Potentials in electron volts of Some Light Atoms Calculated in the LSD, LDA, and HF Approximations[a]

Atom	LSD	LDA	HF	Exptl.
H	13.4	12.0	—	13.6
He	24.5	26.4	—	24.6
Li	5.7	5.4	5.3	5.4
Be	9.1	—	8.0	9.3
B	8.8	—	7.9	8.3
C	12.1	—	10.8	11.3
N	15.3	—	14.0	14.5
O	14.2	16.5	11.9	13.6
F	18.4	—	16.2	17.4
Ne	22.6	22.5	19.8	21.6
Na	5.6	5.3	4.9	5.1
Ar	16.2	16.1	14.8	15.8
K	4.7	4.5	4.0	4.3

Note: LSD = local spin-density method; LDA = local-density approximation; HF = Hartree–Fock.

[a] Adapted from Gunnarsson and Lundqvist (1976).

Table 8.2 LSD Spectroscopic Constants for Diatomic Molecules[a]

	r_e (bohrs)		D_e (eV)		ω_e (cm^{-1})	
	Expt.	LSD	Expt.	LSD	Expt.	LSD
H_2	1.40	1.45	4.8	4.9	4400	4190
Li_2	5.05	5.12	1.1	1.0	350	330
B_2	3.00	3.03	3.0	3.9	1050	1030
C_2	2.35	2.35	6.3	7.3	1860	1880
N_2	2.07	2.07	9.9	11.6	2360	2380
O_2	2.28	2.27	5.2	7.6	1580	1620
F_2	2.68	2.61	1.7	3.4	890	1060
Na_2	5.82	5.67	0.8	0.9	160	160
Al_2	4.66	4.64	1.8	2.0	350	350
Si_2	4.24	4.29	3.1	4.0	510	490
P_2	3.58	3.57	5.1	6.2	780	780
S_2	3.57	3.57	4.4	5.9	730	720
Cl_2	3.76	3.74	2.5	3.6	560	570

[a] From Becke (1986a).

calculated from LSD, LDA, and Hartree–Fock methods with experimental values; see Table 8.1. The ionization potentials were calculated as the energy difference between the neutral atom and the positively charged ion, both computed by the same method. For an interesting study of negative ions, see Fischer, Lagowski and Vosko (1987).

Gunnarsson and Lundqvist (1976), calculating the potential energy curve for the H_2 molecule, clearly showed that LSD gives the proper dissociation limit for diatomic molecules, while LDA fails, paralleling the performance of the unrestricted-HF method as compared with the restricted-HF method. In Table 8.2 are listed the elegant LSD results of Becke (1986a), who solved the diatomic Kohn–Sham equations numerically. LSD bond lengths and vibrational frequencies are good; LSD dissociation energies are overestimated.

Another interesting example is the bonding in the transition-metal dimers, Mo–Mo and Cr–Cr. The major difficulty with the theoretical prediction of bonding in such molecules is the extremely large effect of electron correlation—both restricted-Hartree–Fock and unrestricted-Hartree–Fock methods give unbound molecules. The LSD results (Delley, Freeman, and Ellis 1983, Bernholc and Holzwarth 1983) agree well with experiment and surpass the more demanding generalized valence-bond calculations (Goodgame and Goddard 1982). Other studies of metal-metal bonds include the works of Dunlap (1983), Ziegler, Tschinke, and Becke (1987a), Lee, Bylander and Kleinman (1988), and

Dhar and Kestner (1988); for metal-hydrogen and metal-methyl bonds, see Ziegler, Tschinke, and Becke (1987b).

A recent review by Jones (1987) offers many more examples of LSD calculations on molecular structures.

In complicated cases, the details of the actual calculational methods of course are important. Gaussian or other basis sets can be employed (for example, Kitaura, Satoko, and Morokuma 1979).

8.3 Self-interaction correction

The success of the local-density approximation is impressive, given its crude presumption that the local exchange-correlation energy per particle of a uniform electron gas can be carried over to atomic and molecular systems, where density gradients are large. But one should not be overoptimistic about the LSD calculations. The fact is, as was stated over 20 years ago by Tong and Sham (1966), that the local spin-density approximation underestimates the exchange energy E_x by at least 10%, while overestimating the correlation energy E_c by a factor of 2 or more, with the absolute error in E_x approximately three times that for E_c (Vosko and Wilk 1983). In many cases, although the error in E_x is partially compensated, the total energy is a further away from the experimental value than is the Hartree–Fock energy. The success of LSD for calculating evergy differences, for example ionization energies and binding energies, must be largely due to a cancellation of errors.

To improve upon the local spin-density approximation, we first consider the problem of self-interaction of electrons in the approximate functionals. An electron interacts with other electrons in the molecule via the Coulomb potential; it does not interact with itself. This is manifest in the wave-function approach through the Hamiltonian (1.1.2) in which the electron–electron interaction term excludes self-interaction. The Hartree–Fock approximation offers a clear example. The total electronic energy of (1.3.2) contains no self-interaction contribution, because the self-interaction part of the Coulomb energy cancels that of the exchange part; see (1.3.6).

This requirement that there be no self-interaction, so easily met in the wave-function approach, cannot be satisfied by an arbitrary approximation in density-functional theory without special effort. The LSD approximation contains spurious self-interaction, as does the Thomas–Fermi theory.

In TF theory, the electron–electron interaction energy is approximated by (6.1.5),

$$V_{ee}^{TF}[\rho] = J[\rho] = \frac{1}{2} \iint \frac{\rho(\mathbf{r}_1)\rho(\mathbf{r}_2)}{r_{12}} d\mathbf{r}_1 \, d\mathbf{r}_2 \qquad (8.3.1)$$

The existence of self-interaction contribution in this formula becomes clear when we apply it to a one-electron system with wave function $\phi(\mathbf{r})$, yielding

$$J[|\phi(\mathbf{r})|^2] \neq 0 \tag{8.3.2}$$

The exact potential-energy functional must give zero for one electron, however;

$$V_{ee}[|\phi(\mathbf{r})|^2] = 0 \tag{8.3.3}$$

Fermi and Amaldi (1934) recognized this deficiency in TF theory and proposed the simple self-interaction-corrected formula

$$V_{ee}^{FA}[\rho] = \frac{N-1}{2N} \iint \frac{\rho(\mathbf{r}_1)\rho(\mathbf{r}_2)}{r_{12}} d\mathbf{r}_1 d\mathbf{r}_2 \tag{8.3.4}$$

This formula is self-interaction-free in the sense that it satisfies (8.3.3).

Self-interaction is also present in the $X\alpha$ approximation of (7.4.10); the correction for this was suggested by Lindgren (1971). We follow the development of Perdew and Zunger (1981), who first addressed this problem in the context of a general E_{xc}.

In approximate density-functional theory, including LDA, one approximates the exact functional

$$V_{ee}[\rho^\alpha, \rho^\beta] = J[\rho^\alpha + \rho^\beta] + E_{xc}[\rho^\alpha, \rho^\beta] \tag{8.3.5}$$

by

$$\tilde{V}_{ee}[\rho^\alpha, \rho^\beta] = J[\rho^\alpha + \rho^\beta] + \tilde{E}_{xc}[\rho^\alpha, \rho^\beta] \tag{8.3.6}$$

where tildes indicate approximate functionals. The local spin-density approximation E_{xc}^{LDA} is an example of $\tilde{E}_{xc}$. The necessary requirement (8.3.3) to exclude self-interactions can be rewritten as

$$V_{ee}[\rho_i^\alpha, 0] = J[\rho_i^\alpha] + E_{xc}[\rho_i^\alpha, 0] = 0 \tag{8.3.7}$$

with a similar formula for ρ_i^β, where ρ_i^α and ρ_i^β are the single-particle densities for the ith orbital with spin α and β, respectively. Using the definition (8.2.12) for the exchange-energy functional and (8.2.11) for the correlation-energy functional, (8.3.7) can be put in a more detailed form,

$$J[\rho_i^\alpha] + E_x[\rho_i^\alpha, 0] = 0 \tag{8.3.8}$$

and

$$E_c[\rho_i^\alpha, 0] = 0 \tag{8.3.9}$$

Equation (8.3.9) states that correlation does not exist for one-electron systems.

To eliminate self-interaction in an approximate $\tilde{E}_{xc}$, Perdew and Zunger (1981) proposed the self-interaction-corrected (SIC) version of a given approximation, $\tilde{E}_{xc}$,

$$E_{xc}^{SIC}[\rho^\alpha, \rho^\beta] = \tilde{E}_{xc}[\rho^\alpha, \rho^\beta] - \sum_{i\sigma} (J[\rho_i^\sigma] + \tilde{E}_{xc}[\rho_i^\sigma, 0]) \tag{8.3.10}$$

Note that the SIC procedure does not change the exact functional, by virtue of (8.3.7). The Fermi–Amaldi formula (8.3.4) may be regarded as an example of the SIC; to obtain it let $\bar{E}_{xc} = 0$ and $\rho^\alpha = \rho/N$ in (8.3.10). The SIC exchange-correlation potential becomes

$$v_{xc}^{i\sigma,\text{SIC}}(\mathbf{r}) = \frac{\delta E_{xc}^{\text{SIC}}}{\delta \rho_i^\sigma(\mathbf{r})} = \frac{\delta E_{xc}[\rho^\alpha, \rho^\beta]}{\delta \rho_i^\sigma(\mathbf{r})} - \int \frac{\rho_i^\sigma(\mathbf{r}')}{|\mathbf{r}-\mathbf{r}'|} d\mathbf{r}' - \frac{\delta E_{xc}[\rho_i^\sigma, 0]}{\delta \rho_i^\sigma} \quad (8.3.11)$$

The SIC one-electron equations, obtained by the procedure (8.1.16)–(8.1.19) with E_{xc}^{SIC} for E_{xc}, become

$$\left[-\frac{1}{2}\nabla^2 + v(\mathbf{r}) + \beta_e b(\mathbf{r}) + \int \frac{\rho(\mathbf{r}')}{|\mathbf{r}-\mathbf{r}'|} d\mathbf{r}' + v_{xc}^{i\alpha,\text{SIC}}(\mathbf{r})\right]\phi_{i\alpha}^{\text{SIC}}(\mathbf{r}) = \varepsilon_{i\alpha}^{\text{SIC}} \phi_{i\alpha}^{\text{SIC}}(\mathbf{r})$$

(8.3.12)

$$\left[-\frac{1}{2}\nabla^2 + v(\mathbf{r}) - \beta_e b(\mathbf{r}) + \int \frac{\rho(\mathbf{r}')}{|\mathbf{r}-\mathbf{r}'|} d\mathbf{r}' + v_{xc}^{j\beta,\text{SIC}}(\mathbf{r})\right]\phi_{j\beta}^{\text{SIC}}(\mathbf{r}) = \varepsilon_{j\beta}^{\text{SIC}} \phi_{j\beta}^{\text{SIC}}(\mathbf{r})$$

Note that unlike the spin-polarized Kohn–Sham equations, the one-electron equation for SIC orbitals have different potentials for different orbitals. Such an orbital dependence of the potential causes the orbitals to be nonorthogonal. Additional effort must be taken to obtain orthogonal SIC orbitals, but this produces very small effects on the total energy (Perdew and Zunger 1981). The SIC improves the LSD approximation considerably, as shown in Tables 8.3 and Table 8.4. Gunnarsson and Jones (1981) argued that the SIC improvement over LSD is mainly for the most tightly bound core electrons.

For a generalized exchange-only SIC method, see Manoli and Whitehead (1988).

One might wonder whether the SIC equations belong to the density-functional approach, since they differ from the spin Kohn–Sham equa-

Table 8.3 LSD Calculation of Exchange Energies of Atoms (in electron volts)[a]

Atom	LSD	LSD–SIC	Hartree–Fock
H	−6.9	−8.5	−8.5
He	−23.2	−27.9	−27.9
Ne	−297.6	−337.8	−329.5
Ar	−755.8	−842.4	−821.3
Kr	−2407.5	−2632.0	−2561.9

[a] After Perdew and Zunger (1981).

Table 8.4 LSD Calculations of Correlation Energies of Atoms (in electron volts)[a]

Atom	LSD	LSD–SIC	Exptl.
H	−0.6	−0.0	−0.0
He	−3.0	−1.5	−1.1
Be	−6.0	−3.1	−2.6
Ne	−19.9	−11.4	−10.4
Mg	−23.9	−13.6	−11.6
Ar	−38.4	−22.3	−19.9

[a] After Perdew and Zunger (1981).

tions. In fact, they do: the orbitals used in defining E_{xc}^{SIC} are uniquely determined from the spin density by way of (8.1.13). The Kohn–Sham method with a local potential independent of orbital is only one form of the density-functional theory, perhaps the most natural and probably the most popular form. As still another, equally valid method, in the next section we will encounter nonlocal potentials, where exchange energy is exactly treated. We shall also later consider self-interaction corrections to approximate correlation-energy functionals.

The orthogonalized Hartree method described in §1.3 may be viewed as a primitive self-interaction-corrected density-functional method. Self-interaction corrections also are included in the $X\alpha$ method "with theoretically determined α parameter" (Gáspár 1974, Gáspár and Nagy 1982, 1985, 1986, 1987) and in "the Ξa method" (Gopinathan, Whitehead and Bogdanovic 1976; Gopinathan 1977; Tseng, Hong, and Whitehead 1980; Rooney, Tseng, and Whitehead 1980; Tseng and Whitehead 1981a,b; Vaidehi and Gopinathan 1984).

For typical applications of SIC methods, see Heaton, Harrison, and Lin (1983), Pederson, Heaton and Lin (1984, 1985), Harrison (1983, 1987), and Cortona (1988).

8.4 The Hartree–Fock–Kohn–Sham method

Since the exchange-energy functional can be written explicitly in terms of the set of spin orbitals, (8.2.12), which in turn are determined by the spin density via the constrained-search minimization of the kinetic energy (8.1.13), one can include the exchange effects exactly, leaving it only to approximate the correlation energy, a small fraction of E_{xc}. This idea was originally suggested by Kohn and Sham (1965), but has been used only recently.

With the explicit exchange-energy functional (8.2.12) used in (8.1.16), the total energy can be expressed in terms of the set of spin orbitals,

$$E[\rho^\alpha, \rho^\beta] = \sum_{i\sigma} n_{i\sigma} \int d\mathbf{r} \phi_{i\sigma}^*(\mathbf{r})(-\tfrac{1}{2}\nabla^2)\phi_{i\sigma}(\mathbf{r}) + J[\rho]$$
$$-\frac{1}{2} \iint \frac{1}{r_{12}} [|\rho_1^{\alpha\alpha}(\mathbf{r}_1, \mathbf{r}_2)|^2 + |\rho_1^{\beta\beta}(\mathbf{r}_1, \mathbf{r}_2)|^2] \, d\mathbf{r}_1 \, d\mathbf{r}_2$$
$$+ E_c[\rho^\alpha, \rho^\beta] + \int d\mathbf{r}[(v(\mathbf{r}) + \beta_e(b(\mathbf{r}))\rho^\alpha(\mathbf{r})$$
$$+ (v(\mathbf{r}) - \beta_e b(\mathbf{r}))\rho^\beta(\mathbf{r})] \quad (8.4.1)$$

where the first-order density matrix $\rho_1^{\sigma\sigma}(\mathbf{r}_1, \mathbf{r}_2)$ is given by (8.2.13). As in conventional Hartree–Fock theory, the minimization of $E[\rho^\alpha, \rho^\beta]$ can now be carried out through the orbitals with the orthonormalization constraints

$$\int \phi_{i\sigma}^*(\mathbf{r}) \phi_{j\sigma}(\mathbf{r}) \, d\mathbf{r} = \delta_{ij} \quad (8.4.2)$$

There result the Euler–Lagrange equations

$$-\tfrac{1}{2}\nabla^2 \phi_{i\sigma}(\mathbf{r}) + \int v_{\text{eff}}^\sigma(\mathbf{r}, \mathbf{r}') \phi_{i\sigma}(\mathbf{r}') \, d\mathbf{r}' = \frac{\varepsilon_{i\sigma}' \phi(\mathbf{r})}{n_{i\sigma}} = \varepsilon_{i\sigma} \phi_{i\sigma}(\mathbf{r}) \quad (8.4.3)$$

where $\varepsilon_{i\sigma}'$ are the Lagrangian multipliers for the constraints (8.4.2) and the *nonlocal* effective potentials are

$$v_{\text{eff}}^\alpha(\mathbf{r}, \mathbf{r}') = \left[v(\mathbf{r}) + \beta_e b(\mathbf{r}) + \int \frac{\rho(\mathbf{r}'')}{|\mathbf{r} - \mathbf{r}''|} d\mathbf{r}'' + \frac{\delta E_c}{\delta \rho^\alpha(\mathbf{r})} \right] \delta(\mathbf{r} - \mathbf{r}') - \frac{\rho_1^{\alpha\alpha}(\mathbf{r}, \mathbf{r}')}{|\mathbf{r} - \mathbf{r}'|}$$

$$v_{\text{eff}}^\beta(\mathbf{r}, \mathbf{r}') = \left[v(\mathbf{r}) - \beta_e b(\mathbf{r}) + \int \frac{\rho(\mathbf{r}'')}{|\mathbf{r} - \mathbf{r}''|} d\mathbf{r}'' + \frac{\delta E_c}{\delta \rho^\beta(\mathbf{r})} \right] \delta(\mathbf{r} - \mathbf{r}') - \frac{\rho_1^{\beta\beta}(\mathbf{r}, \mathbf{r}')}{|\mathbf{r} - \mathbf{r}'|}$$

$$(8.4.4)$$

The spin density and the corresponding number of electrons are given by

$$\rho^\sigma(\mathbf{r}) = \sum_i |\phi_{i\sigma}(\mathbf{r})|^2, \quad N^\sigma = \int \rho^\sigma(\mathbf{r}) \, d\mathbf{r}. \quad (8.4.5)$$

To determine the ground state in a given case, the numbers of electrons with α spin and β spin are both varied to minimize the total energy, under the constraint

$$N = N^\alpha + N^\beta \tag{8.4.6}$$

Equation (8.4.3) differs from the Kohn–Sham equation (8.1.19) in having a nonlocal exchange potential (the last term in each of (8.4.4)) that is exact and explicit; equation (8.4.3) also differs from the Hartree–Fock equation (1.3.31) in including correlation effects through a local correlation potential. If $E_c[\rho^\alpha, \rho^\beta]$ is known, (8.4.3) provides a route to the exact electron density and exact electronic energy. There are various names in the literature for this method; we call it the *Hartree–Fock–Kohn–Sham* (HFKS) method.

The conventional Hartree–Fock approximation can be regarded as a density-functional approach in the HFKS scheme with correlation completely neglected, but not in the KS scheme. Instead of the exact *nonlocal* exchange potential in HFKS equations, the KS equations use an effective *local* potential that is not known and has to be approximated. Another trade of accuracy for simplicity! There is an integral equation relating the exact nonlocal HFKS exchange potential to the effective local KS potential (Talman and Shadwick 1976—see §8.7 below).

The correlation-energy functional has to be given some specific approximate form for the HFKS method to be employed in an actual calculation. One can take the local spin-density approximation of (8.2.25), but, as shown in Table 8.6 in §8.7 below, it overestimates the true value by a great deal. To overcome this defect, Stoll, Pavlidou, and Preuss (1978) suggested that the LSD should only be used in atoms and molecules for correlation between electrons of different spins, giving the formula

$$E_c^{SPP}[\rho^\alpha, \rho^\beta] = E_c^{LSD}[\rho^\alpha, \rho^\beta] - E_c^{LSD}[\rho^\alpha, 0] - E_c^{LSD}[\rho^\beta, 0]$$
$$= \int \rho \varepsilon_c(\rho, \xi) \, d\mathbf{r} - \int \rho^\alpha \varepsilon_c(\rho^\alpha, 1) \, d\mathbf{r} - \int \rho^\beta \varepsilon_c(\rho^\beta, 1) \, d\mathbf{r} \tag{8.4.7}$$

where the symmetry $E_c^{LSD}[\rho^\alpha, 0] = E_c^{LSD}[0, \rho^\alpha]$ has been used. Note that this approximation excludes the electronic self-interactions in the sense that is gives zero if there is only one electron in the system,

$$E_c^{SPP}[|\phi_i(\mathbf{r})|^2, 0] = 0 \tag{8.4.8}$$

For follow-up, see Stoll, Golka and Preuss (1980).

Equation (8.4.7) gives atomic correlation energies with errors of only about 10%. Kemister and Nordholm (1985) performed the HFKS self-consistent calculation with the correlation energy E_c^{SPP} for a series of diatomic molecules. Their calculations were based on the conventional *ab initio* quantum-chemistry program (HF) modified by including the correlation effects. Other HFKS calculations are by Baroni and Turcel (1983) and Baroni (1984).

Vosko and Wilk (VW) (1983) proposed another improvement over LSD by taking

$$E_c^{VW}[\rho^\alpha, \rho^\beta] = E_c^{LSD}[\rho^\alpha, \rho^\beta] - (N^\alpha E_c^{LSD}[\rho^\alpha/N^\alpha, 0] + N^\beta E_c^{LSD}[\rho^\beta/N^\beta, 0])$$

$$= \int \rho(\mathbf{r})\varepsilon_c(\rho, \xi)\, d\mathbf{r} - \int \rho^\alpha \varepsilon_c(\rho^\alpha/N^\alpha, 1)\, d\mathbf{r}$$

$$- \int \rho^\beta \varepsilon_c(\rho^\beta/N^\beta, 1)\, d\mathbf{r} \qquad (8.4.9)$$

The rationale for this approximation is that it gives the correct result at the two limits: for $N=1$, $E_c^{VW}=0$, and for $N \to \infty$ and slowly-varying densities, $E_c^{VW} = E_c^{LSD}$. E_c^{VW} compares favorably in accuracy with E_c^{SPP} and also with the results of Perdew and Zunger (1981).

Jones and Gunnarsson (1985) argue against using exact treatment of exchange plus a local-density description of correlation, by giving evidence that the nature of the one-electron orbitals involved (s, p, d, etc.) must be considered in treating correlation.

8.5 The exchange-correlation-energy functional via the exchange-correlation hole

In the previous sections we have introduced the LSD approximation to the exchange-correlation energy functional, E_{xc}, shown how the spurious self interaction contained in LSD can be corrected, and established the HFKS method in which exchange energy is exactly included and only correlation energy needs to be approximated. We continue to study E_{xc} via the exchange-correlation hole in this section and via wave-vector analysis in the next. Both approaches are based on an adiabatic connection that will first be described. They offer insight into the local-density approximation, and they also lead to considerable numerical improvement.

The adiabatic connection has been developed by Harris and Jones (1974), Gunnarsson and Lundqvist (1976), Langreth and Perdew (1977), and Harris (1984). To motivate it, recall the expression for E_{xc} from (7.1.4),

$$E_{xc}[\rho] = (V_{ee}[\rho] - J[\rho]) + (T[\rho] - T_s[\rho]) \qquad (8.5.1)$$

For simplicity, total electron density (as opposed to spin density) is used as the basic variable in the discussion. As it stands, this expression for E_{xc} is a sum of two unrelated contributions: the nonclassical part of electron–electron interaction energy and the difference between the Kohn–Sham noninteracting kinetic energy and the exact kinetic energy.

8.5 THE KOHN–SHAM METHOD: ELABORATION

The first is associated with the second-order reduced density matrix, the second with the first-order reduced density matrix. The adiabatic connection gives rise to a single coherent expression for $E_{xc}[\rho]$.

Consider the functional, defined by constrained search for a given N-representable N-electron density $\rho(\mathbf{r})$ (Levy and Perdew 1985b).

$$F_\lambda[\rho(\mathbf{r})] = \underset{\Psi \to \rho}{\text{Min}} \langle \Psi | \hat{T} + \lambda \hat{V}_{ee} | \Psi \rangle$$

$$= \langle \Psi_\rho^\lambda | \hat{T} + \lambda \hat{V}_{ee} | \Psi_\rho^\lambda \rangle \quad (8.5.2)$$

where Ψ_ρ^λ is the N-electron wave function that minimizes $\langle \hat{T} + \lambda \hat{V}_{ee} \rangle$ and yields the density ρ. Such a minimum always exists (Lieb 1982). The parameter λ characterizes the strength of the electron–electron interaction. Comparing (8.5.2) with (3.4.5) and (7.3.2), we recognize

$$F_1[\rho] = F[\rho] = T[\rho] + V_{ee}[\rho] \quad (8.5.3)$$

and

$$F_0[\rho] = \tilde{T}_s[\rho] \quad (8.5.4)$$

Now define

$$\tilde{E}_{xc}[\rho] = V_{ee}[\rho] - J[\rho] + T[\rho] - \tilde{T}_s[\rho]$$

$$= F_1[\rho] - F_0[\rho] - J[\rho]$$

$$= \int_0^1 d\lambda \frac{\partial F_\lambda[\rho]}{\partial \lambda} - J[\rho] \quad (8.5.5)$$

$\tilde{E}_{xc}[\rho]$ equals $E_{xc}[\rho]$ of (8.5.1) if ρ is noninteracting v-representable, for in that case $\tilde{T}_s[\rho] = T_s[\rho]$ [see (7.3.4)]. In most Kohn–Sham calculations the noninteracting v-representability of the physical density ρ is assumed, and so $\tilde{E}_{xc}[\rho]$ can be legitimately assumed to be $E_{xc}[\rho]$.

We want to evaluate $\tilde{E}_{xc}[\rho]$ from (8.5.5), for which we need $\partial F_\lambda[\rho]/\partial \lambda$. We first examine the requirement on Ψ_ρ^λ in (8.5.2). The variational search in (8.5.2) can be carried out, just like the procedure (7.3.6)–(7.3.9) (see also Freed and Levy 1982), except under certain unusual limitations described by Englisch and Englisch (1983). The constraint that $\psi \to \rho(\mathbf{r})$ can be expressed as

$$\rho(\mathbf{r}) = \langle \Psi | \hat{\rho}(\mathbf{r}) | \Psi \rangle = \langle \Psi | \sum_i^n \delta(\mathbf{r} - \mathbf{r}_i) | \Psi \rangle \quad (8.5.6)$$

It is necessary that Ψ_ρ^λ make stationary the functional

$$\langle \Psi | \hat{T} + \lambda \hat{V}_{ee} | \Psi \rangle + \int \left(v_\lambda(\mathbf{r}) - \frac{E_\lambda}{N} \right) \langle \Psi | \hat{\rho}(\mathbf{r}) | \Psi \rangle \, d\mathbf{r}$$

$$= \langle \Psi | \hat{T} + \lambda \hat{V}_{ee} + \sum_i^N v_\lambda(r_i) - E_\lambda | \Psi \rangle \quad (8.5.7)$$

where $(v_\lambda(\mathbf{r}) - E_\lambda/N)$ is the Lagrangian multiplier for the constraint (8.5.6). Therefore, Ψ_ρ^λ has to be an eigenstate of a Hamiltonian $\hat{H}_\lambda$,

$$\hat{H}_\lambda |\Psi_\rho^\lambda\rangle = \left(\hat{T} + \lambda \hat{V}_{ee} + \sum_i^N v_\lambda(r_i) \right) |\Psi_\rho^\lambda\rangle = E_\lambda |\Psi_\rho^\lambda\rangle \qquad (8.5.8)$$

with eigenvalue E_λ. Hence we have, from the Hellmann–Feynmann theorem,

$$\frac{\partial E_\lambda}{\partial \lambda} = \langle \Psi_\rho^\lambda | \frac{\partial \hat{H}_\lambda}{\partial \lambda} | \Psi_\rho^\lambda \rangle$$

$$= \langle \Psi_\rho^\lambda | \hat{V}_{ee} | \Psi_\rho^\lambda \rangle + \langle \Psi_\rho^\lambda | \frac{\partial}{\partial \lambda} \sum_i^N v_\lambda(\mathbf{r}_i) | \Psi_\rho^\lambda \rangle$$

$$= \langle \Psi_\rho^\lambda | \hat{V}_{ee} | \Psi_\rho^\lambda \rangle + \int \rho(\mathbf{r}) \frac{\partial}{\partial \lambda} v_\lambda(\mathbf{r}) \, d\mathbf{r}, \qquad (8.5.9)$$

which, upon comparison with

$$E_\lambda = \langle \Psi_\rho^\lambda | \hat{T} + \lambda \hat{V}_{ee} + \sum_i^N v_\lambda(\mathbf{r}_i) | \Psi_\rho^\lambda \rangle \qquad (8.5.10)$$

and (8.5.2), leads to the simple formula

$$\frac{\partial F_\lambda[\rho]}{\partial \lambda} = \langle \Psi_\rho^\lambda | \hat{V}_{ee} | \Psi_\rho^\lambda \rangle \qquad (8.5.11)$$

Finally, inserting this expression into (8.5.5), we obtain

$$\bar{E}_{xc}[\rho] = \int_0^1 d\lambda \, \langle \Psi_\rho^\lambda | \hat{V}_{ee} | \Psi_\rho^\lambda \rangle - J[\rho]$$

$$= \iint \frac{1}{r_{12}} \bar{\rho}_2(\mathbf{r}_1, \mathbf{r}_2) \, d\mathbf{r}_1 \, d\mathbf{r}_2 - J[\rho]$$

$$= \frac{1}{2} \iint \frac{1}{r_{12}} \rho(\mathbf{r}_1) \rho(\mathbf{r}_2) \bar{h}(\mathbf{r}_1 \mathbf{r}_2) \, d\mathbf{r}_1 \, d\mathbf{r}_2$$

$$= \frac{1}{2} \iint \frac{1}{r_{12}} \rho(\mathbf{r}_1) \bar{\rho}_{xc}(\mathbf{r}_1, \mathbf{r}_2) \, d\mathbf{r}_1 \, d\mathbf{r}_2 \qquad (8.5.12)$$

where the average *exchange-correlation hole* $\bar{\rho}_{xc}$ is given by [see (2.4.13)]

$$\bar{\rho}_{xc}(\mathbf{r}_1, \mathbf{r}_2) = \rho(\mathbf{r}_2) \bar{h}(\mathbf{r}_1, \mathbf{r}_2) \qquad (8.5.13)$$

and the average pair correlation function $\bar{h}$ is given by [compare (2.4.11)]

$$\int_0^1 d\lambda \, \rho_2^\lambda(\mathbf{r}_1, \mathbf{r}_2) = \bar{\rho}_2(\mathbf{r}_1, \mathbf{r}_2) = \tfrac{1}{2} \rho(\mathbf{r}_1) \rho(\mathbf{r}_2)[1 + \bar{h}(\mathbf{r}_1, \mathbf{r}_2)] \qquad (8.5.14)$$

Here $\rho_2^\lambda(\mathbf{r}_1, \mathbf{r}_2)$ is the diagonal part of the second-order density matrix derived from Ψ_ρ^λ through (2.4.3). Since $\rho_2(\mathbf{r}_1, \mathbf{r}_2)$ is symmetric [see (2.3.11)], so is the average pair-correlation function:

$$\bar{h}(\mathbf{r}_1, \mathbf{r}_2) = \bar{h}(\mathbf{r}_2, \mathbf{r}_1) \tag{8.5.15}$$

Equation (8.5.12) is an important result, interpreting the exchange-correlation energy as the classical Coulomb interaction between the electron density $\rho(\mathbf{r})$ and a charge $\bar{\rho}_{xc}$, the exchange-correlation hole averaged over λ. In this derivation of the adiabatic connection based on the constrained-search formulation, the key point is the use of the $F_\lambda[\rho]$ that connects a noninteracting system ($\lambda = 0$) to an interacting system ($\lambda = 1$).

Two important formulas result from (8.5.12) (Gunnarsson and Lundqvist 1976). First is the sum rule

$$\int \bar{\rho}_{xc}(\mathbf{r}_1, \mathbf{r}_2) \, d\mathbf{r}_2 = \int \rho(\mathbf{r}_2) \bar{h}(\mathbf{r}_1, \mathbf{r}_2) \, d\mathbf{r}_2 = -1 \tag{8.5.16}$$

which follows immediately from (2.4.14). Thus $\bar{\rho}_{xc}(\mathbf{r}_1, \mathbf{r}_2)$ represents a hole around $\mathbf{r}_1$ with unit positive charge. Note that (8.5.16) is an equality for every $\mathbf{r}_1$. Such a local condition, trivially satisfied in a wave-function approach [see for example (2.5.29)], can be a very stringent test for density-functional theory, where one usually has to approximate $\bar{\rho}_{xc}(\mathbf{r}_1, \mathbf{r}_2)$ in terms of the electron density. The second consequence of (8.5.12) is that E_{xc} only depends on a certain spherically averaged behavior of $\rho_{xc}(\mathbf{r}_1, \mathbf{r}_2)$, namely

$$E_{xc}[\rho] = \tfrac{1}{2} \int d\mathbf{r} \rho(\mathbf{r}) \int_0^\infty 4\pi s \, ds \rho_{xc}^{SA}(\mathbf{r}, s) \tag{8.5.17}$$

where the spherical average exchange-correlation hole is given by

$$\rho_{xc}^{SA}(\mathbf{r}, s) = \frac{1}{4\pi} \int_\Omega \bar{\rho}_{xc}(\mathbf{r}, \mathbf{r}') \, d\mathbf{r}', \qquad \Omega: |\mathbf{r} - \mathbf{r}'| = s \tag{8.5.18}$$

where Ω is the integration domain. The sum rule (8.5.16) can now be written as

$$4\pi \int s^2 \, ds \rho_{xc}^{SA}(\mathbf{r}, s) = -1 \tag{8.5.19}$$

We can decompose ρ_{xc} into exchange and correlation contributions. From the definition of the exchange-energy functional, (8.2.12), we can define the *exchange hole* for spin-compensated electron density as

$$\rho_x(\mathbf{r}_1, \mathbf{r}_2) = -\frac{1}{2} \frac{|\rho_1(\mathbf{r}_1, \mathbf{r}_2)|^2}{\rho(\mathbf{r}_1)} \tag{8.5.20}$$

where $\rho_1(\mathbf{r}_1, \mathbf{r}_2)$ is the first-order density matrix that yields $\rho(\mathbf{r}_1)$ and has minimum kinetic energy [see (7.3.1)]. Thus, the exchange energy is given by

$$E_x[\rho] = \frac{1}{2} \int\int \frac{1}{r_{12}} \rho(\mathbf{r}_1) \rho_x(\mathbf{r}_1, \mathbf{r}_2) \, d\mathbf{r}_1 \, d\mathbf{r}_2 \qquad (8.5.21)$$

and the exchange hole satisfies

$$\int \rho_x(\mathbf{r}_1, \mathbf{r}_2) \, d\mathbf{r}_2 = -1 \qquad (8.5.22)$$

If we define the *correlation hole* ρ_c by

$$\bar{\rho}_{xc}(\mathbf{r}_1, \mathbf{r}_2) = \rho_x(\mathbf{r}_1, \mathbf{r}_2) + \rho_c(\mathbf{r}_1, \mathbf{r}_2) \qquad (8.5.23)$$

then (8.5.16) and (8.5.22) lead to

$$\int \rho_c(\mathbf{r}_1, \mathbf{r}_2) \, d\mathbf{r}_2 = 0 \qquad (8.5.24)$$

and the correlation energy is

$$E_c[\rho] = \frac{1}{2} \int\int \frac{1}{r_{12}} \rho(\mathbf{r}_1) \rho_c(\mathbf{r}_1, \mathbf{r}_2) \, d\mathbf{r}_1 \, d\mathbf{r}_2 \qquad (8.5.25)$$

Thus, the exchange energy equals the Coulomb interaction energy of the electrons with a charge distribution containing one unit charge, while the correlation energy comes from the interaction of the electrons with a neutral charge distribution.

Following Gunnarsson and Lundqvist (1976), now consider the local-density approximation,

$$E_{xc}^{\text{LDA}}[\rho] = \int \rho(\mathbf{r}) \varepsilon_{xc}(\rho(\mathbf{r})) \, d\mathbf{r} \qquad (8.5.26)$$

This formula corresponds to

$$\rho_{xc}^{\text{LDA}}(\mathbf{r}_1, \mathbf{r}_2) = \rho(\mathbf{r}_1) \bar{h}_0(|\mathbf{r}_1 - \mathbf{r}_2|; \rho(\mathbf{r}_1)) \qquad (8.5.27)$$

where $\bar{h}_0(|\mathbf{r}_1 - \mathbf{r}_2|; \rho)$ is the λ-averaged pair-correlation function for an interacting uniform electron gas of density ρ. Note that the density factor in (8.5.27) is $\rho(\mathbf{r}_1)$, not the $\rho(\mathbf{r}_2)$ that appears in the exact formula of (8.5.13). ρ_{xc}^{LDA} determines the LDA exchange-correlation energy density through

$$\varepsilon_{xc}(\rho(\mathbf{r}_1)) = \frac{1}{2} \int \frac{1}{r_{12}} \rho_{xc}^{\text{LDA}}(\mathbf{r}_1, \mathbf{r}_2) \, d\mathbf{r}_2 \qquad (8.5.28)$$

Equation (8.5.27) sheds light on the surprising success of the LDA. First,

ρ_{xc}^{LDA} is spherically symmetric in the sense of (8.5.18). This is not true for the exact ρ_{xc}, but presents no shortcoming in the exchange-correlation energy E_{xc}, for according to (8.5.17) it only needs the spherically averaged hole. Second, ρ_{xc}^{LDA} obeys the sum rule (8.5.16),

$$\int \rho_{xc}^{\text{LDA}}(\mathbf{r}_1, \mathbf{r}_2) \, d\mathbf{r}_2 = -1 \tag{8.5.29}$$

which is true because for every $\mathbf{r}_1$, ρ_{xc}^{LDA} is the exact exchange-correlation hole of a homogeneous electron gas with density $\rho(\mathbf{r}_1)$. Thus, the LDA describes the total charge of ρ_{xc} correctly. Cancellation of error can be expected. However, the symmetry requirement of (8.5.15) is not met by $\bar{h}_0(|\mathbf{r}_1 - \mathbf{r}_2|; \rho(\mathbf{r}_1))$.

Figures 8.1 and 8.2 compare the exact and the LDA exchange holes for the neon atom (Gunnarsson, Jonsen, and Lundqvist 1979). Clearly, the LDA gives a much better description for the spherically averaged hole than for the hole itself. For a recent discussion of some of these considerations, see Sahni, Bohnen, and Harbola (1988).

The local-density approximation is only one way of using the pair correlation $\bar{h}_0$ to obtain E_{xc}. There are other ways that lead to improvement over the LDA. In the average-density (AD) approximation (Gunnarsson, Jonson, and Lundqvist 1976), one uses

$$\rho_{xc}^{\text{AD}}(\mathbf{r}_1, \mathbf{r}_2) = \bar{\rho}(\mathbf{r}_1)\bar{h}_0(|\mathbf{r}_1 - \mathbf{r}_2|; \bar{\rho}(\mathbf{r}_1)) \tag{8.5.30}$$

where the average density is given by

$$\bar{\rho}(\mathbf{r}) = \int w(\mathbf{r} - \mathbf{r}'; \bar{\rho}(\mathbf{r}))\rho(\mathbf{r}') \, d\mathbf{r}' \tag{8.5.31}$$

The corresponding exchange-correlation energy is equal to

$$E_{xc}^{\text{AD}}[\rho] = \int \rho(\mathbf{r})\varepsilon_{xc}(\bar{\rho}(\mathbf{r})) \, d\mathbf{r} \tag{8.5.32}$$

The density-dependent weighting factor $w(\mathbf{r}; \rho)$ is here constructed so that ρ_{xc}^{AD} becomes exact when the density is almost constant everywhere. A table of w can be found in the paper of Gunnarsson, Jonson, and Lundqvist (1979). ρ_{xc}^{AD} is spherically symmetric around $\mathbf{r}_1$ and also satisfies the sum rule (8.5.16). Note that the LDA corresponds to a weighting factor that is a Dirac delta function; other choices may be made to incorporate some additional requirement; for example, see Alvarellos, Tarazona, and Chacón (1986).

Another approximation, due to Gunnarsson, Jonson, and Lundqvist (1977) and Alonso and Girifalco (1977, 1978b), defines the weighted-

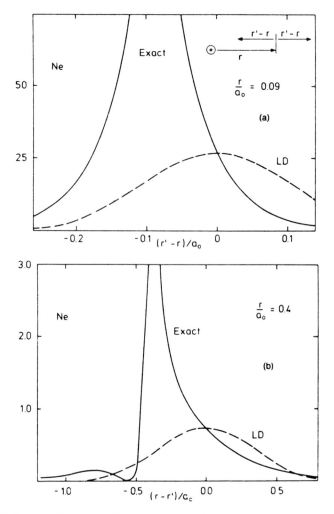

Figure 8.1 Exchange hole $\rho_x(\mathbf{r}, \mathbf{r}')$ for a neon atom. The full curves shown exact results and broken curves show the results in the LD approximation. The curves in (a) and (b) are for two different values of r. (From Gunnarsson, Jonson, and Lundqvist 1979.)

density (WD) method; namely,

$$\rho_{xc}^{\rm WD}(\mathbf{r}_1, \mathbf{r}_2) = \rho(\mathbf{r}_2)\bar{h}_0(|\mathbf{r}_1 - \mathbf{r}_2|; \bar{\rho}(\mathbf{r}_1)) \qquad (8.5.33)$$

where $\bar{\rho}(\mathbf{r}_1)$ is determined by the sum rule

$$\int \rho_{xc}^{\rm WD}(\mathbf{r}_1, \mathbf{r}_2)\, d\mathbf{r}_2 = \int \rho(\mathbf{r}_2)\bar{h}_0(|\mathbf{r}_1 - \mathbf{r}_2|; \bar{\rho}(\mathbf{r}_1))\, d\mathbf{r}_2 = -1 \quad (8.5.34)$$

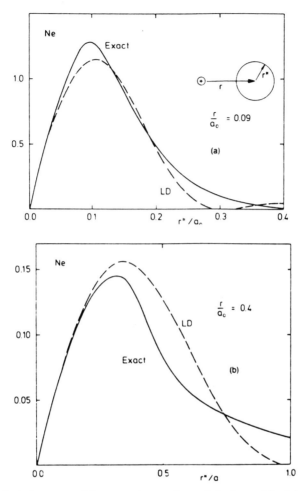

Figure 8.2 Spherical average of the neon atom exchange hole times r'' for (a) $r = 0.09$ a.u. and (b) $r = 0.4$ a.u. The full curves give the exact results and the dashed curves are obtained in the LD approximation. (From Gunnarsson, Jonson, and Lundqvist 1979.)

The WD exchange-correlation energy is given by inserting ρ_{xc}^{WD} into (8.5.12). Comparing (8.5.13) with (8.5.33), we see that ρ_{xc}^{WD} has the correct prefactor $\rho(\mathbf{r}_2)$, which is not the case with ρ_{xc}^{LDA} or ρ_{xc}^{AD}.

The modified WD scheme (MWD) (Gunnarsson and Jones 1980) approximates the pair-correlation function by the approximate analytic function

$$h^{MWD}(|\mathbf{r}_1 - \mathbf{r}_2|; \rho) = -A \exp\left[-B^5/|\mathbf{r}_1 - \mathbf{r}_2|^5\right] \qquad (8.5.35)$$

where the parameters A and B are dependent on ρ and are determined by the sum rule

$$\rho \int h^{\mathrm{MWD}}(\mathbf{r}; \rho) \, d\mathbf{r} = -1 \quad (8.5.36)$$

and the requirement that MWD gives the correct exchange-correlation energy for a homogeneous system,

$$\frac{\rho}{2} \int \frac{1}{r} h^{\mathrm{MWD}}(\mathbf{r}; \rho) \, d\mathbf{r} = \varepsilon_{xc}(\rho) \quad (8.5.37)$$

Other choices of analytic form for h^{MWD} can be made (Pedroza 1986).

The AD, WD, and MWD approximate $E_{xc}[\rho]$ all are *nonlocal* functionals of the electron density. They improve the local-density approximation significantly. Both AD and MWD reduce the error in E_{xc}^{LD}, while WD improves the exchange energy part but not the correlation-energy part of E_{xc}^{LD}.

8.6 The exchange-correlation-energy functional via wave-vector analysis

We continue the discussion on exchange-correlation energy via what is called the *wave-vector analysis*. The idea is to decompose the exchange-correlation energy into components of different wavelength.

The starting place is the result of the adiabatic connection of (8.5.12),

$$E_{xc}[\rho] = \iint \frac{1}{r_{12}} \bar{\rho}_2(\mathbf{r}_1, \mathbf{r}_2) \, d\mathbf{r}_1 \, d\mathbf{r}_2 - J[\rho]$$

$$= \iint \frac{1}{r_{12}} \left[\bar{\rho}_2(\mathbf{r}_1, \mathbf{r}_2) - \frac{1}{2} \rho(\mathbf{r}_1) \rho(\mathbf{r}_2) \right] d\mathbf{r}_1 \, d\mathbf{r}_2 \quad (8.6.1)$$

From Equation (C.43) of Appendix C, we can write the diagonal elements of the second-order reduced density matrix as

$$\rho_2^\lambda(\mathbf{r}_1, \mathbf{r}_2) = \langle \Psi_\rho^\lambda | \sum_{i<j}^N \sum^N \delta(\mathbf{r}_i' - \mathbf{r}_1) \, \delta(\mathbf{r}_j' - \mathbf{r}_2) | \Psi_\rho^\lambda \rangle$$

$$= \frac{1}{2} \langle \Psi_\rho^\lambda | \sum_i^N \sum_j^N \delta(\mathbf{r}_i' - \mathbf{r}_1) \, \delta(\mathbf{r}_j' - \mathbf{r}_2)$$

$$- \sum_i^N \delta(\mathbf{r}_i' - \mathbf{r}_1) \, \delta(\mathbf{r}_i' - \mathbf{r}_2) | \Psi_\rho^\lambda \rangle$$

$$= \frac{1}{2} \{ \Psi_\rho^\lambda | \hat{\rho}(\mathbf{r}_1) \hat{\rho}(\mathbf{r}_2) - \hat{\rho}(\mathbf{r}_1) \, \delta(\mathbf{r}_1 - \mathbf{r}_2) | \Psi_\rho^\lambda \rangle \quad (8.6.2)$$

where we have used (3.7.14). Averaging (8.6.2) over λ from zero to one,

8.6 THE KOHN–SHAM METHOD: ELABORATION

we obtain

$$\bar{\rho}(\mathbf{r}_1, \mathbf{r}_2) = \int_0^1 d\lambda \frac{1}{2} [\langle \Psi_\rho^\lambda | \hat{\rho}(\mathbf{r}_1) \hat{\rho}(\mathbf{r}_2) | \Psi_\rho^\lambda \rangle - \rho(\mathbf{r}_1) \delta(\mathbf{r}_1 - \mathbf{r}_2)] \quad (8.6.3)$$

and (8.6.1) becomes

$$E_{xc}[\rho] = \frac{1}{2} \iint \frac{1}{r_{12}} d\mathbf{r}_1 \, d\mathbf{r}_2 \left\{ \int_0^1 d\lambda NS_\lambda(\mathbf{r}_1 \mathbf{r}_2) - \rho(\mathbf{r}_1) \delta(\mathbf{r}_1 - \mathbf{r}_2) \right\} \quad (8.6.4)$$

where S_λ is the density fluctuation

$$NS_\lambda(\mathbf{r}_1, \mathbf{r}_2) = \langle \Psi_\rho^\lambda | \hat{\rho}(\mathbf{r}_1) \hat{\rho}(\mathbf{r}_2) | \Psi_\rho^\lambda \rangle - \rho(\mathbf{r}_1)\rho(\mathbf{r}_2)$$
$$= \langle \Psi_\rho^\lambda | [\hat{\rho}(\mathbf{r}_1) - \rho(\mathbf{r}_1)][\hat{\rho}(\mathbf{r}_2) - \rho(\mathbf{r}_2)] | \Psi_\rho^\lambda \rangle \quad (8.6.5)$$

Now we introduce the transformations

$$\frac{1}{r_{12}} = \frac{1}{(2\pi)^3} \int \frac{4\pi}{|\mathbf{k}|^2} e^{i\mathbf{K} \cdot (\mathbf{r}_1 - \mathbf{r}_2)} \, d\mathbf{k} \quad (8.6.6)$$

$$\delta(\mathbf{r}_1 - \mathbf{r}_2) = \frac{1}{(2\pi)^3} \int e^{i\mathbf{K} \cdot (\mathbf{r}_1 - \mathbf{r}_2)} \, d\mathbf{k} \quad (8.6.7)$$

$$S_\lambda(\mathbf{r}_1, \mathbf{r}_2) = \frac{1}{(2\pi)^6} \int d\mathbf{k}_1 \int d\mathbf{k}_2 \, e^{i(\mathbf{k}_1 \cdot \mathbf{r}_1 - \mathbf{k}_2 \cdot \mathbf{r}_2)} S_\lambda(\mathbf{k}_1, \mathbf{k}_2) \quad (8.6.8)$$

$$S_\lambda(\mathbf{k}_1, \mathbf{k}_2) = \int d\mathbf{r}_1 \int d\mathbf{r}_2 \, e^{-i\mathbf{k}_1 \cdot \mathbf{r}_1 + i\mathbf{k}_2 \cdot \mathbf{r}_2} S_\lambda(\mathbf{r}_1, \mathbf{r}_2) \quad (8.6.9)$$

On insertion of (8.6.6)–(8.6.9), the exchange-correlation energy of (8.6.4) becomes

$$E_{xc}[\rho] = \frac{1}{2} \int \frac{d\mathbf{k}}{(2\pi)^3} \frac{4\pi}{|\mathbf{k}|^2} N \int_0^1 d\lambda [S_\lambda(\mathbf{k}) - 1] \quad (8.6.10)$$

where

$$S_\lambda(\mathbf{k}) = S_\lambda(\mathbf{k}, \mathbf{k}) \quad (8.6.11)$$

Define

$$E_{xc}(\mathbf{k}) = \frac{1}{2} \int_0^1 d\lambda \frac{4\pi}{|\mathbf{k}|^2} N[S_\lambda(\mathbf{k}) - 1] \quad (8.6.12)$$

Then we obtain the wave-vector decomposition of $E_{xc}[\rho]$,

$$E_{xc}[\rho] = \int \frac{d\mathbf{k}}{(2\pi)^3} E_{xc}(\mathbf{k}) \quad (8.6.13)$$

This result is due to Langreth and Perdew (1977); see also Rasolt, Malmstrom, and Geldart (1979).

The local-density approximation to E_{xc} may be decomposed into wave-vector components in the same way (Langreth and Perdew 1977); namely,

$$E_{xc}^{LDA} = \int \frac{d\mathbf{k}}{(2\pi)^2} \int d\mathbf{r} \frac{1}{2} \int_0^1 d\lambda \frac{4\pi}{|\mathbf{k}|^2} \rho(\mathbf{r})[S_\lambda^{LDA}(\mathbf{k}; \rho(\mathbf{r})) - 1] \quad (8.6.14)$$

where $S_\lambda^{LDA}(\mathbf{k}; \rho(\mathbf{r}))$ is the quantity $S_\lambda(\mathbf{k})$ for a homogeneous electron gas with density $\rho(\mathbf{r})$. Langreth and Perdew argued that the local density approximation (8.6.14) is correct for $E_{xc}(\mathbf{k})$ at large k, but fails badly at small k. However, this failure is rather unimportant because of the factor k^2 in the volume element. This in part explains the surprising success of the local-density approximation.

Langreth and Mehl (1981, 1983), based on the work of Langreth and Perdew (1980), wrote an identity for the exchange-correlation functional of (8.6.12) (as summarized by Pedroza 1986),

$$E_{xc}(\mathbf{k}, \rho(\mathbf{r})) = E_{xc}^{LDA}(\mathbf{k}, \rho(\mathbf{r})) + \frac{1}{2\pi k^2} \int d\mathbf{r}\, k_F^2 Z_{xc}(k_F, q; k) q^2 \quad (8.6.15)$$

where $E_{xc}^{LDA}(k, \rho(\mathbf{r}))$ is defined by (8.6.13) and (8.6.14), k_F is the local Fermi wave vector

$$k_F = [3\pi^2 \rho(\mathbf{r})]^{1/3} \quad (8.6.16)$$

$$q(\mathbf{r}) = \frac{1}{6} \frac{|\nabla \rho(\mathbf{r})|}{\rho(\mathbf{r})} \quad (8.6.17)$$

and (8.6.15) is just a definition for $Z_{xc}(k_f, q; k)$. Guided by exact results for small k, Langreth and Mehl proposed the approximation

$$Z_{xc}(k_F, q; k) = Z_x(k_F, q; k) + Z_c(k_F, q; k) \quad (8.6.18)$$

with

$$Z_x(k_F, q; k) \cong Z_x(k_F, 0, k) \quad (8.6.19)$$

and

$$Z_c(k_F, q; k) \cong Z_c(k_F, 0; k)\eta(k - k_c) \quad (8.6.20)$$

where η is the step function, $k_c = 6fq$, and f is a parameter chosen as 0.15. They further used the results of random-phase approximation to $Z_x(k_F, 0; k)$ and $Z_c(k_F, 0; k)$ obtained from previous calculations (Langreth and Perdew 1980). The final result (in atomic units) is

$$E_{xc}^{LM}[\rho] = E_{xc}^{LDA} + 2.14 \times 10^{-3} \int d\mathbf{r} \frac{|\nabla \rho|^2}{\rho^{4/3}} [2e^{-F} - \tfrac{7}{9}] \quad (8.6.21)$$

where
$$F = 0.26 \frac{|\nabla \rho|}{\rho^{7/6}}$$

This formula for exchange-correlation is the *Langreth–Mehl approximation*. It substantially improves upon LDA results and is comparable to or better than the self-interaction correction methods; see the original paper for the results of atomic calculations, see Savin, Wedig, Preuss, and Stoll (1984) for the results of molecular calculations, and see Mohammed and Sahni (1985) for the results for metallic surfaces. Further elucidation of this approximation can be found in Hu and Langreth (1985, 1986).

The Langreth–Mehl correlation-energy functional has been further improved by Perdew (1986) by naturally separating exchange from correlation and going beyond the random-phase approximation. For a numerical comparison, see Table 8.6 in the next section. For recent discussion, see Langreth and Vosko (1987, 1988) and Rasolt and Geldart (1988).

8.7 Other studies of the exchange-correlation-energy functional

It has been pointed out by many researchers that the major source of error in the local-density approximation is the exchange energy (Tong and Sham 1966, Jones and Gunnarsson 1985, Gunnarsson and Jones 1985, Perdew 1985a, Wang and Overhauser 1986; also see §7.4). The conventional gradient expansion does not provide satisfactory correction to the local-density approximation (see §6.9). Alternatives need to be found.

By imposing the conditions for the correct exchange hole on the approximate hole given by the gradient expansion, Perdew (1985a) proposed a model which leaves only a 1% error in exchange energy. This model has also been further simplified (Perdew and Yue 1986) to the so-called generalized gradient approximation (GGA):

$$E_x^{GGA}[\rho] = -\frac{3}{4}\left(\frac{3}{\pi}\right)^{1/3} \int d\mathbf{r}\, \rho^{4/3} F(s) \qquad (8.7.1)$$

with
$$s = \frac{|\nabla \rho(r)|}{(2k_F \rho)} \qquad (8.7.2)$$

$$k_F = (3\pi^2 \rho)^{1/3} \qquad (8.7.3)$$

and
$$F(s) = (1 + 1.296 s^2 + 14 s^4 + 0.2 s^6)^{1/15} \qquad (8.7.4)$$

Table 8.5 Exchange Energies E_x of Neutral Atoms Calculated from Nonrelativistic Hartree–Fock Densities[a,b]

Atom	LSD	IM	GGA	Exact
H	−0.268	−0.308	−0.311	−0.3125
He	−0.884	−1.015	−1.033	−1.026
Li	−1.538	−1.747	−1.789	−1.781
Be	−2.31	−2.60	−2.68	−2.67
Ne	−11.03	−11.82	−12.22	−12.11
Ar	−27.86	−29.39	−30.29	−30.18
Zn	−65.63	−68.25	−69.93	−69.7
Kr	−88.6	−91.8	−93.8	−93.9
Xe	−170.6	−175.6	−178.6	−179.1

[a] From Perdew and Yue (1986).
[b] LSD is the local spin density formula of (8.2.16), IM is the Langreth–Mehl formula of (8.6.21), GGA is the Perdew–Yue formula (8.7.1).

The LDA corresponds to $F(s) = 1$; other forms of $F(s)$ have been suggested (Becke 1986b). Table 8.5 compares results from various models: the local spin-density approximation, the Langreth–Mehl formula of §8.6, and the GGA. In unrelated work, DePristo and Kress (1987) constructed a rational-function approximation to the exchange-energy functional.

A different approach to the exchange-energy functional was taken by Talman and Shadwick (1976). They constructed the optimum local potential for the Hartree–Fock approximation, which can be interpreted as the exact Kohn–Sham potential for exchange alone (Sahni, Gruenebaum, and Perdew 1982, Langreth and Mehl 1983, Perdew 1985b). Given an expression of the total energy in terms of N orbitals, for example $E_{HF}[\{\psi_i\}]$ of (1.3.2), Talman and Shadwick (1976) produced a method of finding the single-electron Hamiltonian $-\frac{1}{2}\nabla^2 + v(\mathbf{r})$ whose eigenstates ψ_i minimize the total energy. For the case of the Hartree–Fock functional, the resulting variational equation is

$$\frac{\delta E_{HF}}{\delta v(\mathbf{r})} = \sum_i \int d\mathbf{r}' \frac{\delta E_{HF}}{\delta \psi_i^*(\mathbf{r}')} \frac{\delta \psi_i^*(\mathbf{r}')}{\delta v(\mathbf{r})} = 0 \qquad (8.7.5)$$

in which $\delta E_{HF}/\delta \psi_i^*(\mathbf{r}')$ is given by $\hat{F}\psi_i$ of (1.3.8) and $\delta \psi_i^*(\mathbf{r}')/\delta v(\mathbf{r})$ can be obtained from first-order perturbation theory. Equation (8.7.5) can be solved to obtain the optimum local potential $v(\mathbf{r})$. The exchange component is automatically contained in this potential.

Sham and coworkers have carried out a series of studies on the exchange-correlation-energy functional (Sham 1985, Sham and Schlüter

Table 8.6 Errors in Correlation Energies Determined from Various Density-functional Formulas[a]

Species	Experimental Correlation Energy	Colle–Salvetti[b]	LYP[c]	VWN[d]	SPP[e]	IM[f]	Perdew[g]
He	0.0420	−0.0004	0.0017	0.07	0.017	0.004	0.003
Li^+	0.0435	0.0003	0.004	0.092	0.026	−0.13	0.0035
Be	0.094	−0.0014	0.001	0.131	0.022	−0.006	0.001
B^+	0.111	−0.005	−0.004	0.142	0.02	−0.033	−0.008
Ne	0.387	−0.012	−0.004	0.359	−0.001	−0.009	0.008
Li	0.045	0.005	0.008	0.106	0.027	0.006	0.009
B	0.125	0.003	0.003	0.165	0.022	−0.008	0.003
C	0.156	0.005	0.005	0.203	0.02	−0.006	0.009
N	0.189	−0.001	0.004	0.241	0.015	−0.002	0.017
H_2O	0.372	−0.036	0.034	0.292	−0.028	0.011	−0.007
CH_4	0.293	−0.003	0.001	0.3	0.014	0.079	0.035

[a] Tablulated E_{corr}(exptl) − E_{corr}(calc) for the various methods, in atomic units.
[b] Colle and Salvetti (1975).
[c] Lee, Yang, and Parr (1987).
[d] Savin, Stoll, and Preuss (1986). Method of Vosko, Wilk and Nusair (1980).
[e] Savin, Stoll, and Preuss (1986). Method of Stoll, Pavlidou, and Preuss (1978).
[f] Method of Langreth and Mehl (1984).
[g] Method of Perdew (1986).

1985, Lannoo, Schlüter, and Sham 1985). A review is available by Sham (1986). In this work, the exact Kohn–Sham local potential is related to the Dyson self-energy by a generalization of (8.7.5) to include electron correlation (Sham 1985).

In a very interesting paper, Colle and Salvetti (1975) constructed an approximate correlation-energy formula in terms of the Hartree–Fock second-order density matrix. Their results, and those obtained later by the same method (McWeeny 1976, Amaral and McWeeny 1983, Amaral 1985), are very promising—an average error of about 2% for the atoms and molecules studied. Lee, Yang, and Parr (1988) further turned the Colle–Salvetti correlation-energy formula into an explicit functional of the electron density. Table 8.6 includes a comparison of various approximations to the correlation energy. For additional developments of the Colle–Salvetti ideas, see Cohen, Frishberg, Lee, and Massa (1986), and Cohen, Santhanam, and Frishberg (1980). See Gritsenko and Zhidomirov (1987) for approximations based on an effective pair distribution function.

One expects successively more accurate correlation and exchange-correlation functionals to be developed in the future.

There have been many other interesting studies of the exchange and correlation energies, for instance Leite, Ferreira and Pereira (1972), Keller and Gázquez (1979), Payne and Fujiwara (1979), San-Fabián and Moscardó (1984), von Barth and Pedroza (1985), Ossicini and Bertoni (1985), Sharma and Thakkar (1985), Carroll, Bader, and Vosko (1987), Gritsenko and Zhidomirov (1987), Fuentealba, Stoll, and Savin (1988), and Becke (1988a,b).

9
EXTENSIONS

9.1 Finite-temperature Kohn–Sham theory

In the last two chapters, we have demonstrated how one can rigorously treat the kinetic-energy functional by use of the auxiliary Kohn–Sham orbitals and thereby replace a problem of N interacting electrons with a problem of N noninteracting electrons moving in an effective local potential. The Kohn–Sham theory, in combination with some approximation for the exchange-correlation energy functional, becomes a powerful and practical tool for calculations on many-electron systems. We now present the finite-temperature extension.

In §3.6, it was shown that in a grand canonical ensemble at a given temperature θ and chemical potential μ, the grand potential is a functional of the electron density and reaches its minimum at the equilibrium electron density. This was stated in (3.6.11), which we now rewrite as

$$\Omega^0 = \underset{\rho}{\text{Min}}\, \Omega[\rho]$$
$$= \text{Min}\left\{ T[\rho] + V_{ee}[\rho] - \theta S[\rho] + \int \rho(\mathbf{r})(v(\mathbf{r}) - \mu)\, d\mathbf{r} \right\} \quad (9.1.1)$$

where we have restricted the interaction energy to be electron–electron interaction energy, since we are interested in electronic systems. Note that $\Omega[\rho]$ depends on θ, as in (3.6.12). If we knew explicit expressions for the kinetic energy $T[\rho]$, the potential energy $V_{ee}[\rho]$, and the entropy $S[\rho]$, we would be able to obtain the equilibrium electron density and Ω^0 from (9.1.1). The finite-temperature Kohn–Sham theory (Kohn and Sham 1965, Kohn and Vashishta 1983) invokes an indirect method for treating $T[\rho] - \theta S[\rho]$ together.

Define $A_s[\rho]$ as the Helmholtz free energy of a reference system of noninteracting electrons with density $\rho(\mathbf{r})$ at temperature θ, where $\rho(\mathbf{r})$ is a density for the interacting system of interest. Thus, in terms of the kinetic energy $T_s[\rho]$ and entropy $S_s[\rho]$ of such a system,

$$A_s[\rho] = T_s[\rho] - \theta S_s[\rho] \quad (9.1.2)$$

Then

$$\Omega[\rho] = A_s[\rho] + \int \rho(\mathbf{r})(v(\mathbf{r}) - \mu)\, d\mathbf{r} + J[\rho] + F_{xc}[\rho] \quad (9.1.3)$$

where $F_{xc}[\rho]$ is the exchange-correlation contribution to the free energy,

$$F[\rho] = T[\rho] + V_{ee}[\rho] - \theta S[\rho]$$
$$= A_s[\rho] + J[\rho] + F_{xc}[\rho] \quad (9.1.4)$$

In other words,

$$F_{xc}[\rho] = (T[\rho] - \theta S[\rho]) - (T_s[\rho] - \theta S_s[\rho]) + V_{ee}[\rho] - J[\rho] \quad (9.1.5)$$

corresponding to (7.1.14).

We first describe the statistical mechanics of the noninteracting system. The grand potential becomes

$$\Omega_s[\rho] = A_s[\rho] + \int \rho(\mathbf{r})(v(\mathbf{r}) - \mu)\, d\mathbf{r} \quad (9.1.6)$$

and the equilibrium ensemble density operator of (2.7.14) is simply

$$\hat{\Gamma}^0 = \frac{e^{-\beta(\hat{H}_s - \mu \hat{N})}}{\operatorname{Tr} e^{-\beta(\hat{H}_s - \mu \hat{N})}} \quad (9.1.7)$$

where the Hamiltonian is

$$\hat{H}_s = \int dx_1\, \hat{\psi}^+(x_1)(-\tfrac{1}{2}\nabla^2)\hat{\psi}(x_1) + \int dx_1\, \hat{\psi}^+(x_1)\hat{\psi}(x_1)v(\mathbf{r}_1) \quad (9.1.8)$$

We use the second-quantization form of the operators, equations (C.37) and (C.38) of Appendix C, because the Fock space is involved in the grand canonical ensemble. The associated equilibrium-state electron density is given by

$$\rho(\mathbf{r}) = \operatorname{Tr}[\hat{\rho}(\mathbf{r})\hat{\Gamma}^0]$$
$$= \sum_i^\infty f(\varepsilon_i - \mu)\, |\phi_i(\mathbf{r})|^2 \quad (9.1.9)$$

where $f(\varepsilon_i - \mu)$ is the Fermi function

$$f_i = f(\varepsilon_i - \mu) = [1 + e^{\beta(\varepsilon_i - \mu)}]^{-1} \quad (9.1.10)$$

The $\phi_i(\mathbf{r})$ and ε_i are the eigenfunctions and eigenvalues of

$$[-\tfrac{1}{2}\nabla^2 + v(\mathbf{r})]\phi_i(\mathbf{r}) = \varepsilon_i \phi_i(\mathbf{r}) \quad (9.1.11)$$

Equation (9.1.9) can be understood in an intuitive way: Because the electrons are noninteracting, they occupy the single-particle eigenstates according to the Fermi–Dirac statistics. This can be rigorously proved (see §7.5b of Blaizot and Ripka 1986 and also see McQuarrie 1976).

The kinetic energy is given by

$$T_s = \sum_i^\infty f_i \int \phi_i(\mathbf{r})(-\tfrac{1}{2}\nabla^2)\phi_i(\mathbf{r})\,d\mathbf{r}$$

$$= \sum_i^\infty \varepsilon_i f_i - \int v(\mathbf{r})\rho(\mathbf{r})\,d\mathbf{r} \quad (9.1.12)$$

and the entropy by

$$S_s = -k_B \sum_i^\infty \{f_i \ln f_i + (1-f_i)\ln(1-f_i)\} \quad (9.1.13)$$

Now consider the interacting system. The variational principle (9.1.1), combined with (9.1.3), leads to the Euler–Lagrange equation

$$\frac{\delta A_s[\rho]}{\delta \rho(\mathbf{r})} + v(\mathbf{r}) - \mu + \int \frac{\rho(\mathbf{r}')}{|\mathbf{r}-\mathbf{r}'|}\,d\mathbf{r}' + \frac{\delta F_{xc}[\rho]}{\delta \rho(\mathbf{r})} = 0 \quad (9.1.14)$$

This has the same form as the Euler equation for a system of noninteracting electrons moving in a local potential $v_{\text{eff}}(\mathbf{r})$, defined by

$$v_{\text{eff}}(\mathbf{r}) = v(\mathbf{r}) + \int \frac{\rho(\mathbf{r}')}{|\mathbf{r}-\mathbf{r}'|}\,d\mathbf{r}' + \frac{\delta F_{xc}[\rho]}{\delta \rho(\mathbf{r})} \quad (9.1.15)$$

With this definition, (9.1.14) takes independent-particle form, and solution proceeds as just described. We obtain eigenfunctions and eigenvalues from the finite-temperature Kohn–Sham equations,

$$[-\tfrac{1}{2}\nabla^2 + v_{\text{eff}}(\mathbf{r})]\phi_i(\mathbf{r}) = \varepsilon_i \phi_i(\mathbf{r}) \quad (9.1.16)$$

and therefore the electron density from

$$\rho(\mathbf{r}) = \sum_i^\infty f_i |\phi_i(\mathbf{r})|^2 \quad (9.1.17)$$

Equations (9.1.15)–(9.1.17) have to be solved self-consistently. The resulting electron density can be used to determine the grand potential, with the kinetic energy T_s and entropy S_s given by (9.1.12) and (9.1.13). The derivation also can be accomplished through arguments similar to those in §7.2 and §7.3.

Note that the Kohn–Sham theory itself does not provide $F_{xc}[\rho]$, which must be obtained somehow. The adiabatic-connection expression for $F_{xc}[\rho]$ is given by Perdew (1986). In practice, approximations are made, usually the local density approximation

$$F_{xc}^{\text{LDA}}[\rho] = \int \rho(\mathbf{r}) f_{xc}(\rho(\mathbf{r}))\,d\mathbf{r} \quad (9.1.18)$$

where $f_{xc}(\rho)$ is the exchange-correlation contribution per particle to the free energy of a homogeneous electron gas of density ρ at temperature θ. for data on $f_{xc}(\rho)$, see Perrot and Dharma-wardana (1984) and Kanhere, Panat, Rajagopal, and Callaway (1986). This approximation has been successfully applied to the study of the metal–insulator transition in doped semiconductors (Ghazali and Hugon 1978) and of plasmas (Dharma–Wardana 1987).

As in the ground-state formulation, the free energy $A_s[\rho]$ can also be approximated directly. There is the finite-temperature Thomas–Fermi theory, for which see the review by Brack (1985). Recently, there has been developed an integral formulation of the finite-temperature Kohn–Sham theory, using Feynman path integrals (Yang 1988b). This is an extension of the zero-temperature formulation described in §7.5. The essential feature is the use of multidimensional integrals to rigorously replace the infinite set of Kohn–Sham orbitals.

9.2 Excited States

We now consider the difficult problem of the excited states of N-electron systems. Is there a workable density-functional theory for them? This is an important and contemporary research area.

In wave-function theory, excited states are much more difficult to handle than ground states (except for states that have symmetries different from the ground-state symmetry). The basic difficulty is that an excited-state wave function must be orthogonal to the ground-state wave function, and indeed also to all wave functions of excited states below it. This is a constraint that is at best cumbersome to impose in an approximate method. We can expect no better in density-functional theory.

The finite-temperature Kohn–Sham density-functional theory described in §3.6 and §9.1 is in effect an excited-state theory. At a given θ the equilibrium state is a mixture of ground and excited states; the equilibrium density is a corresponding average of ground and excited-state densities; the finite-temperature density-functional method deals with that density as the basic variable. As we vary the temperature, we vary the mix of states, and so we might hope to be able somehow to invert the whole process to extract density and energies for individual excited states.

One might ask whether, by any chance, the Euler equation of ground-state density-functional theory has solutions, other than the ground-state solution, corresponding to true excited states. One could be optimistic about this, since the ground-state density determines the Hamiltonian and hence all of the eigenstates, ground plus excited.

However, there is the orthogonality difficulty mentioned above. Perdew and Levy (1985) show that, indeed, nonground-state solutions of the ground-state Euler equation, if there are any, do correspond to some excited states. However, v-representability becomes a knotty problem here, and one certainly would not expect a general solution for excited states to be obtained in this way.

For the lowest excited state of a given symmetry, one knows that the conventional wave-function variational procedures are easy: one just restricts trial functions in the variational procedure to functions of a given symmetry and by energy minimization reaches the lowest state of that symmetry. So too with the density-functional theory, though there is the unhappy fact that the Hohenberg–Kohn functional $F[\rho]$ will in general differ from symmetry to symmetry. (Analogously, a boson $F[\rho]$ will be drastically different from a fermion $F[\rho]$.) Basic for this "subspace theory of excited states" are the papers of Gunnarsson and Lundqvist (1976), von Barth (1979), Englisch and Englisch (1983), and Kutzler and Painter (1987).

A next question is, given a particular excited state of interest, whether there might exist a variational principle whereby its density and energy could be found, and in which its density is the basic variable. Such a principle does in fact exist, based on a principle proved years ago in wave-function theory by MacDonald (1934). Define

$$I_2[\Psi, U] = \int \Psi^*(\hat{H} - U)^2 \Psi \, d\mathbf{x} \tag{9.2.1}$$

where U is a real constant. MacDonald proved that if E_b is that bound-state eigenvalue of $\hat{H}$ that minimizes $(E - U)^2$, then

$$I_2[\Psi, U] \geq (E_b - U)^2 \tag{9.2.2}$$

for all normalized Ψ of appropriate symmetry. The density-functional analog, proved by Valone and Capitani (1981), is readily established using constrained search. Namely, define the density functional

$$R_2[\rho, U] = \operatorname*{Min}_{\Psi \to \rho} I_2[\Psi, U] \tag{9.2.3}$$

Then

$$R_2[\rho, U] \geq (E_b - U)^2 \tag{9.2.4}$$

and we have the desired principle. Valone and Capitani showed that there is no v-representability problem in this method, because the search in (9.2.3) can be extended to ensembles. They further proved the generalization to the general m using $(\hat{H} - U)^m$, again following MacDonald, and an extension to density matrices. Unfortunately, implementation of (9.2.4) requires quadratures involving $\hat{H}^2$, and the functional R_2 does not have the favorable universality properties of the ground-state Hohenberg–Kohn functional. Indeed, Lieb (1985) has proved that there exists no *universal* variational density functional procedure yielding an *individual* excited state.

We now come to the excited-state method that appears to have promise for substantial application, deriving from a proposal by Theophilou (1979). Theophilou began by proving the lovely theorem that if $\phi_1, \phi_2, \ldots, \phi_m, \ldots$, are orthonormal trial functions for the m lowest eigenstates of Hamiltonian H, having exact eigenvalues $E_1, E_2, \ldots, E_m, \ldots$, then

$$\sum_{i=1}^{m} \langle \phi_i | \hat{H} | \phi_i \rangle \geq \sum_{i=1}^{m} E_i \tag{9.2.5}$$

To prove this, first augment the set of ϕ_i by additional functions to give a complete orthonormal set $\{\phi_i\}$, and expand these ϕ_i in the exact orthonormal eigenfunctions of $\hat{H}$, ψ_j. Then

$$\phi_i = \sum_{j=1}^{\infty} C_{ij} \psi_j \tag{9.2.6}$$

and the **C** matrix is unitary:

$$\sum_{i=1}^{\infty} |C_{ij}|^2 = \sum_{j=1}^{\infty} |C_{ij}|^2 = 1 \tag{9.2.7}$$

If we define

$$P_j = \sum_{i=1}^{m} |C_{ij}|^2 \tag{9.2.8}$$

we will have

$$0 \leq P_i < 1 \tag{9.2.9}$$

and

$$\sum_{i=1}^{\infty} P_i = \sum_{i=1}^{\infty} \sum_{j=1}^{\infty} |C_{ij}|^2 = \sum_{j=1}^{m} \sum_{i=1}^{\infty} |C_{ij}|^2 = \sum_{j=1}^{m} (1) = m \tag{9.2.10}$$

Now we find

$$\sum_{i=1}^{m} [\langle \phi_i | \hat{H} | \phi_i \rangle - E_i] = \sum_{i=1}^{m} \left[\sum_{j=1}^{\infty} |C_{ij}|^2 E_j - E_i \right]$$

$$= \sum_{j=1}^{\infty} \sum_{i=1}^{m} |C_{ij}|^2 E_j - \sum_{j=1}^{m} E_j$$

$$= \sum_{j=1}^{\infty} P_j E_j - \sum_{j=1}^{m} E_j$$

$$= \sum_{j=1}^{m} (P_j - 1)E_j + \sum_{j=m+1}^{\infty} P_j E_j$$

$$= \sum_{j=1}^{m} (1 - P_j)(E_m - E_j) + \sum_{j=m+1}^{\infty} P_j(E_j - E_m) \geq 0 \quad (9.2.11)$$

This proves (9.2.5). The result also follows from MacDonald (1933).

Based on (9.2.5), constrained search gives a density variational principle for the average of the first m eigenstates, which we may call $\bar{E}(m)$ (see Pathak 1984.) With $\bar{\rho} = (1/m) \sum_{i=1}^{m} \rho_i$,

$$\bar{E}(m) = \underset{\bar{\rho}}{\text{Min}} \left\{ \underset{\{\phi_i\} \to \bar{\rho}}{\text{Min}} \left[\frac{1}{m} \sum_{i=1}^{m} \langle \phi_i | \hat{H} | \phi_i \rangle \right] \right\} \quad (9.2.12)$$

An extension in Kohn–Sham form was given by Hadjisavvas and Theophilou (1985). Katriel (1980) restates the result as just the usual ground-state result for a suitably defined supersystem.

Kohn, Gross, and Oliveira (Kohn 1986, Gross, Oliveira and Kohn 1988a,b) generalize the Theophilou theorem to ensembles with nonequal fractions of the various states, and thereby achieve merging with the finite-temperature theory of Mermin (1965). Their theorem is

$$\sum_{i=1}^{M} w_i \langle \phi_i | \hat{H} | \phi_i \rangle \geq \sum_{i=1}^{M} w_i E_i \quad (9.2.13)$$

provided that

$$w_1 \geq w_2 \geq \cdots \geq w_M > 0 \quad (9.2.14)$$

The proof follows from (9.2.5) once one notices that $[w_{m+1} \equiv 0]$

$$\sum_{i=1}^{M} w_i [\langle \phi_i | \hat{H} | \phi_i \rangle - E_i] = \sum_{m=1}^{M} (w_m - w_{m+1}) \sum_{i=1}^{m} [\langle \phi_i | \hat{H} | \phi_i \rangle - E_i] \quad (9.2.15)$$

For the density-functional procedure corresponding to (9.2.13), define the average density

$$\bar{\rho} = \frac{1}{M} \sum_{i=1}^{M} w_i \rho_i \quad (9.2.16)$$

and the density functional for the energy (as a function of the given parameters $w_1, w_2, \ldots$),

$$\bar{E}[\bar{\rho}] = \underset{\{\phi_i\} \to \bar{\rho}}{\text{Min}} \left[\frac{1}{M} \sum_{i=1}^{m} w_i \langle \phi_i | \hat{H} | \phi_i \rangle \right] \quad (9.2.17)$$

Furthermore, with suitable definitions of entropy and free energy, one can obtain a whole computational scheme in the form of Mermin's theory, and corresponding Kohn–Sham equations. First calculations on excited states of helium atoms are very encouraging (Oliveira, Gross, and Kohn 1988a).

The preceding assumes that all states are nondegenerate. For full details when degeneracies are present, which is of course very common for excited states, see Gross, Oliveira, and Kohn (1988a). For another interesting approach to excited states, see Fritsche (1986). For a fairly general discussion, see Englisch, Fieseler and Haufe (1988).

9.3 Time-dependent systems

Without completely going through what has been done in the subject, we now discuss a topic in the density-functional theory that clearly will be important for future applications of the theory: time-dependent density-functional theory. The subject is in its infancy. For details, see Gross and Dreizler (1985) and Bartolotti (1986), but see also Xu and Rajagopal (1985) and Dhara and Ghosh (1987).

Aside from an earlier paper by Ying (1974), the first important contribution to the time-dependent theory is the work of Peukert (1978), who invoked an action principle to study time-dependent phenomena but did not claim to be able to invert the mapping $v(\mathbf{r}, t) \to \rho(\mathbf{r}, t)$, which, if it were possible, would provide the extension of the Hohenberg–Kohn theory to the time-dependent domain.

As is well known, for a time-dependent situation the wave function $\Psi(\mathbf{x}, t)$, where $\mathbf{x}$ stands for space and spin coordinates of all particles, in general must be allowed to be complex. This means that it will be defined by specification of two real functions, $\Phi(\mathbf{x}, t)$ and $S(\mathbf{x}, t)$, in the polar representation of Ψ:

$$\Psi(\mathbf{x}, t) = \Phi(\mathbf{x}, t) \exp\left[iS(\mathbf{x}, t)\right] \qquad (9.3.1)$$

A density-functional theory should include a demonstration that $\rho(\mathbf{r}, t)$ determines both Φ and S, to the extent necessary to determine all observables.

Bartolotti (1981, 1982b), and independently Deb and Ghosh (1982), treated the special important case in which the external potential $v(\mathbf{r}, t)$ has a periodic dependence on time, producing periodic $\Psi(\mathbf{x}, t)$, $\Phi(\mathbf{x}, t)$, and $\rho(\mathbf{r}, t)$—a steady state. Their solution also requires that the applied frequency is not large enough to cause a transition from the ground state to an excited state. Deb and Ghosh used a proof by contradiction, in the original manner of Hohenberg and Kohn; Bartolotti used constrained search, in the manner of Levy and Lieb. We do not give details, for which one may refer to the original papers. Suffice it that under the conditions stated, $\rho(\mathbf{r}, t)$ does determine $v(\mathbf{r}, t)$ and all properties, and there is a stationary principle of Hohenberg–Kohn type. If each Kohn–Sham orbital is written in the polar form

$$\psi_k(\mathbf{r}, t) = \phi_k(\mathbf{r}, t) \exp[is_k(\mathbf{r}, t)] \tag{9.3.2}$$

there result coupled differential equations for the ϕ_k and s_k, the time-dependent Kohn–Sham equations

$$[-\tfrac{1}{2}\nabla^2 + \tfrac{1}{2}\nabla s_k \cdot \nabla s_k + v_{\text{eff}}]\phi_k = \varepsilon_k \phi_k \tag{9.3.3}$$

$$\frac{\partial \phi_k^2}{\partial t} = \nabla \cdot [\phi_k^2 \nabla s_k] \tag{9.3.4}$$

where each $\varepsilon_k = \varepsilon_k(\mathbf{r}, t)$ is the sum of a normal Lagrange multiplier for the normalization constraint and a space-time-dependent Lagrange multiplier for the charge-current conservation constraint

$$\frac{\partial \phi_k^2}{\partial t} = -\nabla \cdot \mathbf{j}_k \tag{9.3.5}$$

where $\mathbf{j}_k$ is the Kohn–Sham single-particle current vector

$$\mathbf{j}_k = \tfrac{1}{2}[\psi_k \hat{\mathbf{p}}^* \psi_k^* + \psi_k^* \hat{\mathbf{p}} \psi_k] = -\phi_k^2 \nabla^2 s_k \tag{9.3.6}$$

The effective potential v_{eff} in (9.3.3) is the Kohn–Sham effective potential of (7.1.16), including the time-dependent perturbation. Bartolotti (1986) reports static polarizabilities of helium atoms calculated using these formulas. A less detailed way to summarize these equations is to write

$$[-\tfrac{1}{2}\nabla^2 + v_{\text{eff}}]\psi_k = i\frac{\partial \psi_k}{\partial t} \tag{9.3.7}$$

but this suppresses the several constraints that must be satisfied.

The restriction to periodic perturbations was removed by Runge and Gross (1984), who established the $\rho(\mathbf{r}, t) \to v(\mathbf{r}, t)$ mapping, to the extent needed, under more general conditions. We say "to the extent needed" here because $v(\mathbf{r}, t)$ is determined only up to an arbitrary time-dependent function, which is of no consequence for time-independent observables. The condition imposed by Runge and Gross is that $v(\mathbf{r}, t)$ possess a Taylor series about $v(\mathbf{r}, t_0)$. In addition, criticism by Xu and Rajagopal (1985) and reply by Dhara and Ghosh (1987) revealed the additional necessary conditions that $v(\mathbf{r}, t)$ and $\rho(\mathbf{r}, t)$ vanish at the boundary of the system, $v(\mathbf{r}, t)$ at least as fast as $\rho(\mathbf{r}, t)$. A whole exact density-functional theory is then valid, which Runge and Gross (1984) elucidated, as well as a Kohn–Sham version. The Kohn–Sham equations are of the form (9.3.7), with v_{eff} suitably defined. The proof of Runge and Gross uses reduction *ad absurdum* in the style of Hohenberg and Kohn; a corresponding proof using constrained search is given by Kohl and Dreizler (1986). Kohl and Dreizler make various other general remarks. The extension to time-dependent ensembles is provided by Li and Tong (1985).

9.4 Dynamic linear response

The time-dependent density-functional theory of the previous section has already found extensive applications in determining the dynamic linear-response properties of many-electron systems. Thus, density-functional theory provides an efficient way to calculate the dynamic polarizability, which is required for describing various optical properties (linear and nonlinear) of matter as well as for investigating long-range interactions of atoms and molecules (Zangwill and Soven 1980, Stott and Zaremba 1980a). For a description of the conventional wave-function approaches to this problem, which are much more complicated, see Dalgarno (1962) and Miller and Bederson (1978). In the following, we essentially follow Zangwill and Soven (1980).

Consider a many-electron system in the ground state for the external potential $v(\mathbf{r})$. Under a small time-dependent perturbation $\delta v(\mathbf{r}, t)$, there is a density response $\delta\rho(\mathbf{r}, t)$. Then the associated Fourier components $\delta v(\mathbf{r}, \omega)$ and $\delta\rho(\mathbf{r}, \omega)$ are related by the equation

$$\delta\rho(\mathbf{r}, \omega) = \int d\mathbf{r}' \, \chi(\mathbf{r}, \mathbf{r}'; \omega) \, \delta v(\mathbf{r}', \omega) \tag{9.4.1}$$

Here $\chi(\mathbf{r}, \mathbf{r}', \omega)$ is the exact density–density response function, which has a spectral representation

$$\chi(\mathbf{r}, \mathbf{r}'; \omega) = \sum_m \left\{ \frac{\langle 0| \hat{\rho}(\mathbf{r}) |m\rangle \langle m| \hat{\rho}(\mathbf{r}') |0\rangle}{\omega - (E_m - E_0) + i\gamma} - \frac{\langle 0| \hat{\rho}(\mathbf{r}') |m\rangle \langle m| \hat{\rho}(\mathbf{r}) |0\rangle}{\omega + (E_m - E_0) + i\gamma} \right\} \tag{9.4.2}$$

where $|0\rangle$ is the exact many-electron ground state with energy E_0, $\{|m\rangle\}$ is the complete set of many-electron eigenstates with energies $\{E_m\}$, $\hat{\rho}(\mathbf{r})$ is the density operator, and γ is a positive infinitesimal. This representation is not computationally tractable, because it presumes complete knowledge of many-electron wave functions and energies.

Since the ground-state electron density is given by the sum of the squares of the Kohn–Sham orbitals, the time-dependent density response can be obtained through the following self-consistent equations:

$$\delta\rho(\mathbf{r}, \omega) = \int d\mathbf{r}' \chi_s(\mathbf{r}, \mathbf{r}'; \omega) \, \delta v_{\text{eff}}(\mathbf{r}', \omega) \tag{9.4.3}$$

and

$$\delta v_{\text{eff}}(\mathbf{r}, \omega) = \delta v(\mathbf{r}, \omega) + \int \frac{\delta \rho(\mathbf{r}', \omega)}{|\mathbf{r} - \mathbf{r}'|} d\mathbf{r}' + \delta v_{xc}(\mathbf{r}, \omega) \qquad (9.4.4)$$

where $\chi_s(\mathbf{r}, \mathbf{r}', \omega)$ is the density–density response function of the Kohn–Sham noninteracting ground state having the same density $\rho(\mathbf{r})$, and

$$\delta v_{xc}(\mathbf{r}, \omega) = \int d\mathbf{r}' f_{xc}(\mathbf{r}, \mathbf{r}'; \omega) \, \delta \rho(\mathbf{r}', \omega) \qquad (9.4.5)$$

where $f_{xc}(\mathbf{r}, \mathbf{r}'; \omega)$ depends on the ground-state density $\rho(\mathbf{r})$. An approximation to f_{xc} employing the corresponding static formula for the homogeneous electron gas was originally used, but a frequency-dependent version recently has been developed (Gross and Kohn 1985).

The validity of these self-consistent equations hinges on the assumption that the density $\rho(\mathbf{r}) + \delta \rho(\mathbf{r}, t)$ is noninteracting v-representable; that is, that it can be reproduced by a system of noninteracting electrons in an appropriate single-particle potential $v_{\text{eff}}(\mathbf{r}) + \delta v_{\text{eff}}(\mathbf{r}, t)$. See the discussion of this point by Mearns and Kohn (1987).

The response function for the noninteracting ground state has a much simpler spectral representation. Using (9.4.2) for the noninteracting system, one obtains (Zangwill and Soven 1980)

$$\chi_s(\mathbf{r}, \mathbf{r}'; \omega) = 2 \sum_{k}^{\varepsilon_k < \varepsilon_F} \phi_k^*(\mathbf{r}) \phi_k(\mathbf{r}') G^+(\mathbf{r}, \mathbf{r}'; \varepsilon_k + \omega)$$

$$+ 2 \sum_{k}^{\varepsilon_k < \varepsilon_F} \phi_k(\mathbf{r}) \phi_k^*(\mathbf{r}') G^-(\mathbf{r}', \mathbf{r}; \varepsilon_k - \omega) \qquad (9.4.6)$$

where $\{\phi_k, \varepsilon_k\}$ are the eigenstates and eigenenergies of $\hat{h}_{\text{eff}}$, the Kohn–Sham Hamiltonian of (7.2.6), ε_F is the Fermi energy (which is greater than the highest occupied eigenvalue but smaller that the lowest unoccupied eigenvalue), and G^+/G^- is the one-particle retarded/advanced Green's function

$$G^{\pm}(\mathbf{r}, \mathbf{r}'; \omega) = \langle \mathbf{r} | \frac{1}{\omega - \hat{h}_{\text{eff}} \pm i\gamma} | \mathbf{r}' \rangle$$

$$= \sum_j \frac{\phi_j(\mathbf{r}) \phi_j^*(\mathbf{r}')}{\omega - \varepsilon_j \pm i\gamma} \qquad (9.4.7)$$

The factor 2 in (9.4.6) takes care of the summation over electron spin states, as these formulas are restricted to spin-compensated situations (closed-shell systems). The Green's functions are usually obtained by solving the associated differential equations, not by the infinite summation of (9.4.7).

Let $\delta v(\mathbf{r}, \omega) = (1/2)z\varepsilon_0 \cos \omega t$, where ε_0 is the magnitude of an external electric field in the z direction. Then the solution of (9.4.3) and (9.4.4) gives the dynamic polarizability, which is the ratio of the induced dipole moment to the external field strength,

$$\alpha(\omega) = -\frac{2}{\varepsilon_0} \int z \, \delta\rho(\mathbf{r}, \omega) \, d\mathbf{r}$$

$$= -e^2 \int z\chi(\mathbf{r}, \mathbf{r}'; \omega)z' \, d\mathbf{r} \, d\mathbf{r}' \qquad (9.4.8)$$

The imaginary part of $\alpha(\omega)$ gives the Golden Rule formula for the photoabsorption cross section

$$\sigma(\omega) = 4\pi(\omega/c) \, \text{Im} \, \alpha(\omega) \qquad (9.4.9)$$

The static polarizability is given by $\alpha(0)$. See Table 9.1 for a comparison of calculated values of $\alpha(0)$ with experimental values for noble-gas atoms. The independent-particle approximation amounts to replacing χ in (9.4.8) by χ_s of (9.4.6) for the corresponding noninteracting system.

The density-functional theory of dynamic linear response has been applied to small molecules by Levine and Soven (1984) and to solids by Baroni, Giannozzi, and Testa (1987), and has been further extended to nonlinear response (Senatore and Subbaswamy 1987) and to finite-temperature systems (Grimaldi, Grimaldi-Lecourt, and Dharma-wardana 1985). Variational-perturbation methods have been used with the time-dependent Kohn–Sham theory to obtain dynamic polarizability (Ghosh and Deb 1982c, Bartolotti 1984). Also, an integral formulation has been proposed by Yang (1988c), in which χ_s is given explicitly in terms of the Kohn–Sham effective potential via path integrals, and orbitals are not needed. This may provide a useful way to calculate the linear dynamic response for large molecules.

Table 9.1 Static Polarizabilities[a] of the Noble-gas atoms; α_0 is the Result of an Independent-particle Calculation; α^{SCF} is the Self-consistent-Field Value

	α_0	α^{SCF}	α_{expt}
Neon	0.50	0.43	0.40 ± 0.01
Argon	2.60	1.74	1.64 ± 0.01
Krypton	4.11	2.60	2.48 ± 0.01
Xenon	6.89	4.12	4.04 ± 0.02

[a] From Zangwill and Soven (1980)

9.5 Density-matrix-functional theory

Given the existence of the density-functional theory for a many-electron system, in which all properties are obtained starting from a variational principle that has the electron density $\rho(\mathbf{r})$ as the basic variable, there must be a corresponding density-matrix-functional theory, in which the basic variable is the first-order reduced density matrix of (2.3.3), $\gamma_1(\mathbf{x}', \mathbf{x})$, or the first-order spinless density matrix of (2.4.1), $\rho_1(\mathbf{r}', \mathbf{r})$. In this section, we consider both exact and approximate versions of such a one-matrix theory.

The Hartree–Fock theory is the prototype first-order density-matrix theory. Indeed, in §2.5 we wrote the Hartree–Fock theory explicitly as a density-matrix theory; see (2.5.7)–(2.5.9). In view of a strong theorem of Lieb (1981a) one can in fact forget (2.5.7) and the energy formula of (2.5.9) will still minimize to the Hartree–Fock energy. So we have, in effect, an approximate but well-defined and well-understood density-matrix-functional theory.

But we should like to do better. In principle there is an exact density-matrix theory, as we now show. The first-order density matrix determines the density,

$$\rho(\mathbf{r}) = \rho_1(\mathbf{r}, \mathbf{r}) = \int \gamma_1(\mathbf{x}', \mathbf{x})|_{\mathbf{x}=\mathbf{x}'} \, ds \qquad (9.5.1)$$

Therefore, by the Hohenberg–Kohn theorem it determines all properties including the energy,

$$E = E[\gamma_1] = E[\rho_1] \qquad (9.5.2)$$

Constrained search gives the explicit form of the functional,

$$E[\gamma_1] = \operatorname*{Min}_{\gamma_N \to \gamma_1} \operatorname{tr}(\hat{\gamma}_N \hat{H}) \qquad (9.5.3)$$

(Here we could follow Valone (1980a) and Lieb (1982) in letting the search extend over all trial N-electron ensembles; that is, over all trial density operators of the form (2.2.7).) Equation (9.5.3) was first given by Levy (1979), though with the search only over pure-state wave functions.

In what was the first study of exact one-matrix functional theory, Donnelly and Parr (1978) wrote the variational principle as

$$\delta\{E[\gamma_1] - \mu N[\gamma_1]\} = 0 \qquad (9.5.4)$$

which stands for

$$E_0 = \operatorname*{Min}_{\gamma_1} E[\gamma_1] \qquad (9.5.5)$$

In (9.5.4), μ is the chemical potential, and can be shown to be numerically equal to the chemical potential of Chapter 4. There are delicate problems with (9.5.4) (Valone 1980b), but it appears to be essentially correct (see later). If we parametrize γ_1 in terms of natural spin orbitals ψ_i and occupation numbers n_i, equation (2.3.14), we obtain from (9.5.4)

$$\mu = \left(\frac{\partial E}{\partial n_i}\right)_{\psi_i, n_j \neq n_i} \quad \text{for all } i \tag{9.5.6}$$

on the assumption that

$$0 < n_i < 1 \tag{9.5.7}$$

If there is any natural orbital among the complete set of natural orbitals for which (9.5.7) is not true, then (9.5.6) does not hold; for such an orbital with $n_i = 0$, $(\partial E/\partial n_i) > \mu$; for such an orbital with $n_i = 1$, $(\partial E/\partial n_i) \leq \mu$. It is a conjecture, but probably true, that (9.5.7) holds for all orbitals in an atom or molecule. These results about the chemical potential go back to Gilbert (1975). For a more recent discussion, see Zumbach and Maschke (1985). This situation has an analogy in thermodynamics. Recall that the chemical potential of a component is the same in two phases if the component is present in each phase; if a component is absent from some phase, its chemical potential there is higher than it is elsewhere. In the Hartree–Fock model, occupation numbers are 0 or 1; this is like a thermodynamic system in which impermeable barriers prevent transfer between phases, for which chemical potentials bear no necessary relation from one phase to the next.

The result of considering variations of the natural orbitals themselves in (9.5.4), remembering to impose orthonormalization conditions on the orbitals, is a set of coupled differential equations for the natural orbitals (Donnelly and Parr 1978). These equations are very different from the Kohn–Sham equations (7.2.7) and from the Janak equations (7.6.9). For some pertinent analysis and comparison, see Donnelly (1979) and Valone (1980a). For a quite different slant, see Ludena and Sierraalta (1985).

This one-matrix theory possesses a remarkable simplifying and challenging feature that should be emphasized (Levy 1987a). Namely, since from (2.3.27) the first-order density matrix determines all components of the energy explicitly except $V_{ee}[\gamma]$, the constrained search in (9.5.3) may be restricted to search for the γ_N that minimizes V_{ee}:

$$V_{ee}[\gamma_1] = \underset{\gamma_N \to \gamma_1}{\text{Min}} \ \text{tr} \ (\hat{\gamma}_N \hat{V}_{ee}) \tag{9.5.8}$$

Or, since $\gamma_N \to \gamma_2 \to \gamma_1$, we could even write

$$V_{ee}[\gamma_1] = \underset{\gamma_2 \to \gamma_1}{\text{Min}} \iint \frac{1}{r_{12}} \gamma_2(\mathbf{x}_1 \mathbf{x}_2, \mathbf{x}_1 \mathbf{x}_2) \, d\mathbf{x}_1 \, d\mathbf{x}_2 \tag{9.5.9}$$

where the trial γ_2 would of course have to satisfy N-representability conditions (Harriman 1978). Equation (9.5.9) offers the enticing possibility of finding $V_{ee}[\gamma_1]$ to good accuracy by inspired guess of the dependence on parameters in a trial γ_2 (Valone 1980b).

9.6 Nonelectronic and multicomponent systems

So far, the density-functional theory has been presented as a theory describing some number of electrons moving in a given external potential $v(\mathbf{r})$. The techniques of density-functional theory are, however, much more general. The particles need not be electrons and there can be more than one kind of particle.

First, suppose that the particles are not electrons, but that the system comprises N_A particles of species A, as for example nucleons in an atomic nucleus. Let the charge on a particle be Z_A and its mass be m_A. Assume that there is some universal law of interaction between the particles, say $u(\mathbf{r}_1, \mathbf{r}_2) = u(r_{12})$, and suppose that a single-particle external electrostatic potential $v(\mathbf{r})$ acts on them. The Hamiltonian then will be

$$\hat{H} = -\sum_i^{N_A} \frac{1}{2m_A} \nabla_i^2 + \sum_{i<j}^{N_A} u(r_{ij}) + Z_A \sum_i^{N_A} v(\mathbf{r}_i) \tag{9.6.1}$$

The corresponding ground-state wave function $\Psi(\mathbf{x}_1, \mathbf{x}_2, \ldots, \mathbf{x}_N)$ will determine the single-particle density

$$\rho_A(\mathbf{r}_1) = N_A \int \cdots \int |\Psi|^2 \, ds_1 \, ds_2 \cdots ds_{N_A} \, d\mathbf{r}_2 \, d\mathbf{r}_3 \cdots d\mathbf{r}_{N_A} \tag{9.6.2}$$

The whole Hohenberg–Kohn–Levy theory then goes through easily: ρ_A determines v and all ground-state properties including the energy, and there is an energy functional $E_A[\rho]$ that is stationary for the true ground-state density. That is, the ground-state energy E_0 is determined by the minimum principle (cf. §3.4):

$$E_0 = \underset{\rho_A}{\text{Min}} \, E_A[\rho_A] = \underset{\Psi \to \rho_A}{\text{Min}} \, \langle \Psi | \hat{H} | \Psi \rangle$$

$$= \underset{\rho_A}{\text{Min}} \left\{ F_A[\rho_A] + Z_A \int v(\mathbf{r}) \rho(\mathbf{r}) \, d\mathbf{r} \right\} \tag{9.6.3}$$

where

$$F_A[\rho_A] = \underset{\Psi \to \rho_A}{\text{Min}} \, \langle \Psi | -\sum_i^{N_A} \frac{1}{2m_A} \nabla_i^2 + \sum_{i<j}^{N_A} u(r_{ij}) | \Psi \rangle \tag{9.6.4}$$

An entire density-functional computational apparatus is thereby provided, just as for electrons (and generally presenting the same problems as for electrons). Note that there is a dependence on m_A and Z_A, and that the form of $u(r_{ij})$ will also affect the results in a given case. If the particles are bosons and not fermions, the corresponding Kohn–Sham theory is simpler than for electrons: the Kohn–Sham equations become a single differential equation for density (Levy, Perdew, and Sahni 1984).

Suppose now that there are several different kinds of particles; for simplicity, suppose two kinds: N_A particles of mass m_A and charge Z_A, N_B particles of mass m_B and charge Z_B. Let the interaction potentials be u_{AA}, u_{AB}, and u_{BB}; let the external potential acting on A particles by $v_A(\mathbf{r})$, that acting on B particles $v_B(\mathbf{r})$. The Hamiltonian then is

$$\hat{H} = \sum_i^{N_A} -\frac{1}{2m_A}\nabla_i^2 + \sum_k^{N_B} -\frac{1}{2m_B}\nabla_k^2 + \sum_{i<j}^{N_A} u_{AA}(r_{ij}) + \sum_i^{N_A}\sum_k^{N_B} u_{AB}(r_{ik})$$

$$+ \sum_{k<l}^{N_B} u_{BB}(r_{kl}) + \sum_i^{N_A} Z_A v_A(\mathbf{r}_i) + \sum_k^{N_B} Z_B v_B(\mathbf{r}_k) \quad (9.6.5)$$

With ground-state wave function $\Psi(x^{N_A}, x^{N_B})$, the single-particle densities ρ_A, ρ_B, are given by

$$\rho_A(\mathbf{r}_1) = N_A \iint \cdots \int |\Psi|^2 \, ds_1 \, ds_2 \cdots ds_N \, d\mathbf{r}_2 \, d\mathbf{r}_3 \cdots d\mathbf{r}_N$$

$$\rho_B(\mathbf{r}_{N_A+1}) = N_B \iint \cdots \int |\Psi|^2 \, ds_1 \, ds_2 \cdots ds_N \, d\mathbf{r}_1 \cdots d\mathbf{r}_{N_A} \, d\mathbf{r}_{N_A+2} \cdots d\mathbf{r}_N \quad (9.6.6)$$

and so on, where we have let coordinates 1 to N_A refer to A, $N_A + 1$ to N (where $N = N_A + N_B$) refer to B. Together, the densities ρ_A, ρ_B determine the external potentials v_A and v_B and all ground-state properties, and again there is a constrained-search procedure for obtaining the ground-state wave functions, densities, and energy. Namely,

$$E_0 = \underset{\rho_A,\rho_B}{\text{Min}}\, E[\rho] = \underset{\Psi \to \rho_A,\rho_B}{\text{Min}}\, \langle \Psi | \hat{H} | \Psi \rangle$$

$$= \underset{\rho_A,\rho_B}{\text{Min}} \left\{ F[\rho_A, \rho_B] + Z_A \int v_A(\mathbf{r}) \rho_A(\mathbf{r}) \, d\mathbf{r} + Z_B \int v_B(\mathbf{r}) \rho_B(\mathbf{r}) \, d\mathbf{r} \right\} \quad (9.6.7)$$

where

$$F[\rho_A, \rho_B] = \underset{\Psi \to \rho_A,\rho_B}{\text{Min}}\, \langle \Psi | \sum_i -\frac{1}{2m_i}\nabla_i^2 + \sum_{i<j}^{N_A} u_{AA}(r_{ij})$$

$$+ \sum_i^{N_A}\sum_k^{N_B} u_{AB}(r_{ik}) + \sum_{k<l}^{N_B} u_{BB}(r_{kl}) | \Psi \rangle \quad (9.6.8)$$

This F lacks universality in the sense that it depends on the specific forms of u_{AA}, u_{AB}, u_{BB}, and on m_A, m_B; the energy formula (9.6.7) naturally also depends on Z_A, Z_B. But again we have a full density-functional procedure. The generalization to three or more kinds of particles is elementary.

For detailed proofs of these results, which are simple extensions of §§3.1–3.4, see Capitani, Nalewajski, and Parr (1982); see also Englisch and Englisch (1984a).

One application of these ideas is to atoms or molecules in the non-Born–Oppenheimer approximation (Capitani, Nalewajski, and Parr 1982). Such a species has several components: electrons plus as many different kinds of nuclei as are present. The ground state of a molecule in a large box can be thought of, in fact, as a homogeneous (!) fluid of many components, governed by density-functional theory. (There is very large correlation between particles, of course.) From this point of view, every type of atom will have its own escaping tendency from a molecule; see the interesting paper by Lohr (1984) in which there is introduced the "protonic counterpart of electronegativity".

There is no problem extending these results to finite temperature. It is the density-functional theory in this generalized form that is the basis for the manifold applications of density-functional theory in macroscopic chemical physics, already briefly alluded to in §3.7. Whenever the interactions are known, the density-functional theory will apply, in quantum or classical form as appropriate, irrespective of whether the system is homogeneous. For classical applications, see, for example, Rowlinson and Widom (1982), Chandler, McCoy, and Singer (1986a,b), Curtin and Ashcroft (1986), Curtin (1987), Curtin and Runge (1987), Haymet (1987), and Ding, Chandler, Smithline, and Haymet (1987).

10
ASPECTS OF ATOMS AND MOLECULES

10.1 Remarks on the problem of chemical binding

We have now completed the presentation of the basic principles of the density-functional theory of the electronic structure of atoms, molecules, and solids, including the time-dependent version of the theory and its extension to excited states. We have also given a number of illustrative applications of the theory. The reader should now understand the essentials of the theory, its limits, and the lines on which it most likely will be developing in the future.

In this and the next (the last) chapter, we survey selected additional applications of the theory. We also identify and briefly characterize various alternative viewpoints that can be taken on the theory.

In the present chapter, we discuss what density-functional theory has to say about the problem that above all other problems characterizes the science of chemistry itself—the nature of the chemical bond. We begin by first considering the problem of the binding effects between rare-gas atoms at intermediate distances. We proceed to the question of what is an atom in an ordinary molecule, the HSAB principle from the atoms-in-molecules standpoint, bond-charge models of the chemical bond, and semiempirical electronegativity-equalization schemes.

First a cautionary remark. One should not expect anything radically new. To establish the nature of chemical bonding from the wave-function picture was not easy; it took the valence-bond and molecular-orbital descriptions of H_2, and these both had to be validated by numerical calculation. Try to derive chemical bonding from $\hat{H}\psi = E\psi$, the Pauli principle, and nothing else, and you will meet difficulties! To give a general density-functional description of bonding may be awkward (at best) unless one again introduces orbitals. The Kohn–Sham method does that, and, indeed, the Kohn–Sham method provides a whole picture of bonding essentially equivalent to the wave-function picture. That may be most of the story, practically speaking. Nevertheless, in the present chapter we probe for more.

10.2 Interatomic forces

Before we get into the subtleties of chemical binding in molecules, let us consider the problem of calculating the tiny binding forces that exist between neutral closed-shell atoms at close range. This is a problem for which no simple orbital theory had been successful, and yet Gordon and Kim (1972) succeeded in providing a simple density-functional theory that explained the basic facts semiquantitatively.

Consider two neutral rare-gas atoms A and B, at a distance apart such that the electron density of the pair is essentially the sum of densities of A and B separately, and yet these densities overlap. That is, assume

$$\rho_{AB} = \rho_A + \rho_B \quad (10.2.1)$$

although

$$\int \rho_A(\mathbf{r})\rho_B(\mathbf{r})\,d\mathbf{r} \neq 0 \quad (10.2.2)$$

This overlap will be responsible for an interaction over and above the $1/R^6$ long-range interaction due to the dispersion forces; the problem is to see whether a quantitative theory of the intermediate range can be given within the additive-density assumption of (10.2.1). The densities ρ_A and ρ_B can be accurately calculated using conventional Hartree–Fock theory. What Gordon and Kim did was to compute the total energy $E_{AB}[\rho_{AB}]$ from approximate density-functional theory, and similarly $E_A[\rho_A]$ plus $E_B[\rho_B]$, and subtract. Note that this is in the spirit of the calculations referred to at the beginning of §6.4, and that it involves the common-sense idea that errors may be expected to cancel when one uses the same method to calculate E_A and E_B as one does for E_{AB}.

Gordon and Kim (1972) proposed using the Thomas–Fermi formula for kinetic energy, (6.1.19), the Dirac formula for exchange energy, (6.1.20), and a uniform-gas formula for the correlation energy that involved equations (E.19) and (E.20) of Appendix E and an interpolation formula between them, $E_c[\rho] = \int \rho(\mathbf{r})\varepsilon_c(\rho)\,d\mathbf{r}$. The interaction energy then becomes, at each R,

$$\begin{aligned}
\Delta E &= E_{AB}(R) - E_{AB}(\infty) = E_{AB}(R) - E_A - E_B \\
&= C_F \int \{\rho_{AB}^{5/3}(\mathbf{r}) - \rho_A^{5/3}(\mathbf{r}) - \rho_B^{5/3}(\mathbf{r})\}\,d\mathbf{r} + \left\{ -\int \left(\frac{Z_A}{r_A} + \frac{Z_B}{r_B}\right)\rho_{AB}(\mathbf{r})\,d\mathbf{r} \right. \\
&\quad \left. + \int \frac{Z_A}{r_A}\rho_A(\mathbf{r})\,d\mathbf{r} + \int \frac{Z_B}{r_B}\rho_B(\mathbf{r}) \right\} + \left\{ \frac{Z_A Z_B}{R} + J[\rho_{AB}] - J[\rho_A] - J[\rho_B] \right\} \\
&\quad - C_x \int \{\rho_{AB}^{4/3}(\mathbf{r}) - \rho_A^{4/3}(\mathbf{r}) - \rho_B^{4/3}(\mathbf{r})\}\,d\mathbf{r} + \{E_c[\rho_{AB}] - E_c[\rho_A] - E_c[\rho_B]\}
\end{aligned}$$

$$(10.2.3)$$

Table 10.1 Potential Parameters for Rare-gas Pair Interactions[a]

Pair	r_m calc. (exptl.)[b]	ε calc (exptl.)[c]
He–He	2.49 (2.96)	62.5 (16.5)
Ne–Ne	2.99 (3.09)	56 (63)
Ar–Ar	3.62 (3.70)	175 (195)
Kr–Kr	3.89 (3.87–4.08)	248 (270–340)
Xe–Xe	4.15 (4.5)	417 (342–360)
He–Ne	2.73 (2.79–3.65)	59 (23–32)
He–Ar	3.11 (3.37–3.56)	92 (34–54)
He–Kr	3.28 (3.54–3.81)	100 (33–58)
He–Xe	3.50 (3.76–3.93)	106 (50–69)
Ne–Ar	3.42 (3.38–3.58)	79 (80–111)
Ne–Kr	3.60 (3.53–3.67)	81 (80–121)
Ne–Xe	3.87 (3.70–4.04)	74 (79–116)
Ar–Kr	3.78 (3.80–3.96)	203 (214–238)
Ar–Xe	3.92 (4.01–4.29)	245 (236–262)
Kr–Xe	4.03 (4.21–4.60)	313 (223–276)

[a] From Gordon and Kim (1972), where references for experimental data are cited).
[b] Position of potential minimum in angstroms.
[c] Depth of potential minimum in 10^{-16} erg.

There is considerable cancellation within each of the terms in curly brackets, and also between the two sets of purely electrostatic contributions; indeed, if it were not for (10.2.2) there would be a complete cancellation at intermediate to large R. Table 10.1 gives some of the results obtained when Hartree–Fock densities are inserted into this formula. They are remarkably good, considering the crudity of the approximations made. Corresponding results on ion–rare gas interactions may be found in Kim and Gordon (1974c). A scaling procedure improves the results for light atoms (Waldman and Gordon 1979).

This method in fact goes back to 1955 (Massey and Sida 1955, Gaydaenko and Nikulin 1970, Nikulin and Tsarev 1975). It admits of many refinements, extensions, and applications. There are the extensions to include dispersion forces (Kim and Gordon 1974c, Cohen and Pack 1974, Rae 1975, Harris 1975, 1984), external fields (Heller, Harris, and Gelbart 1975), and open-shell/closed-shell interactions (Clugston and Gordon 1977a), extensions to describe ion–rare gas interactions more accurately (Gianturco 1976), and extensions to determine anisotropic interactions (Kim 1978, Lee and Kim 1979, Ree and Winter 1980). An application to ligand-field splitting in transition-metal halides is very interesting (Clugston and Gordon 1977b). For an early application to rare-gas atoms adsorbed on surfaces, see Freeman (1975); application to molecular solids also has been taken up (LeSar and Gordon 1982).

Rae (1973) very early pointed out the need for a self-interaction correction. In a definitive paper, Wood and Pyper (1981) tested alternative versions of the theory and recommended the version due to Lloyd and Pugh (1977). However, see the suggestion of Brual and Rothstein (1978), which Wood and Pyper do not mention. For a recent detailed comparative study, see Spackman (1986).

The decomposition of (10.2.1) clearly violates the Hellmann–Feynman and virial theorems if ρ_{AB} is not determined variationally (Harris and Heller 1975), but the results are not much affected if one does a full Hartree–Fock reminimization on the whole system (Green, Garrison, and Lester 1975). Furthermore, there being a density-variational principle, first-order errors in electron density will produce only second-order errors in energy (Heller, Harris, and Gelbart 1975).

Finally, we should mention that there is an appealing different ansatz from (10.2.1), which also makes use only of the Hartree–Fock descriptions of atoms (Kolos and Radzio 1978, Radzio-Andzelm 1981). If one assumes

$$\Psi_{AB} = \mathscr{A}[\Psi_A \Psi_B] \tag{10.2.4}$$

where $\mathscr{A}$ is the antisymmetrization operator and Ψ_A and Ψ_B are atom-A and atom-B Hartree–Fock wave functions, one does not obtain (10.2.1) but a modified form involving overlap integrals between orbitals on different atoms, namely,

$$\rho_{AB} = 2 \sum_{\mu\nu} \psi_\mu (\Delta^{-1})_{\mu\nu} \psi_\nu \tag{10.2.5}$$

where the ψ_μ and ψ_ν are Hartree–Fock orbitals on the two constituents and Δ is the overlap integral matrix. One can now apply (10.2.3) just as before (or modifications of it). The results for He_2 are substantially improved.

10.3 Atoms in molecules

We now come to real molecules, for which the individual atoms have to a certain degree, if not completely, lost their identity. Is there any point in talking about the *atom in a molecule*? A computational chemist might say no: the computer program for molecules will give molecular properties; why bother with the artificial construct of "atom" in such a case? But this utterly ignores what chemistry is all about, for practicing chemists never for an instant forget which atoms or functional groups are in molecules with which they are dealing. Chemistry is the science of why particular atoms behave in particular ways, and why also do particular functional groups. Combination does not destroy atomic identity; it only perturbs it slightly, or a little more. An atom here is not the same as an atom there, but it is almost so.

We are therefore compelled to ask, what is an atom in a molecule? How can we find it in the Schrödinger equation? Alternatively, how can we find it in density-functional theory? The problem is a profound one.

The definitive study from the wave-function viewpoint is the work of Moffitt (1951). Given a wave function for some state of a molecule, built from atomic orbitals in some specific way, one can identify contributing states for the component atoms. A typical atom will appear as a negative ion, a neutral species, and/or a positive ion, and for each of these there often will enter both ground and excited states. A weakness of this description is that it is not unique. Putting a complete set of orbitals on each atom in a molecule gives a highly overcomplete set of functions; the essential ambiguity in the description is inescapable.

A unique and appealing definition of atoms in a molecule has been given, comprehensively discussed, and widely applied by Bader and coworkers (Bader and Nguyen-Dong 1981). The idea is very simple, and there are many properties of this partitioning that endow it with a special, fundamental interest. What Bader proposes is to define atoms by a space-partitioning of the total molecular electron density, with dividing surfaces between individual atomic regions defined as the surfaces on which the normal component of the density gradient is zero. For simplicity, consider a diatomic molecule AB with ground-state electron density ρ^0_{AB}. then the Bader partitioning is

$$\rho^0_{AB} = \rho_A + \rho_B \tag{10.3.1}$$

with ρ_A and ρ_B disjoint densities for atoms A and B, the surface between them being defined as indicated. A and B are not atoms in ground states, but in perturbed "valence" states. There are substantial advantages to this partitioning, as Bader has discussed (Bader 1986), including a certain atomic virial theorem. But there also are disadvantages: overlapping between atoms is forbidden, transferability is necessarily limited, and the chemical bond itself seems to have disappeared into thin air.

We shall now describe a density-functional theory solution of this same problem (Parr, Donnelly, Levy, and Palke 1978, Parr 1984). It provides a related but different definition of atoms in molecules.

Again consider a diatomic molecule AB. In its ground state at some internuclear distance R (not necessarily the equilibrium distance), it has electronic energy E_{AB}, electron density ρ^0_{AB}, and chemical potential μ_{AB}; the external potential is

$$v_{AB} = -\frac{Z_A}{r_A} - \frac{Z_B}{r_B} = v^0_A + v^0_B \tag{10.3.2}$$

where

$$v_A^0 = -\frac{Z_A}{r_A}, \quad v_B^0 = -\frac{Z_B}{r_B} \tag{10.3.3}$$

and r_A and r_B are distances of the electron from nucleus A and nucleus B, respectively. The isolated atoms A and B in their ground states have energies E_A^0, E_B^0, numbers of electrons N_A^0 and N_B^0, electron densities ρ_A^0 and ρ_B^0, chemical potentials μ_A^0 and μ_B^0, external potentials v_A^0 and v_B^0. The chemical potentials μ_A^0 and μ_B^0 may be different.

Clearly, while ρ_{AB}^0 is roughly $\rho_A^0 + \rho_B^0$, it is not precisely so if any kind of interaction is present (compare the discussion of rare-gas dimers in the previous section). We must imagine distortions of each atom as the atoms come together. They are no longer isolated; each becomes an open system (as emphasized by Bader), exchanging electrons and energy with the other to produce the ground state of the molecule AB. At equilibrium, we will have the basic equality

$$\mu_A = \mu_B = \mu_{AB} \tag{10.3.4}$$

We also assume that

$$\rho_{AB}^0 = \rho_A + \rho_B \tag{10.3.5}$$

where ρ_A integrates to N_A, not necessarily to N_A^0, ρ_B integrates to N_B, not necessarily to N_B^0; ρ_{AB} integrates to $N_A + N_B = N_A^0 + N_B^0$. In contrast with Bader's definition, ρ_A and ρ_B are not assumed to be disjoint.

The additive density assumption of (10.3.5) is optimistic. Should one wish to add a term integrating to zero (one way to do it) or integrating to some finite "bond charge" (another way to do it), that of course would be feasible and might ultimately prove to be desirable.

The change of atom A from its isolated ground state, density ρ_A^0, to its state in the molecule (its "valence state"), density ρ_A, can be regarded as a change of one ground state to another owing to change in an external field v_A^0 to v_A, say, plus possibly a change of particle number, N_A^0 to n_A; it is therefore covered by density-functional theory. There is a promotion energy for A, and another for B,

$$\Delta E_A^{(p)} = \int [\rho_A(\mathbf{r}) - \rho_A^0(\mathbf{r})] v_A^0(\mathbf{r}) \, d\mathbf{r} + F[\rho_A] - F[\rho_A^0]$$

$$\Delta E_B^{(p)} = \int [\rho_B(\mathbf{r}) - \rho_B^0(\mathbf{r})] v_B^0(\mathbf{r}) \, d\mathbf{r} + F[\rho_B] - F[\rho_B^0] \tag{10.3.6}$$

and a total promotion energy

$$\Delta E^{(p)} = \Delta E_A^{(p)} + \Delta E_B^{(p)} \tag{10.3.7}$$

We may complete the definition of the atoms A and B in the molecule

AB by requiring that of all decompositions of ρ_{AB} into ρ_A and ρ_B satisfying (10.3.4) and (10.3.5) that one is preferred that minimizes the total promotion energy:

$$\Delta E^{(p)}[\rho_A, \rho_B] = \text{minimum} \quad (10.3.8)$$

In very rare circumstances, $\Delta E^{(p)}$ could be negative and a modification of this condition would be required. Each of $\Delta E_A^{(p)}$ and $\Delta E_B^{(p)}$ would be positive if no change of particle number took place. When there is a charge transfer, one could be negative but the other then would almost always be more positive [see (4.1.15)]. The best approach then would probably be to minimize the promotion energy before charge transfer, to states defined by N_A^0, N_B^0, and v_A, v_B.

Calculations exploring such a procedure have been made by Palke (1980) and by Guse (1981). Rychlewski and Parr (1986) succeeded in accurately implementing it for the case of H_2, using an equivalent wave-function procedure that will work for any homonuclear diatomic molecule with an even number of electrons. The hydrogen atom in the molecule turns out to be distorted in the manner one would expect, its promotion energy at the equilibrium distance being 0.045 a.u. For a related calculation, see Adams and Clayton (1986); for an attempt in the same direction based on the first-order density matrix, see Li and Parr (1986).

Equation (10.3.4) may appear paradoxical when applied to the molecule AB at infinite internuclear distance, for then one has atoms with their own electronegativities, $\mu_A^0 = -\frac{1}{2}(I_A + A_A)$ and $\mu_B^0 = -\frac{1}{2}(I_B + A_B)$. These are in general not equal. However, when the two atoms constitute one equilibrium system, whatever distance they are apart, an added electron goes to whichever atom has the maximum electron affinity, and an electron taken away is taken most easily from whichever atom has the lower ionization potential. Thus, the equilibrium chemical potential of AB at infinite separation is (Perdew, Parr, Levy, and Balduz 1982)

$$\mu_{AB}^0 = -\tfrac{1}{2}(I_{\min} + A_{\max}) \quad (10.3.9)$$

In any particular problem, it should ordinarily be clear which of μ_A^0, μ_B^0, and μ_{AB}^0 will apply.

These ideas require intensive further study. The Bader partitioning is a secure, valuable picture, but partitionings with overlapping atoms also have advantages.

10.4 More on the HSAB principle

We now continue the discussion, begun in §5.2, of the role of the electronegativity and hardness in chemical bonding interactions and the

HSAB principle of Pearson (1963). We have the quantities η, $\eta(\mathbf{r},\mathbf{r}')$, $\eta(\mathbf{r})$, S, $s(\mathbf{r},\mathbf{r}')$, and $s(\mathbf{r})$, respectively defined by (5.3.1), (5.3.10), (5.5.14), (5.3.2), (5.5.11), and (5.5.1). Do they in fact lead to the rule "hard likes hard and soft likes soft"? There is as yet no convincing argument available. Nevertheless, we try to go beyond §§5.2 and 5.3.

The discussion has been carried on in the literature since 1963 (Pearson 1973, Parr and Pearson 1983). As has been long argued, the principle "soft likes soft" has to do with high polarizabilities enhancing covalent bond formation. The principle "hard likes hard" relates to factors enhancing purely electrostatic interactions. But having hardness and softness precisely defined, we may reasonably hope to find an explicit formulation valid to the extent that bond formation from separate atoms is a second-order effect. Validity would be to second order using the quantities indicated because, as was emphasized in Chapter 5, these quantities suffice to quantify energy changes to second order only.

The crudest model considers only the charge transfer between atoms induced by an electronegativity (chemical potential) difference. This model has been displayed countless times in the literature and textbooks (see for example, Huheey 1978) and was described in §5.2. The number of electrons transferred is given by (5.2.6) and binding energy enhancement by (5.2.7); these are greater if hardnesses of the partners are both small. This, roughly, is why soft likes soft, but this model is oversimplified. Note, for example, that no electronic-energy dependence on internuclear separation can be predicted in this way.

More completely, as A approaches B, not only do N_A and N_B change but so also do the effective external potentials v_A and v_B acting on the electrons in A and B, respectively. The following second-order atoms-in-molecules argument is the basis for the analysis of this effect by Nalewajski (1984a). Let ΔN_A and ΔN_B be the changes in going from the separate atom A, having chemical potential μ_A^0, to the atom A in the molecule, having chemical potential μ_A, and similarly with B. Letting $\Delta N = \Delta N_A = -\Delta N_B$, we will have, using appropriate formulas from Chapter 5,

$$\mu_A = \mu_A^0 + \left(\frac{\partial \mu_A}{\partial N_A}\right)_{v_A} \Delta N_A + \int \left[\frac{\delta \mu}{\delta v_A(\mathbf{r})}\right]_{n_A} \Delta v_A(\mathbf{r})\, d\mathbf{r}$$

$$= \mu_A^0 + 2\eta_A \Delta N + \int f_A(\mathbf{r}) \Delta v_A(\mathbf{r})\, d\mathbf{r}$$

$$\mu_B = \mu_B^0 - 2\eta_B \Delta N + \int f_B(\mathbf{r}) \Delta v_B(\mathbf{r})\, d\mathbf{r} \qquad (10.4.1)$$

where η_A, η_B are atom-A and atom-B hardnesses and f_A and f_B are their

Fukui functions. The equilibrium condition $\mu_A = \mu_B$ then yields the first-order formula

$$\Delta N = \frac{(\mu_B^0 - \mu_A^0) + \int f_B(\mathbf{r})\, \Delta v_B(\mathbf{r})\, d\mathbf{r} - \int f_A(\mathbf{r})\, \Delta v_A(\mathbf{r})\, d\mathbf{r}}{2(\eta_A + \eta_B)} \qquad (10.4.2)$$

This is an improved version of (5.2.6). Note that the terms involving the Fukui functions now moderate the chemical-potential difference in driving the charge transfer. Nalewajski and Koniski (1987) have given an electrostatic model of (10.4.2).

We are interested in the energy change to second order. Using

$$\Delta \rho(\mathbf{r}) = f(\mathbf{r})\, \Delta N + \int \left[\frac{\delta \rho(\mathbf{r})}{\delta v(\mathbf{r}')}\right]_N \Delta v(\mathbf{r}')\, d\mathbf{r}' \qquad (10.4.3)$$

from (5.1.1) we find

$$\Delta E_A = \mu_A^0\, \Delta N_A + \int \rho_A(\mathbf{r})\, \Delta v_A(\mathbf{r})\, d\mathbf{r}$$
$$+ \frac{1}{2}\left\{ 2\eta_A (\Delta N_A)^2 + 2\, \Delta N_A \int f(\mathbf{r})\, \Delta v_A(\mathbf{r})\, d\mathbf{r} \right.$$
$$\left. + \iint \left[\frac{\delta \rho_A(\mathbf{r})}{\delta v_A(\mathbf{r}')}\right]_{N_A} \Delta v_A(\mathbf{r})\, \Delta v_A(\mathbf{r}')\, d\mathbf{r}\, d\mathbf{r}' \right\} \qquad (10.4.4)$$

We have a similar expression for ΔE_B, and hence a formula for the total energy change, $\Delta E = \Delta E_A + \Delta E_B$. Minimizing this with respect to ΔN then gives (10.4.2) above, and then, upon back-substitution of this into ΔE, the energy change to second order,

$$\Delta E = \left\{ \int \rho_A(\mathbf{r})\, \Delta v_A(\mathbf{r})\, d\mathbf{r} + \int \rho_B(\mathbf{r})\, \Delta v_B(\mathbf{r})\, d\mathbf{r} \right\} - [(\eta_A + \eta_B)(\Delta N)^2]$$
$$+ \left\{ \iint \left[\frac{\delta \rho_A(\mathbf{r})}{\delta v_A(\mathbf{r}')}\right]_N \Delta v_A(\mathbf{r})\, \Delta v_A(\mathbf{r}')\, d\mathbf{r}\, d\mathbf{r}' \right.$$
$$\left. + \iint \left[\frac{\delta \rho_B(\mathbf{r})}{\delta v_B(\mathbf{r}')}\right]_N \Delta v_B(\mathbf{r})\, \Delta v_B(\mathbf{r}')\, d\mathbf{r}\, d\mathbf{r}' \right\} \qquad (10.4.5)$$

where ΔN is given by (10.4.2). Note that this reduces to (5.2.7) when $\Delta v_A = \Delta v_B = 0$.

Equation (10.4.5) has three components, each enclosed in brackets. the first (combined with the nucleus–nucleus repulsion ΔV_{nn}) is the electrostatic contribution, $\Delta E_{\text{electrostatic}}$; the second is the contribution associated with the flow of electrons from B to A, $\Delta E_{\text{covalent}}$; and the third is the polarization contribution, $\Delta E_{\text{polarization}}$. We discuss each of these briefly.

For a standard acid–base reaction, a large second term signals substantial covalent bonding (Pearson 1967) and so we label it

$$\Delta E_{\text{covalent}} = -(\eta_A + \eta_B)(\Delta N)^2$$

$$= -\frac{\left[\mu_B^0 - \mu_A^0 + \int f_B(\mathbf{r})\,\Delta v_B(\mathbf{r})\,d\mathbf{r} - \int f_A(\mathbf{r})\,\Delta v_A(\mathbf{r})\,d\mathbf{r}\right]^2}{4(\eta_A + \eta_B)} \quad (10.4.6)$$

This term is always negative, and the larger it is in absolute value the greater is the energy lowering. When $\eta_A + \eta_B$ is small, the soft-likes-soft principle follows. For typical soft–soft interactions, the frontier terms in the numerator may overwhelm the electronegativity difference, but it still is not entirely clear how the directional preference in chemical reactions (frontier control) comes into play.

When A and B are hard, typically with small size and high charge, the bonding is mainly ionic and the dominant term in ΔE is the Hellmann–Feynman electrostatic contribution

$$\Delta W_{\text{electrostatic}} = \int \rho_A(\mathbf{r})\,\Delta v_A(\mathbf{r})\,d\mathbf{r} + \int \rho_B(\mathbf{r})\,\Delta v_B(\mathbf{r})\,d\mathbf{r} + \Delta V_{nn} \quad (10.4.7)$$

One may then say that there is "charge control" (Klopman 1968). Note that this contribution is first order. The remaining term in the energy change is the polarization energy

$$\Delta E_{\text{polarization}} = \iint \left[\frac{\delta\rho_A(\mathbf{r})}{\delta v_A(\mathbf{r}')}\right]_{N_A} \Delta v_A(\mathbf{r})\,\Delta v_A(\mathbf{r}')\,d\mathbf{r}\,d\mathbf{r}'$$

$$+ \iint \left[\frac{\delta\rho_B(\mathbf{r})}{\delta v_B(\mathbf{r}')}\right]_{N_B} \Delta v_B(\mathbf{r})\,\Delta v_B(\mathbf{r}')\,d\mathbf{r}\,d\mathbf{r}' \quad (10.4.8)$$

It follows from (1.5.9) that this term is always negative. It can be connected with softness by use of (5.5.22); it will be big for reactions involving soft species.

A rather different viewpoint was taken by Berkowitz (1987). Consider A and B at some large distance apart, still much like ground-state A and B, but already in equilibrium so that their chemical potentials are equal:

$$\mu_A = \mu_B \quad (10.4.9)$$

As the reaction coordinate changes by a small amount, this condition will remain satisfied, but there may be a flow of electrons from B to A and a change in external potential. We will have

$$d\mu_A = 2\eta_A \, dN_A + \int f_A(\mathbf{r}) \, dv_A(\mathbf{r}) \, d\mathbf{r}$$
$$d\mu_B = 2\eta_B \, dN_B + \int f_B(\mathbf{r}) \, dv_B(\mathbf{r}) \, d\mathbf{r}$$
(10.4.10)

But $dN_A = -dN_B = dN$ and $d\mu_A = d\mu_B$. Consequently, we have a sort of Clapeyron equation for this change along a reaction coordinate:

$$dN = \frac{\int f_B(\mathbf{r}) \, dv_B(\mathbf{r}) \, d\mathbf{r} - \int f_A(\mathbf{r}) \, dv_A(\mathbf{r}) \, d\mathbf{r}}{2\eta_A + 2\eta_B} \qquad (10.4.11)$$

Starting from this formula, Berkowitz argues that for some small change along the reaction coordinate, as parameterized by a change $dv(\mathbf{r})$ in the potential due to nuclei, the induced electron flow is greater the softer are A and B (small $\eta_A + \eta_B$), the larger is the frontier quantity $f_B - f_A$ (difference of local softnesses), and the larger is the overlap between Fukui functions.

Of course, energy effects are associated with electron transfer not only from atom to atom, but also from an atom into a bond region. A more detailed model is needed of these effects, that hopefully will encompass the whole spectrum of soft–soft to hard–hard interactions. Such a model should have the capacity to predict bond lengths. The semiempirical model of the next section and the semiempirical calculational schemes of the section following represent attempts at such refinement. One should also note the recent work of Nalewajski and Koniski (1987), wherein the off-diagonal elements of the hardness kernel $\eta(\mathbf{r}, \mathbf{r}')$ are given special attention.

There very well may be a hard-likes-hard and soft-likes-soft rule operating where all quantities are local. Local hardness and local softness not being simple reciprocals of each other [cf. (5.5.15)], this would provide a two-parameter description potentially of the greatest importance for biomolecules. But much work is needed to establish whether this speculation has merit.

For an essentially homogeneous system divided into two parts, A and B, we see from (10.4.2) that the flow of electrons from B to A could nevertheless be extremely large if the hardness η were extremely small. This situation, of very large or infinite softness, may characterize the superconducting state (Tachibana and Parr 1989).

10.5 Modeling the chemical bond: The bond-charge model

In the previous section, we addressed the HSAB principle using density-functional concepts, invoking a model resolution of a molecule into constituent atoms. Full understanding will certainly require consideration of the chemical bond itself. With that in mind, we here describe a bond-charge model of chemical bonds.

Following Borkman and Parr (1968), consider a neutral homonuclear diatomic molecule characterized by an equilibrium internuclear distance R_e, a harmonic force constant

$$k_e = \left(\frac{\partial^2 W}{\partial R^2}\right)_{R_e} \tag{10.5.1}$$

cubic and quartic force constants

$$l_e = \left(\frac{\partial^3 W}{\partial R^3}\right)_{R_e}, \quad m_e = \left(\frac{\partial^4 W}{\partial R^4}\right)_{R_e} \tag{10.5.2}$$

and dissociation energy

$$D_e = W(\infty) - W(R_e) \tag{10.5.3}$$

where W is the total electronic energy as a function of R including nuclear–nuclear repulsion—see (1.1.9). Near $R = R_e$, suppose that we can represent $W(R)$ by the form

$$W(R) = W_0 + \frac{W_1}{R} + \frac{W_2}{R^2} \tag{10.5.4}$$

where W_0, W_1, and W_2 are empirical constants for each molecule. Though this form is not highly accurate, it is fairly so. It predicts (as is readily verified) (Borkman and Parr 1968)

$$R_e l_e = -6k_e, \quad R_e^2 m_e = 36k_e, \quad \frac{k_e m_e}{l_e^2} = 1 \tag{10.5.5}$$

Empirically, one finds modest deviations from these formulas (Calder and Ruedenberg 1968), on average -9.7 in place of -6 in the first formula, 77 in place of 36 in the second formula, and 0.82 instead of 1 in the third formula. This is much better than a power-series expansion in $(R - R_e)$, of course, which necessarily diverges for $R \geq R_e$ (since it diverges at $R = 0$) and for which the coefficients of successive terms are unrelated. Equation (10.5.4) is in fact just the first term in an expansion of the vibrational potential in the variable $[(1/R) - (1/R_e)]$. For details and extensions of all this, including applications to vibrational potential functions of polyatomic molecules and solids, see Parr and Brown (1968),

Martin (1968), Simons and Parr (1971), Brown and Parr (1971), Simons (1972), Simons, Parr, and Finlan (1973), Simons and Finlan (1974), Pasternak, Anderson, and Leech (1977, 1978), and Gázquez, Ray, and Parr (1978).

The special interest of (10.5.4) is that it demands, quite literally, a particular modeling. The virial relations of (1.6.11) and (1.6.12) immediately give

$$T(R) = -W_0 + \frac{W_2}{R^2} \qquad (10.5.6)$$

and

$$V(R) = 2W_0 + \frac{W_1}{R} \qquad (10.5.7)$$

The coefficients W_1 and W_2 are of most interest (but see Politzer (1970) for discussion of W_0). In the electronic kinetic energy, the $1/R^2$ dependence signifies a particle-in-a box behavior or the uncertainty principle operating; the $1/R$ dependence implies Coulomb attractions or repulsions. Suppose that there is an electronic charge (the "bond charge") $-qe$ located at the bond center, and net charge $+(q/2)e$ on each of the atoms. Suppose that the bond charge has a kinetic energy as if it were a charge moving in a one-dimensional box of length vR, where v is a second parameter. Then we find

$$W_2 = \left(\frac{\hbar^2 \pi}{2mv^2}\right)q, \qquad W_1 = -\frac{7}{4}q^2 e^2 \qquad (10.5.8)$$

or, in terms of experimentally accessible R_e and k_e,

$$q = \left(\frac{4R_e^2 k_e}{7e^2}\right)^{1/2}, \qquad v = \left(\frac{qh^2}{4mR_e^4 k_e}\right)^{1/2} \qquad (10.5.9)$$

Some q and v values determined from these formulas are given in Table 10.2. They are quite reasonable as effective measures of bond population and delocalization length. The parameter q, indeed, could be taken as an empirical measure of bond order (Politzer 1969a,b argues for q/R_e^2 instead). Note that R_e, k_e, l_e, and m_e can be predicted if q and v are both known. Note that v is essentially constant for a given row of the periodic table.

For a heteropolar molecule, one can determine the location of q in several ways, for example using known electronegativities of the constituent atoms. Perhaps most interesting is the model obtained using electronegativities defined from within the model itself (Ray, Samuels, and Parr 1979). The basic idea is due to Pasternak (1977), following very much the spirit of Gordy (1946). A bond charge located a distance r from

Table 10.2 Bond-charge-model Parameters q and v for Ground States of Homonuclear Diatomic Molecules[a,b]

Molecule	q	v
H_2	0.76	2.30
Li_2	1.10	1.01
B_2	1.90	1.00
C_2	2.30	0.99
N_2	2.73	1.00
O_2	2.26	1.05
F_2	1.80	1.07
Na_2	1.11	0.93
Si_2	2.44	0.74
P_2	3.05	0.72
S_2	2.87	0.74
Cl_2	2.52	0.77
K_2	1.21	0.79
Se_2	2.99	0.68
Br_2	2.68	0.70
Te_2	3.18	0.70
I_2	2.83	0.63

[a] After Borkman and Parr (1968). For heteronuclear diatomic values, see Borkman, Simons, and Parr (1969).

[b] See text for definitions of parameters.

a nucleus of net charge Z feels a potential Z/r due to this nucleus; if we are at the covalent radius, this should be an electronegativity measure. For the atoms A and B in the diatomic molecules AA and BB, respectively, we accordingly define electronegativities by the formulas (atomic units)

$$\chi_A = C\frac{q_A}{2r_A}, \quad \chi_B = C\frac{q_B}{2r_B} \qquad (10.5.10)$$

where C is a constant [$C = 0.484, 0.354, 0.273$ for single, double, and triple bonds, respectively, according to Ray, Samuels, and Parr (1979)], q_A and q_B are bond charges, and $2r_A = R_{AA}$ and $2r_B = R_{BB}$; the factor 2 appears because in the homonuclear bond-charge model the net charge on a nucleus is $\frac{1}{2}q$.

Now consider a heteropolar diatomic molecule AB, with bond charge q_{AB} located at a distance r_1 from nucleus A and r_2 from nucleus B, with $r_1 + r_2 = R_{AB} =$ the bond length. Assume $q_{AB} = \frac{1}{2}(q_A + q_B)$; then if the net charge on A is written $\frac{1}{2}(q_A + \Delta)$, the net charge on B will be $\frac{1}{2}(q_B - \Delta)$.

We then have

$$\chi_A \text{ (in AB)} = C\frac{q_A + \Delta}{2r_1}, \quad \chi_B \text{ (in AB)} = C\frac{q_B - \Delta}{2r_2} \quad (10.5.11)$$

But these must be equal to each other and equal to the molecular electronegativity χ_{AB}:

$$\chi_{AB} = \chi_A \text{ (in AB)} = \chi_B \text{ (in AB)} \quad (10.5.12)$$

From this there follows, after simple manipulations,

$$\chi_{AB} = \frac{\chi_A R_{AA} + \chi_B R_{BB}}{2R_{AB}} \approx \frac{\chi_A R_{AA} + \chi_B R_{BB}}{R_{AA} + R_{BB}} \quad (10.5.13)$$

This is in reasonable agreement with experiment if Mulliken electronegativity values are used for all electronegativities. Among other results, $2R_{AB} < R_A + R_B$ is predicted.

An extraordinary implication follows if we compare (10.5.13) with the result of the simple electronegativity equalization calculation of §5.2. Inserting (5.2.6) into (5.2.5) we find

$$\mu_{AB} = \mu_A = \mu_B = \frac{2\eta_B \mu_A^0 + 2\eta_A \mu_B^0}{2\eta_A + 2\eta_B} \quad (10.5.14)$$

where η_A and η_B are absolute hardnesses [(5.3.1)], and hence

$$\mu_{AB} = \frac{S_A \mu_A^0 + S_B \mu_B^0}{S_A + S_B} \quad (10.5.15)$$

where S_A and S_B are softnesses [see (5.3.2)]. Since the chemical potential is the negative of the electronegativity, from comparison between (10.5.13) and (10.5.15) it follows that

$$S_A \propto R_{AA} \quad (10.5.16)$$

or, using (5.3.4),

$$R_{AA} \propto \frac{1}{I_A - A_A} \quad (10.5.17)$$

This prediction is borne out reasonably well, with the proportionality factor being 0.90 (Ray, Samuels, and Parr 1979). Furthermore, going back to the simple bond-charge formulas (10.5.9) and eliminating k_e, we see that for given v, q_e is proportional to the reciprocal of R_e. From (10.5.11) this implies

$$\chi_A \propto \frac{1}{R_{AA}^2} \quad (10.5.18)$$

In the molecule, softness goes as the bond length, electronegativity as its inverse square.

General confirmation for these arguments comes from several sources. Politzer, Parr, and Murphy (1983) have calculated for a number of atoms the distance from the nucleus at which the chemical potential is equal to the classical electrostatic potential [compare (10.5.10)] and found that this distance can be identified with the covalent radius. Balbás, Alonso, and Vega (1986) elaborate and confirm this idea. Komorowski (1987) independently argues for (10.5.16), as do Nalewajski and Koniski (1987); see also Gázquez and Ortiz (1984).

A clean derivation of these results can also be given. Returning to the homonuclear molecule AA, write the energy in the simple bond-charge form as

$$W(R, q) = 2E_A\left(\frac{q}{2}\right) - \frac{Cq^2}{R} + \frac{Dq}{R^2} \tag{10.5.19}$$

where C and D are constants, D now absorbing the parameter v and varying slightly from one row of the periodic table to the next. The leading term is the energy of atom A when it has positive charge $\frac{1}{2}q$,

$$E_A\left(\frac{q}{2}\right) = E_A^0 + \chi_A^0\left(\frac{q}{2}\right) + \eta_A\left(\frac{q}{2}\right)^2 \tag{10.5.20}$$

Then, at each R, we may expect $W(R, q)$ to be a minimum with respect to q (density-functional theory!). That is,

$$\left(\frac{\partial W}{\partial q}\right)_R = 2\left(\frac{dE_A}{dq}\right) + \frac{\partial}{\partial q}\left(-\frac{Cq^2}{R} + \frac{Dq}{R^2}\right) = 0 \tag{10.5.21}$$

But

$$2\left(\frac{dE_A}{dq}\right) = 2\frac{d}{dq}E_A\left(\frac{q}{2}\right) = \frac{d}{dx}E_A(x)\bigg|_{x=q/2} = \chi_A \text{ (in the molecule)} \tag{10.5.22}$$

If, therefore, we define

$$\chi_{\text{bond}} = \frac{\partial}{\partial q}\left(\frac{Cq^2}{R} - \frac{Dq}{R^2}\right) = -\frac{D}{R^2} + \frac{2Cq}{R} \tag{10.5.23}$$

(10.5.21) becomes

$$\chi_A \text{ (in the molecule)} = \chi_{\text{bond}} = \chi_{\text{molecule}} \tag{10.5.24}$$

and we obtain the fundamental formulas

$$\chi_{\text{bond}}^0 = -\frac{D}{R^2}, \quad \eta_{\text{bond}} = -\frac{C}{R} \tag{10.5.25}$$

At the equilibrium $R = R_e$, (10.5.19) gives $q = (2D/CR_e)$. Substituting this into (10.5.23), we find

$$\chi_A \text{ (in the molecule)} = \chi_{\text{bond}} = \frac{3D}{R_e^2} \qquad (10.5.26)$$

This agrees with (10.5.18).

There is an apparent contradiction between setting χ_{molecule} equal to χ_{bond} [in (10.5.24)] and (10.5.21), which at first sight seems to say that χ_{molecule} is equal to zero. The contradiction disappears when one realizes that direct calculation of χ_{molecule} requires changing the total number of electrons in the molecule. What we are doing here is moving electrons from place to place within the molecule, which cannot give the molecular chemical potential. Compare the discussions in §4.4 and §4.5, and also Politzer and Weinstein (1979).

Equations (10.5.25) are fundamental for the considerations of the next section.

10.6 Semiempirical density-functional theory

As has already been said in §10.1, the most sure way to obtain a secure, detailed density-functional picture of the electronic structure of polyatomic molecules is to follow the lead of conventional quantum chemistry and develop the description using orbitals—Kohn–Sham orbitals. One can undertake purely theoretical Kohn–Sham calculations by the methods of Chapters 7 and 8, or some semiempirical implementation of them, and there will be no essential problem.

We do not here go into semiempirical Kohn–Sham theory in any detail; there has been little work in this subject. We merely call attention to the convincing argument in the literature (Lindholm and Lundqvist 1985) indicating that the so-called HAM semiempirical model of electronic structure (Lindholm and Asbrink 1985) can reasonably be viewed as a semiempiricized version of Kohn–Sham theory.

Our concern in the present section is rather the possible semiempirical implementation of the original bare-boned Hohenberg–Kohn theory, in which the basic variable is taken to be the electron density itself. We shall mainly discuss the consequent useful semiempirical electronegativity equalization schemes.

In the results of the previous two sections we already have most of what we need, since a sensible modeling of the electronic charge distribution, involving its energetics, is *a fortiori* a density-functional procedure, by the Hohenberg–Kohn theorem admitting of energy minimization with respect to model parameters. Any ultrasimple modeling must necessarily be handled semiempirically, which is legitimate to the extent that it provides useful insights and predictions. We may be able to get some things out of purely atoms-in-molecules-type modeling (§10.4); to include bonds will be necessary for most purposes (§10.5). Energy minimization must give, in any case, equal chemical potential through the system—chemical potential or electronegativity equalization. So what we are talking about here are electronegativity equalization procedures for polyatomic molecules.

That simple bond-charge models can be viewed as density-functional models was first recognized by Pasternak (1980). The formal reduction of density-functional theory to bond-charge form has been described by Ghosh and Parr (1987a).

At the first level of modeling, one takes a known, fixed molecular structure and number of electrons, assigns electronegativities and hardnesses to each atomic site, and minimizes the total energy with respect to electron transfer, ignoring all electrostatic effects. This is the polyatomic generalization of the atoms-in-molecules model of §5.2 and the beginning of §10.4. Clearly, it is grossly deficient in not possessing the capacity to predict any aspect of molecular structure except atomic charges. Early electronegativity-equalization calculations on polyatomic molecules were of this type (for example, Huheey 1965) or even simpler (Sanderson 1976).

At the next level of modeling, one would include electrostatic effects modeled in some simple way—Coulomb attractions and repulsions between charged atoms, including point dipoles on nuclei as appropriate—but still ignore bonds. This is similar to the model described by (10.4.1)–(10.4.2). Now there begins to be the possibility of predicting bond angles, but not much possibility of predicting bond lengths or force constants. This is the model discovered, developed, and applied, most effectively, by Mortier and coworkers (Mortier, van Genechten, and Gasteiger 1985; Mortier, Ghosh, and Shankar 1986; Yang and Mortier 1986; van Genechten, Mortier, and Geerlings 1987; for a review see Mortier 1987). Table 10.3 shows how very well this method, once calibrated, can predict charge densities in molecules.

Van Genechten, Mortier and Geerlings (1987) ascribe special importance to the final molecular electronegativity after complete equalization, calling it the *intrinsic framework electronegativity*, showing that trends in this quantity can quantitatively predict trends in various properties of solids. A similar result was earlier found by Shankar and Parr (1985), who, prompted by (5.2.6), demonstrated that crystal structures of nontransition elements and of octet and suboctet binary compounds could be classified efficiently using electronegativity difference and hardness sum as the two coordinates in a structure–stability diagram. Much work is needed examining such ideas.

Table 10.3 Comparison of Atomic Charges in the Alanine Dipeptide Molecule Determined by Quantum-chemical Calculation and by an Electronegativity-Equalization Scheme[a]

Atom	STO-3G[b]	EEM[c]
C-1	−0.23	−0.29
C-2	−0.30	0.30
O-3	−0.27	−0.28
H-4	0.09	0.11
H-5	0.08	0.11
H-6	0.09	0.11
N-7	−0.37	−0.38
C-8	0.04	0.04
H-9	0.19	0.19
C-10	0.28	0.25
N-11	−0.36	−0.36
C-12	−0.10	−0.11
H-13	0.07	0.07
C-14	−0.20	−0.23
H-15	0.08	0.10
H-16	0.08	0.11
H-17	0.08	0.10
O-18	−0.24	−0.27
H-19	0.17	0.17
H-20	0.08	0.07
H-21	0.07	0.07
H-22	0.08	0.08

[a] From Mortier, Ghosh and Shankar (1986), which see for details.

[b] Self-consistent-field gross population.

[c] Calculated by the electronegativity-equalization procedure calibrated on small molecules.

At the third level of modeling, one would introduce bond charges located at bond sites (chosen by an appropriate criterion) endowed with hardnesses and electronegativities in accord with the ideas at the end of §10.5. The model of §10.5 is thereby generalized to polyatomic molecules, though there are many delicate issues that need to be resolved for the polyatomic case. For early work that gives grounds for optimism, see Borkman (1969), Borkman and Settle (1971), Brown and Parr (1971) and Gázquez, Ray, and Parr (1978). With such models, one should be able successfully to predict bond lengths, bond angles, force constants, and charge distributions.

At a still higher level of modeling, one would wish to become more sophisticated in the treatment of the interactions between atoms, giving special attention to the Fukui functions and their modeling.

11
MISCELLANY

11.1 Scaling relations

Scaling is an operation on a wave function or density that dilates (or shrinks) all its coordinates while preserving proper normalization. We briefly discussed the scaling of wave functions in §1.6 and one of its applications in density-functional theory in §6.2. Here we address the scaling behavior of the energy as a functional of the density. We deal exclusively with the scaling defined by (1.6.16).

Given a wave function $\Psi(\mathbf{r}^N)$, we define the scaled wave function by

$$\Psi_\lambda(\mathbf{r}^N) = \lambda^{3N/2}\Psi(\lambda\mathbf{r}^N) \qquad (11.1.1)$$

Then the scaled density corresponding to Ψ_λ is

$$\rho_\lambda(\mathbf{r}) = \lambda^3 \rho(\lambda\mathbf{r}) \qquad (11.1.2)$$

An expressed in (1.6.17) and (1.6.18), the kinetic energy and electron–electron interaction energy, as functionals of wave function, scale homogeneously in second and first degree, respectively. Naively, we would expect that they also scale in the same way as functionals of electron density. In fact they do not! This was first shown by Levy and Perdew (1985a).

Manipulating the constrained-search formula for the functional $F[\rho]$, [(3.4.10)], we find

$$F[\rho(\mathbf{r})] = \operatorname*{Min}_{\Psi \to \rho} \int \Psi^*(\mathbf{r}^N)[\hat{T}(\mathbf{r}^N) + \hat{V}_{ee}(\mathbf{r}^N)]\Psi(\mathbf{r}^N)\,d\mathbf{r}^N$$

$$= \operatorname*{Min}_{\Psi \to \rho} \lambda^{3N} \int \Psi^*(\lambda\mathbf{r}^N)[\hat{T}(\lambda\mathbf{r}^N) + \hat{V}_{ee}(\lambda\mathbf{r}^N)]\Psi(\lambda\mathbf{r}^N)\,d\mathbf{r}^N$$

$$= \operatorname*{Min}_{\Psi \to \rho} \lambda^{-2} \int \Psi_\lambda(\mathbf{r}^N)[\hat{T}(\mathbf{r}^N) + \lambda\hat{V}_{ee}(\mathbf{r}^N)]\Psi_\lambda(\mathbf{r}^N)\,d\mathbf{r}^N$$

$$= \operatorname*{Min}_{\Psi_\lambda \to \rho_\lambda} \lambda^{-2}\langle\Psi_\lambda|\hat{T} + \lambda\hat{V}_{ee}|\Psi_\lambda\rangle \qquad (11.1.3)$$

Accordingly, we see that if $\Psi_\rho^{\min}$ gives ρ and minimizes $\langle\hat{T} + \hat{V}_{ee}\rangle$, then

$\lambda^{3N/2}\Psi_\rho^{\min}(\lambda\mathbf{r}^N)$ gives ρ_λ, but minimizes $\lambda^{-2}\langle \hat{T} + \lambda\hat{V}_{ee}\rangle$. Therefore

$$\lambda^{3N/2}\Psi_\rho^{\min}(\lambda\mathbf{r}^N) \neq \Psi_{\rho_\lambda}^{\min}(\mathbf{r}^N) \qquad (11.1.4)$$

where $\Psi_{\rho_\lambda}^{\min}(\mathbf{r}^N)$ gives ρ_λ and minimizes $\langle \hat{T} + \hat{V}_{ee}\rangle$,

$$F[\rho_\lambda] = \langle \Psi_{\rho_\lambda}^{\min}| \hat{T} + \hat{V}_{ee} |\Psi_{\rho_\lambda}^{\min}\rangle \qquad (11.1.5)$$

Because of (11.1.4), $T[\rho_\lambda]$ and $V_{ee}[\rho_\lambda]$ do not scale homogeneously in λ. Based on this result, Levy and Perdew (1985a) further establish the inequalities

$$V_{ee}[\rho_\lambda] < \lambda V_{ee}[\rho], \qquad T[\rho_\lambda] > \lambda^2 T[\rho] \qquad (\lambda < 1) \qquad (11.1.6)$$

and

$$V_{ee}[\rho_\lambda] > \lambda V_{ee}[\rho], \qquad T[\rho_\lambda] < \lambda^2 T[\rho] \qquad (\lambda > 1) \qquad (11.1.7)$$

These contrast with the wave-function equalities (1.6.17) and (1.6.18). The virial theorem itself [(1.6.10)], of course, holds in both wave-function and density-functional theories.

One can proceed in the manner of (11.1.2) to show that the noninteracting Kohn–Sham kinetic energy, either (7.3.1) or (7.3.2), scales homogeneously,

$$T_s[\rho_\lambda] = \lambda^2 T_s[\rho] \qquad (11.1.8)$$

Similarly, for the exchange-energy functional [the spin-compensated version of (8.1.12)]

$$E_x[\rho_\lambda] = \lambda E_x[\rho] \qquad (11.1.9)$$

Equations (11.1.8) and (11.1.9) justify the argument of (6.1.27) and (6.1.29). It can be easily shown that the classical Coulomb potential energy scales homogeneously:

$$J[\rho_\lambda] = \lambda J[\rho] \qquad (11.1.10)$$

The part of $T[\rho]$ that does not scale like λ^2 is the kinetic-energy contribution to the correlation energy, namely,

$$T_c[\rho] = T[\rho] - T_s[\rho] \qquad (11.1.11)$$

But, by (7.3.12),

$$T_c[\rho] > 0 \qquad (11.1.12)$$

Finally, the correlation energy, in the spin-compensated version of (8.2.111),

$$E_c[\rho] = V_{ee}[\rho] - J[\rho] - E_x[\rho] + T_c[\rho] \qquad (11.1.13)$$

also obeys the inequalities (Levy and Perdew 1985)

$$E_c[\rho_\lambda] < \lambda E_c[\rho] \quad (\lambda < 1) \qquad (11.1.14)$$

$$E_c[\rho_\lambda] > \lambda E_c[\rho] \quad (\lambda > 1) \qquad (11.1.15)$$

Scaling inequalities may aid in the search for better approximations to the functionals in density-functional theory. Levy, Yang, and Parr (1985) were able to define a functional of ρ and the scaling parameter λ such that the kinetic and electron–electron interaction energy do scale in the naively expected fashion. Various arguments on scaling relations in density-functional theory can be found in Szasz, Berrios-Pagan, and McGinn (1975), Averill and Painter (1981), Ghosh and Parr (1985), and Sahni and Levy (1986). For a recent review, see Levy (1987b).

11.2 A maximum-entropy approach to density-functional theory

One topic remains to be discussed, the analogy between the density-functional theory of the ground states of inhomogeneous electronic systems and the classical thermodynamics of the equilibrium states of nonhomogeneous macroscopic systems: an analogy only, and only approximate, but in any case provocative.

The chemical potential of density-functional theory is, as we have explained in Chapter 4, closely analogous to the macroscopic chemical potential of thermodynamics and statistical mechanics. The integral-N restriction for atoms and molecules produces the complexities we have discussed, to be sure, but the analogy for the chemical potential is certainly close and useful. Is there more?

Yes, and this we now discuss. In §6.5 we introduced a local quantity $\beta(\mathbf{r})$, which we called a temperature parameter, and we wrote

$$\beta(\mathbf{r}) = 1/kT(\mathbf{r}) \qquad (11.2.1)$$

This may or may not be trite; we here demonstrate that it is not. We follow Ghosh, Berkowitz, and Parr (1984).

At a particular point $\mathbf{r}$ in the electron distribution of an atom or molecule, there is a Kohn–Sham local electronic kinetic-energy density $t_s(\mathbf{r})$ and an electron density $\rho(\mathbf{r})$. We may take, with the argument for the coefficient $-\frac{1}{8}$ for $\nabla^2\rho$ being unequivocal (see §6.5, Berkowitz 1986, and Ghosh and Parr 1986),

$$t_s(\mathbf{r}, \rho) = \sum_i \frac{1}{8}\frac{\nabla\rho_i \, \nabla\rho_i}{\rho_i} - \frac{1}{8}\nabla^2\rho \qquad (11.2.2)$$

where the sum is over the Kohn–Sham orbitals. We now ask what phase-space distribution function $f(\mathbf{r}, \mathbf{p})$ in the sense of the Wigner function of Appendix D would produce this t_s, and this ρ at each point, through

$$\rho(\mathbf{r}) = \int f(\mathbf{r}, \mathbf{p}) \, d\mathbf{p} \tag{11.2.3}$$

and

$$t_s(\mathbf{r}) = \int \frac{p^2}{2} f(\mathbf{r}, \mathbf{p}) \, d\mathbf{p} \tag{11.2.4}$$

Many $f(\mathbf{r}, \mathbf{p})$ could do this; we proceed to choose one that is optimal in a certain information-theoretical sense.

Define the total (global) entropy associated with the density distribution $\rho(\mathbf{r})$ by the formula

$$S = -k \iint f(\mathbf{r}, \mathbf{p}) \ln f(\mathbf{r}, \mathbf{p}) \, d\mathbf{r} \, d\mathbf{p} \tag{11.2.5}$$

Maximize this with respect to the form of $f(\mathbf{r}, \mathbf{p})$, subject to the conditions (11.2.3) and (11.2.4). Letting $\alpha(\mathbf{r})$ and $\beta(\mathbf{r})$ be Lagrange multipliers for (11.2.3) and (11.2.4), respectively, there results the distribution function.

$$f(\mathbf{r}, \mathbf{p}) = e^{-\alpha(\mathbf{r})} e^{-\beta(\mathbf{r}) p^2 / 2} \tag{11.2.6}$$

The normalization condition (11.2.3) gives $\alpha(\mathbf{r})$:

$$\rho(\mathbf{r}) = e^{-\alpha(\mathbf{r})} 4\pi \int_0^\infty e^{-\beta(\mathbf{r}) p^2 / 2} p^2 \, dp$$

$$= e^{-\alpha(\mathbf{r})} [2\pi / \beta(\mathbf{r})]^{3/2} \tag{11.2.7}$$

Equation (11.2.6), consequently, becomes

$$f(\mathbf{r}, \mathbf{p}) = \rho(\mathbf{r}) \left[\frac{\beta(\mathbf{r})}{2\pi} \right]^{3/2} e^{-\beta(\mathbf{r}) p^2 / 2} \tag{11.2.8}$$

or, equivalently, with $T(\mathbf{r})$ defined by (11.2.1),

$$f(\mathbf{r}, \mathbf{p}) = \rho(\mathbf{r}) [2\pi k T(\mathbf{r})]^{-3/2} e^{-p^2 / 2kT(\mathbf{r})} \tag{11.2.9}$$

This is a local Maxwell–Boltzmann distribution law. From (11.2.4) there then follows

$$\beta(\mathbf{r}) = \frac{3}{2} \frac{\rho(\mathbf{r})}{t_s(\mathbf{r})} \tag{11.2.10}$$

Other local quantities follow readily. If one defines the local pressure by (Bartolotti and Parr 1980, Ghosh 1987)

$$p = -\tfrac{1}{3} \operatorname{tr} \boldsymbol{\sigma} \tag{11.2.11}$$

one finds for the Kohn–Sham case the stress tensor

$$\boldsymbol{\sigma}_s(\mathbf{r}) = -\rho \nabla \frac{\delta R_s}{\delta \rho} \qquad (11.2.12)$$

and the local kinetic pressure,

$$p_s(\mathbf{r}) = \rho(\mathbf{r}) kT(\mathbf{r}) \qquad (11.2.13)$$

This is the ideal-gas law. Corrections for nonideality have been examined by Robles (1986). We also have (6.5.18). The local entropy has the Sackur–Tetrode form (Ghosh, Berkowitz, and Parr 1984).

So far, so good; but we should be cautious. The $f(\mathbf{r}, \mathbf{p})$ is not the exact Wigner function, but some approximation to it. Since it reproduces the exact ρ, by the Hohenberg–Kohn theorem it contains all the information about the system. But it remains to be seen whether it is useful for anything except reproducing ρ and t_s.

A very direct, numerical test is provided by calculations of Compton profiles of atoms and molecules from $f(\mathbf{r}, \mathbf{p})$ (Parr, Rupnik, and Ghosh 1986). The momentum-space density analogue of (11.2.3) is the momentum-space formula (see Appendix D)

$$\rho_p(\mathbf{p}) = \int f(\mathbf{r}, \mathbf{p}) \, d\mathbf{r} \qquad (11.2.14)$$

In terms of this, the spherically averaged Compton profile, in the inpulse approximation, is

$$J(q) = \frac{1}{2} \int_{|q|}^{\infty} \rho_p(\mathbf{p}) \frac{1}{p} \, d\mathbf{p} \qquad (11.2.15)$$

This $J(q)$ can be accurately computed and experimentally measured. The approximation (11.2.8) gives the prediction

$$J(q) = (2\pi)^{-1/2} \int \rho(\mathbf{r}) \beta^{1/2}(\mathbf{r}) e^{-\beta(\mathbf{r}) q^2/2} \, d\mathbf{r} \qquad (11.2.16)$$

which can be readily tested. One uses Hartree–Fock local kinetic-energy densities to determine $\beta(\mathbf{r})$, and does the integration numerically. The results are very good, as illustrated in Table 11.1. For molecules, which do not possess spherical symmetry, individual components of the Compton profiles are accessible. Introducing β_x, β_y, β_z as components of local temperature parameter, one obtains the generalization of (11.2.8),

$$f(\mathbf{r}, \mathbf{p}) = (2\pi)^{-3/2} [\beta_x(\mathbf{r}) \beta_y(\mathbf{r}) \beta_z(\mathbf{r})]^{1/2} \rho(\mathbf{r}) \exp\left[-\tfrac{1}{2}(\beta_x p_x^2 + \beta_y p_y^2 + \beta_z p_z^2)\right] \qquad (11.2.17)$$

and hence directional Compton profiles $J_x(q) = \int \rho_p(\mathbf{p}) \, \delta(q - \mathbf{p} \cdot \hat{\mathbf{i}}) \, d\mathbf{p}$,

Table 11.1 Compton Profiles for Argon and Molecular Nitrogen[a]

q	$J(q)$ for Ar	$J(q)$ for N_2	$J_\|(q)$ for N_2	$J_\perp(q)$ for N_2
0.0	5.49 (5.06)	5.74 (5.29)	6.18 (6.08)	5.05 (5.27)
0.2	5.27 (4.96)	5.50 (5.14)	5.85 (5.69)	4.90 (5.13)
0.4	4.70 (4.62)	4.86 (4.70)	5.02 (4.75)	4.50 (4.71)
0.6	3.96 (4.04)	4.03 (4.03)	4.04 (3.72)	3.94 (4.06)
0.8	3.22 (3.33)	3.18 (3.27)	3.14 (2.89)	3.30 (3.28)
1.0	2.59 (2.66)	2.45 (2.56)	2.40 (2.27)	2.67 (2.54)
2.0	1.12 (1.08)	0.72 (0.75)	0.70 (0.79)	0.78 (0.76)
5.0	0.33 (0.36)	0.15 (0.15)	0.13 (0.15)	0.17 (0.15)

[a] From Parr, Rupnik, and Ghosh (1986), where more details are given.
[b] Values in parentheses are from accurate Hartree–Fock calculations.

etc. Numerical results obtained from these formulas are good; see Table 11.1. One concludes that encoding real-space $\rho(\mathbf{r})$ and $t(\mathbf{r})$ information into these approximate $f(\mathbf{r}, \mathbf{p})$ formulas and then proceeding to momentum space loses little information. These $f(\mathbf{r}, \mathbf{p})$ are quantitatively useful. In retrospect, the result is not surprising, because the local kinetic energies of course contain information about momentum of electrons.

But there is also another, very convenient test: Transform this approximate $f(\mathbf{r}, \mathbf{p})$ back to position space to give an approximate first-order density matrix and compute the exchange energy. The formulas required are the inverse of Equation (D.1) of Appendix D,

$$\rho_1(\mathbf{r} + \tfrac{1}{2}\mathbf{s}, \mathbf{r} - \tfrac{1}{2}\mathbf{s}) = \int f(\mathbf{r}, \mathbf{p}) e^{i\mathbf{p} \cdot \mathbf{s}} \, d\mathbf{p} \quad (11.2.18)$$

plus equation (6.5.3) of Chapter 6 for closed-shell Hartree–Fock exchange. We obtain straightforwardly the formulas (Ghosh and Parr 1986)

$$\rho_1(\mathbf{r} + \tfrac{1}{2}\mathbf{s}, \mathbf{r} - \tfrac{1}{2}\mathbf{s}) = \rho(\mathbf{r}) e^{-s^2/2\beta(\mathbf{r})} \quad (11.2.19)$$

and

$$K[\rho] = \frac{\pi}{2} \int \rho^2(\mathbf{r}) \beta(\mathbf{r}) \, d\mathbf{r} \quad (11.2.20)$$

In effect (see Kemister 1986), (11.2.19) is the formula (6.5.12) obtained by a different method; (11.2.20) is the previous (6.5.13). Equation (11.2.20) was shown in §6.5 to be numerically superior to the classic Dirac $\rho^{4/3}$ formula of (6.1.20); see Table 6.3 for comparisons.

As was shown in §6.5, (11.2.19) leads to total-energy functionals of Thomas–Fermi–Dirac form, including such functionals with improved numerical coefficients, as soon as one makes the single additional assumption that $\beta(\mathbf{r})$ is local:

$$\beta(\mathbf{r}) = \beta(\rho(\mathbf{r})) \tag{11.2.21}$$

What this means is that we have thereby generated still another derivation of Thomas–Fermi–Dirac theories: maximize the entropy, as defined by (11.2.5), subject to the conditions (11.2.3) and (11.2.4), and assume locality in $\beta(\mathbf{r})$, (11.2.21). As was stated at the end of §6.5, systematic extension of these ideas is desirable.

This "classical ideal-gas-like" approximate description of an electronic system contains many other interesting relations involving density, temperature, and pressure. These include equations of Orstein–Zernike, Percus-Yevick, and Yvon form (Ghosh and Berkowitz 1985).

11.3 Other topics

Many important researches in density-functional theory have so far not been mentioned in this book. We here briefly comment on a few of them.

Most important, there are a multitude of applications to practical problems in the electronic structure of atoms, molecules, solids, and surfaces, including studies of electron-removal energies and electron affinities (Perdew and Norman 1982, Baroni and Turcel 1983), X-ray structure factors (Massa, Goldberg, Frishberg, Boehme, and La Placa 1985, Cohen, Frishberg, Lee, and Massa 1986, Levy and Goldstein 1987, Kryachko, Petkov, and Stoitsov 1987a), surfaces (Lang and Kohn 1970, Wimmer, Krakauer, Weinert, and Freeman 1981, Fu and Freeman 1988), chemisorption (Feibelman 1985, Weinert, Freeman, and Ohnishi 1986), electrochemistry (Goodisman 1987, Groot and Van der Eerden 1988), embedded atoms (effective medium theory) (Norskov and Lang 1980, Stott and Zaremba 1980b, 1982, Puska, Nieminen, and Manninen 1981, Puska and Nieminen 1984, Kress and DePristo 1987, 1988, Baskes 1987), superconductivity (Tachibana 1988a, Oliveira, Gross and Kohn 1988b), and so on. Discussions of many applications are contained in the various reviews of density-functional theory listed in Appendix G; see especially Williams and von Barth (1983).

The reviews in Appendix G also cover many studies that broaden and deepen density-functional theory itself. We would mention relativistic density-functional theory, which Rajagopal and others have developed (e.g. Ramana and Rajagopal 1983, Engel and Dreizler 1988); density-functional theory in strong fields (Vignale and Rasolt 1987, 1988; Ghosh and Dhara 1988) the large subject of $1/Z$ and N/Z expansions (for example, Englert and Schwinger 1982); the enticing problem of formulating a differential equation for the density (Deb and Ghosh 1983, March 1986a, Hunter 1987, Levy, Perdew, and Sahni 1984, March and Nalewajski 1987, Levy and Ou-Yang 1988, Nalewajski and Kozlowski 1988, Tachibana 1988b). There also is a momentum-space density-functional theory (Henderson 1981; Smith 1984; Gadre and Chakravorty 1987; Das, Ghosh, and Sahni 1988), wherein the functionals are not universal; connections of the subject with the hydrodynamical formulation of quantum mechanics, of which Deb and colleagues have provided beautifully complete and documented discussion (for example, Ghosh and Deb 1982a,b; Deb and Ghosh 1987, Ghosh 1987); and the subject of local force laws (Ghosh 1987, Baltin 1988a,b). There is systematic study of inequalities and stability conditions, on which Nalewajski has made an auspicious beginning (Nalewajski and Parr 1982, Nalewajski and Capitani 1982, Nalewajski 1984c, Nalewajski 1988); a theory for constrained systems (Dederichs, Blügel, Zeller, and Akai 1984) the connection with stochastic mechanics (McClendon 1988); the proposal to combine molecular dynamics with density-functional theory (Car and Parrinello 1985, Payne, Joannopoulos, Alan, Teter, and Vanderbilt 1986, Fois, Selloni, Parrinello, and Car 1988); the formulation using local scaling transformations (Petkov, Stoitsov, and Kryachko 1986, Kryachko, Petkov, and Stoitsov 1987b); and the new ideas of Nordholm (1987) about nonergodic Thomas–Fermi theory.

As we have already indicated in Chapter 8, and elsewhere in the book, there is currently considerable activity on what is unquestionably the most important density-functional problem of all: to establish improved approximations to the exchange-correlation functional $E_{xc}[\rho]$ and/or the correlation functional $E_c[\rho]$. This problem area is of the utmost importance for achieving reliable calculations for large molecules.

11.4 Final remarks

In this exposition we have tried to present accurately the fundamental principles of the density-functional theory of the electronic structure of matter. We have tried to avoid getting bogged down in mathematical details, while not denying that many important mathematical questions remain to be examined. We have tried to describe adequately the main calculational schemes, while also wanting to make clear that the calculational schemes are developing all the time in ways that are not trivial. We have attempted to expound the conceptual advantages of density-functional theory over wave-function theory, while not pretending that all is cut and dried with respect to the most appropriate definitions of the concepts. We have emphasized the formal aspects of the theory because no text is yet available in which they are laid out in full, without wishing to seem oblivious of the fact that it is special applications that will more probably concern many readers. We hope that our bibliography will correct for our many, considerable omissions, but we warn that the field is moving so fast that the bibliography will soon become quite out of date.

The density-functional theory of electronic structure should expand very much in the future, and should become increasingly useful. Its calculational methods are in need of developmental as well as fundamental theoretical work, but they offer a special economy with respect to difficulty as the number of electrons increases. Density-functional concepts offer a compellingly appealing language for discussion of molecular structure and behavior: almost pictorial, almost intuitive, yet quantitative. If the type of analysis in §11.2 should happen to be usefully further developed, about which one can only speculate, the consequences could be very interesting.

We have purposely not put forward most-favored recipes for calculation; the subject is too young for that. Spin-density theory in its Kohn–Sham implementation is a highly useful calculational method, with the LSD approximation for the exchange-correlation functional $E_{xc}[\rho]$. But so also is the Hartree–Fock–Kohn–Sham method, and methods for calculating $E_x[\rho]$ and $E_{xc}[\rho]$ are fast improving.

Dirac's familiar pronouncement long ago, that the laws of chemistry were now known, began, rather than ended, the development of quantum chemistry. The contemporary successful, accurate calculations of molecular properties signal the end of what is only the first phase of the era Dirac began. The next phase, just beginning, comprises the shakedown, codification, and unification of the basic ideas in the science of chemistry itself (as distinct from physics) (Parr 1975). We expect the language of chemistry of the future to be replete with the idiom of density-functional theory.

APPENDIX A
FUNCTIONALS

We first state in simple terms what functionals are (Volterra 1959, Evans 1964). Recall that a *function* is a rule for going from a variable x to a number: $f(x)$. A *functional* is a rule for going from a function to a number: say $F[f]$ where $f(x)$ is a function (such a rule is also referred to as a *mapping*). In a manner of speaking, a functional is a function of which the variable is a function.

The expectation value $\langle \Psi | \hat{H} | \Psi \rangle$ of quantum chemistry is a functional; given Ψ one gets a number from this prescription. Similarly with $\langle \Psi | \Psi \rangle$, so that the usual variational method of quantum chemistry, Equation (1.2.7) of the text, is the search for the minimum (or extremum) of a functional. Indeed, most of what is commonly called the calculus of variations (Arfken 1980, Chapter 17; Courant and Hilbert 1953, Chapter IV, Gelfand and Fomin 1963) is a branch of the calculus of functionals.

The *differential* of a functional is the part of the difference $F[f + \delta f] - F[f]$ that depends on δf linearly. Each $\delta f(x)$ may contribute to this difference, and so we write, for very small δf,

$$\delta F = \int \frac{\delta F}{\delta f(x)} \delta f(x) \, dx, \qquad (A.1)$$

where the quantity $\delta F / \delta f(x)$ is the *functional derivative of F with respect to f* at the point x. Equation (A.1) is the rule for operating on $\delta f(x)$ to give a number δF, and is the extension to continuous variables of the formula for the total differential of a function $F(f_1, f_2, \ldots)$: $dF = \sum_i (\partial F / \partial f_i) \, df_i$.

One way to determine a functional derivative is just to expand $(F[f + \delta f] - F[f])$ in terms of δf, keeping only the first-order term, taking care that the result is put in the form of (A.1). That is to say, $\delta F / \delta f$ is explicitly defined by the process

$$\lim_{\varepsilon \to 0} \left[\frac{F(f + \varepsilon \phi) - F[f]}{\varepsilon} \right] = \left\{ \frac{d}{d\varepsilon} (F[f + \varepsilon \phi]) \right\}_{\varepsilon = 0} = \int \frac{\delta F}{\delta f(x)} \phi(x) \, dx \quad (A.2)$$

where $\phi(x)$ is arbitrary. Taking $\phi(x)$ to be a delta function $\delta(x - x_0)$ reduces this to an explicit formula for $[\delta F / \delta f(x_0)]$. A functional $F[f]$ is *differentiable* if $\delta F/\delta f(x)$ exists in the sense of (A.2). The functional derivative $\delta F/\delta f(x)$ may depend on f values at points other than x; that is, $\delta F/\delta f(x)$ itself, for each value of x, may be a functional of f.

From (A.2), it follows that the functional derivative has properties similar to the ordinary derivative:

$$\frac{\delta}{\delta f(x)}(C_1 F_1 + C_2 F_2) = C_1 \frac{\delta F_1}{\delta f(x)} + C_2 \frac{\delta F_2}{\delta f(x)} \tag{A.3}$$

and

$$\frac{\delta}{\delta f(x)}(F_1 F_2) = \frac{\delta F_1}{\delta f(x)} F_2 + F_1 \frac{\delta F_2}{\delta f(x)}. \tag{A.4}$$

Let us give some examples. Consider first the functional

$$T_{\rm TF}[\rho] = C_F \int \rho^{5/3}(\mathbf{r})\, d\mathbf{r} \tag{A.5}$$

Then

$$T_{\rm TF}[\rho + \delta\rho] = C_F \int [\rho^{5/3}(\mathbf{r}) + \tfrac{5}{3}\rho^{2/3}(\mathbf{r})\,\delta\rho + \cdots]\, d\mathbf{r}$$

$$= T_{\rm TF}[\rho] + \tfrac{5}{3} C_F \int \rho^{2/3}(\mathbf{r})\,\delta\rho(\mathbf{r})\, d\mathbf{r} + \cdots$$

Hence

$$\frac{\delta T_{\rm TF}[\rho]}{\delta \rho(\mathbf{r})} = \tfrac{5}{3} C_F \rho^{2/3}(\mathbf{r}) \tag{A.6}$$

Consider next the functional

$$T_W[\rho] = \frac{1}{8} \int \frac{\nabla \rho(\mathbf{r}) \cdot \nabla \rho(\mathbf{r})}{\rho(\mathbf{r})}\, d\mathbf{r} \tag{A.7}$$

Then to first order,

$$T_W[\rho + \delta\rho] = \frac{1}{8} \int \frac{\nabla(\rho + \delta\rho) \cdot \nabla(\rho + \delta\rho)}{\rho + \delta\rho}\, d\mathbf{r}$$

$$= T_W[\rho] - \frac{1}{4} \int \frac{\nabla^2 \rho}{\rho}\,\delta\rho\, d\mathbf{r} + \frac{1}{8} \int \frac{\nabla \rho \cdot \nabla \rho}{\rho^2}\,\delta\rho\, d\mathbf{r} \tag{A.8}$$

Therefore

$$\frac{\delta T_W[\rho]}{\delta \rho(\mathbf{r})} = -\frac{1}{4}\frac{\nabla^2 \rho}{\rho} + \frac{1}{8}\frac{\nabla \rho \cdot \nabla \rho}{\rho^2} \tag{A.9}$$

Finally consider the functional

$$J[\rho] = \frac{1}{2}\iint \frac{\rho(\mathbf{r}_1)\rho(\mathbf{r}_2)}{r_{12}} d\mathbf{r}_1 \, d\mathbf{r}_2 \tag{A.10}$$

For this we find

$$\frac{\delta J[\rho]}{\delta \rho(\mathbf{r}_1)} = \int \frac{\rho(\mathbf{r}_2)}{r_{12}} d\mathbf{r}_2 \tag{A.11}$$

The following general formula covers many cases of interest. Consider the functional

$$F[\rho] = \int f(x, \rho, \rho^{(1)}, \rho^{(2)}, \ldots, \rho^{(n)}) \, dx \tag{A.12}$$

where $\rho^{(n)}(x) = d^n\rho(x)/dx^n$, and ρ vanishes at the boundary of x. Then (Gelfand and Fomin 1953, p. 42)

$$\frac{\delta F}{\delta \rho(x)} = \frac{\partial f}{\partial \rho} - \frac{d}{dx}\left(\frac{\partial f}{\partial \rho^{(1)}}\right) + \frac{d^2}{dx^2}\left(\frac{\partial f}{\partial \rho^{(2)}}\right) - \cdots + (-1)^n \frac{d^n}{dx^n}\left(\frac{\partial f}{\partial \rho^{(n)}}\right) \tag{A.13}$$

where $\partial f/\partial \rho$, $\partial f/\partial \rho^{(1)}$, ... are partial derivatives: $\partial f/\partial \rho^{(p)}$ is taken holding $\rho^{(m)}$ constant, $m \neq p$. For *local functionals*, $f = f(x, \rho)$, and (A.13) reduces simply to $\delta F/\delta \rho = \partial f/\partial \rho$. Among the examples discussed, only $T_{TF}[\rho]$ of (A.5) is a local functional. For the three-dimensional generalization of (A.13), just replace the operator d/dx with the operator ∇.

As a more involved example of functional differentiation, we give the derivation of (2.5.12) of the text. The Hartree–Fock energy of (2.5.9) as a functional of first-order reduced density matrix can be rewritten as

$$E_{HF}[\gamma_1] = \int \delta(\mathbf{x}_1' - \mathbf{x}_1)[-\tfrac{1}{2}\nabla_1^2 + v(\mathbf{r}_1)]\gamma_1(\mathbf{x}_1', \mathbf{x}_1) \, d\mathbf{x}_1' \, d\mathbf{x}_1$$

$$+ \frac{1}{2}\int \frac{1}{|\mathbf{r}_1 - \mathbf{r}_2|}[\gamma_1(\mathbf{x}_1', \mathbf{x}_1)\gamma(\mathbf{x}_2', \mathbf{x}_2) - \gamma_1(\mathbf{x}_1', \mathbf{x}_2)\gamma_1(\mathbf{x}_2', \mathbf{x}_1)]$$

$$\times \delta(\mathbf{x}_1' - \mathbf{x}_1)\delta(\mathbf{x}_2' - \mathbf{x}_2) \, d\mathbf{x}_1' \, d\mathbf{x}_1 \, d\mathbf{x}_2' \, d\mathbf{x}_2$$

$$= H_1[\gamma_1] + J[\gamma_1] + K[\gamma_1] \tag{A.14}$$

where H_1, J, and K, respectively, denote the one-particle, Coloumb, and exchange contributions to the energy. Following the rule of (A.13), we have

$$\frac{\delta H_1[\gamma_1]}{\delta \gamma_1(\mathbf{x}_1', \mathbf{x}_1)} = [-\tfrac{1}{2}\nabla_1^2 + v(\mathbf{r}_1)]\,\delta(\mathbf{x}_1' - \mathbf{x}_1) \tag{A.15}$$

and
$$\frac{\delta J[\gamma_1]}{\delta \gamma_1(\mathbf{x}_1', \mathbf{x}_1)} = \delta(\mathbf{x}_1' - \mathbf{x}_1) \int \frac{\rho(\mathbf{r}_2)}{r_{12}} d\mathbf{r}_2 \qquad (A.16)$$

For $K[\gamma_1]$, we interchange the dummy integration variables $\mathbf{x}_2$ and $\mathbf{x}_1$; thus

$$K[\gamma_1] = -\frac{1}{2} \int \frac{1}{|\mathbf{r}_1 - \mathbf{r}_2|} \gamma_1(\mathbf{x}_1', \mathbf{x}_1) \gamma_1(\mathbf{x}_2', \mathbf{x}_2) \delta(\mathbf{x}_1' - \mathbf{x}_2)$$
$$\times \delta(\mathbf{x}_2' - \mathbf{x}_1) \, d\mathbf{x}_1' \, d\mathbf{x}_1 \, d\mathbf{x}_2' \, d\mathbf{x}_2 \qquad (A.17)$$

Then

$$\frac{\delta K[\gamma_1]}{\delta \gamma_1(\mathbf{x}_1', \mathbf{x}_1)} = -\int \frac{1}{|\mathbf{r}_1 - \mathbf{r}_2|} \gamma(\mathbf{x}_2', \mathbf{x}_2) \delta(\mathbf{x}_1' - \mathbf{x}_2) \delta(\mathbf{x}_2' - \mathbf{x}_1) \, d\mathbf{x}_2' \, d\mathbf{x}_2$$
$$= -\frac{1}{|\mathbf{r}_1 - \mathbf{r}_1'|} \gamma_1(\mathbf{x}_1, \mathbf{x}_1') \qquad (A.18)$$

Putting (A.15)–(A.17) together, we arrive at (2.5.12).

The meaning of the second functional derivative $\delta^2 F / \delta f(x) \, \delta f(x')$ is that it is the functional derivative of the first functional derivative $\delta F / \delta f(x)$, with respect to $f(x')$, $\delta F / \delta f(x)$ being thought of as depending only parametrically on x. Higher functional derivatives are defined similarly. The order of differentiation is usually unimportant,

$$\frac{\delta^2 F}{\delta f(x) \, \delta f(x')} = \frac{\delta^2 F}{\delta f(x') \, \delta f(x)} \qquad (A.19)$$

with similar identities for higher derivatives.

Let Δf be a finite change in f. A formal "Taylor series" expansion of $F[f + \Delta f]$ can then always be written down:

$$F[f + \Delta f] = F[f] + \sum_{n=1}^{\infty} \frac{1}{n!} \int \int \cdots \int \frac{\delta^{(n)} F}{\delta f(x_1) \, \delta f(x_2) \cdots \delta f(x_n)}$$
$$\times \Delta f(x_1) \, \Delta f(x_2) \cdots \Delta f(x_n) \, dx_1 \, dx_2 \cdots dx_n \qquad (A.20)$$

Here, each functional derivative is evaluated at the initial f. When such a series is terminated, it can be made an accurate representation by having the last functional derivative evaluated not at the initial f but at some $f + \lambda \Delta f$, with λ between 0 and 1 (Volterra 1959, p. 26).

Suppose that F is a functional of $f(x)$: $F = F[f(x)]$. Then for an arbitrary small change δF,

$$\delta F = \int \frac{\delta F}{\delta f(x)} \delta f(x) \, dx \qquad (A.21)$$

If f at each point of x in turn is a functional of g, that is, $f = f[g(x), x]$, we have

$$\delta f(x) = \int \frac{\delta f(x)}{\delta g(x')} \delta g(x') \, dx' \tag{A.22}$$

and hence

$$\delta F = \iint \frac{\delta F}{\delta f(x)} \frac{\delta f(x)}{\delta g(x')} \delta g(x') \, dx \, dx' \tag{A.23}$$

so that

$$\frac{\delta F}{\delta g(x')} = \int \frac{\delta F}{\delta f(x)} \frac{\delta f(x)}{\delta g(x')} \, dx \tag{A.24}$$

This is the basic *chain rule* for functional derivatives.

An important special case of a functional is an ordinary function. Let $F = F(f)$ be some *function* of f. Its functional derivative with respect to f, evaluated at x', must be

$$\frac{\delta F(f(x))}{\delta f(x')} = \frac{dF}{df} \delta(x - x') \tag{A.25}$$

in order that

$$\delta F = \int \frac{\delta F}{\delta f(x')} \delta f(x') \, dx' = \frac{dF}{df} \delta f(x) \tag{A.26}$$

In particular, we may take f to be F itself, obtaining

$$\frac{\delta f(x)}{\delta f(x')} = \delta(x - x') \tag{A.27}$$

For ordinary derivatives, an *inverse* is readily defined: $(dF/df)^{-1} = df/dF$; $(dF/df)(dF/df)^{-1} = 1$. For functional derivatives, because the relation $F = F[f]$ is many-to-one when f is a function and x is a real number, there is no unique inverse. However, we sometimes deal with a function $f(x)$ each value of which is a functional of a function $g(x')$, with each value of $g(x')$ also a functional of $f(x)$. Then we will have

$$\delta f(x) = \int \frac{\delta f(x)}{\delta g(x')} \delta g(x') \, dx'$$

and

$$\delta g(x') = \int \frac{\delta g(x')}{\delta f(x'')} \delta f(x'') \, dx''$$

so that

$$\delta f(x) = \iint \frac{\delta f(x)}{\delta g(x')} \frac{\delta g(x')}{\delta f(x'')} \delta f(x'') \, dx' \, dx'' \quad (A.28)$$

We therefore see that

$$\int \frac{\delta f(x)}{\delta g(x')} \frac{\delta g(x')}{\delta f(x'')} \, dx' = \delta(x - x'') \quad (A.29)$$

If then we define the inverse

$$\left[\frac{\delta f(x)}{\delta g(x')}\right]^{-1} = \frac{\delta g(x')}{\delta f(x)} \quad (A.30)$$

we obtain

$$\int \frac{\delta f(x)}{\delta g(x')} \left[\frac{\delta f(x'')}{\delta g(x')}\right]^{-1} dx' = \delta(x - x'') \quad (A.31)$$

Note the similarity of this result with the situation for a unitary transformation between a set of variables x_j and another set of the same number of variables y_i:

$$y_i = \sum_j A_{ij} x_j, \qquad x_j = \sum_k B_{jk} y_k$$

$$y_i = \sum_j \sum_k A_{ij} B_{jk} y_k = \sum_k (AB)_{ik} y_k = \sum_k \delta_{ik} y_k = y_i$$

$$\therefore (AB)_{ik} = \delta_{ik} = \sum_j A_{ij} B_{jk} \quad (A.32)$$

Thus (A.29) defines the sense in which $\delta f/\delta g$ is the "reciprocal" of $\delta g/\delta f$.

If the variable of F depends on a parameter λ, that is, $F[f(x, \lambda)]$, then

$$\frac{\partial F}{\partial \lambda} = \int \frac{\delta F}{\delta f(x)} \frac{\partial f(x)}{\partial \lambda} \, dx \quad (A.33)$$

which follows from (A.21). For a differentiable function of a functional, $\phi = \phi(F[f(x)])$, we have

$$\frac{\delta \phi}{\delta f(x)} = \frac{d\phi}{dF} \frac{\delta F}{\delta f(x)} \quad (A.34)$$

This rule can be seen to be a special case of (A.24).

A function $f(x_1, x_2, \ldots)$ is said to be homogeneous of degree k in all

x_i if $f(\lambda x_1, \lambda x_2, \dots) = \lambda^k f(x_1, x_2, \dots)$, which means also that

$$\sum_i x_i \frac{\partial f}{\partial x_i} = kf(x_1, x_2, \dots)$$

As a generalization of this concept; a functional $X[f]$ is said to be *homogeneous of degree k in f* if

$$X[\lambda f(x)] = \lambda^k X[f(x)] \quad (A.35)$$

or

$$\int \frac{\delta X}{\delta f(x)} f(x) \, dx = k X[f(x)] \quad (A.36)$$

To prove (A.36), differentiate (A.35) with respect to λ, apply (A.33), and set λ equal to 1. Examples are $T_{\text{TF}}[\rho]$ of (A.5)—homogeneous of degree $\frac{5}{3}$ in ρ; $T_W[\rho]$ of (A.7)—homogeneous of degree 1 in ρ; and $J[\rho]$ of (A.10)—homogeneous of degree 2 in ρ.

The Gibbs free energy of the classical thermodynamics of homogeneous phases is homogeneous of degree 1 in mole number(s), exemplifying all "extensive" properties in the classical thermodynamics of homogeneous phases. The temperature and pressure are homogeneous of degree 0 in mole number(s); these are the so-called "intensive" properties. Unfortunately, such pervasive simple homogeneities are missing from the density-functional theory of inhomogeneous electronic systems.

Finally, we consider the problem of finding the minimum of the functional $F[f]$ over all functions $f(x)$ vanishing at the boundary. Just as in elementary calculus, the necessary condition is

$$\delta F = \int \frac{\delta F}{\delta f(x)} \delta f(x) \, dx = 0 \quad (A.37)$$

for arbitrary increment $\delta f(x)$. We may equivalently require that the solution be a function $f(x)$ satisfying (Choquet-Bruhat, Dewitt-Morette, and Dillard-Bleick 1982, p. 82)

$$\frac{\delta F}{\delta f(x)} = 0 \quad (A.38)$$

For the case $F[f] = \int \mathcal{L}(f, f'; x) \, dx$, (A.38) follows directly from the fundamental lemma of the calculus of variations: If the relation $\int \eta(x) \phi(x) \, dx = 0$, with $\phi(x)$ a continuous function of x, holds for all functions $\eta(x)$ that vanish on the boundary and are continuous together with their first two derivatives, then $\phi(x) = 0$ (Courant and Hilbert 1953, p. 185).

Equation (A.38) is the *Euler–Lagrange equation*, which is a necessary condition for the existence of an *extremal*.

Suppose now that the problem is to minimize $F[f]$ subject to the constraint that

$$G[f] = 0 \qquad (A.39)$$

Often the convenient way to this is to employ *the method of Lagrange multipliers*: Solve the Euler–Lagrange equation for the auxiliary functional $\Omega[f] = F[f] - \lambda G[f]$, that is, solve

$$\frac{\delta \Omega}{\delta f(x)} = \frac{\delta F}{\delta f(x)} - \lambda \frac{\delta G}{\delta f(x)} = 0 \qquad (A.40)$$

disregarding the constraint (A.39). The solution of (A.40) contains the multiplier λ, which must be determined by imposing (A.39) on the solution. The value of λ may have important physical meaning, however, since from (A.40) λ is the ratio of two variations, $\delta F/\delta G$, at the solution point. The quantity λ is sometimes called a *sensitivity coefficient*.

Sometimes a constraint is local (pointwise),

$$g[f, x] = \text{constant} \qquad (A.41)$$

where $g[f, x]$ is simultaneously a functional of f and a function of x. Minimization of $F[f]$ subject to (A.41) can now proceed using a Lagrange multiplier $\lambda(x)$ that is a function of x. Let $\Omega[f] = F[f] - \int \lambda(x) g[f, x] \, dx$, and solve

$$\frac{\delta \Omega}{\partial f(x)} = \frac{\delta F}{\delta f(x)} - \int \lambda(x') \, dx' \, \frac{\delta g[f, x']}{\delta f(x)} = 0 \qquad (A.42)$$

the multiplier $\lambda(x)$ being determined by (A.41). One can view $\lambda(x)$ as an infinite number of Lagrange multipliers each associated with a constraint (A.41) at a point x.

Solutions of Euler–Lagrange equations are *extremals*. For an extremal to be a minimum, it is necessary that the second variation of $F[f]$ be positive at the solution point f; that is,

$$\delta^2 F = \frac{1}{2} \int\!\!\int \frac{\delta^2 F}{\delta f(x) \, \delta f(x')} \bigg|_f \delta f(x) \, \delta f(x') \, dx \, dx' \geq 0 \qquad (A.43)$$

The sufficient conditions for a minimum can be found in Gelfand and Fomin (1963). In practice one can often assume, from the nature of the problem, that a solution of the Euler–Lagrange equation is a minimum without actual verification.

Finally, in problems where $f(x)$ does not necessarily vanish at the boundary, *natural boundary conditions* arise, which must be imposed on $f(x)$ in addition to the Euler–Lagrange equation, to ensure that the differential of the functional δF be zero. For the case $F = \int \mathscr{L}(f, f', x)\, dx$, the natural boundary condition is simply $\partial \mathscr{L} / \partial f'(x) = 0$ at the boundary of x (Courant and Hilbert 1953, p. 208).

Sometimes the Euler–Lagrange equation is simply called the *Euler equation*.

For an introduction to modern functional analysis, see Reed and Simon (1972).

APPENDIX B
CONVEX FUNCTIONS AND FUNCTIONALS

The functions and functionals we deal with often possess unique features based on their convex nature. We begin by defining a function $f(x)$ on a real interval I to be convex if, with $0 \leq \lambda \leq 1$,

$$f(\lambda x_1 + (1-\lambda)x_2) \leq \lambda f(x_1) + (1-\lambda)f(x_2) \tag{B.1}$$

The function $f(x)$ is said to be *strictly convex* if the equality holds only for $x_1 = x_2$. Correspondingly, $f(x)$ is said to be *concave (strictly concave)* if $-f(x)$ is convex (or strictly convex). The nature of a convex function is shown graphically in Figure B1.

For a differentiable function, convexity can be further characterized by the theorem: A differentiable $f(x)$ is convex if and only if

$$f(x_1) - f(x_2) - (x_1 - x_2)f'(x_2) \geq 0 \tag{B.2}$$

for any $x_1, x_2 \in I$; $f(x)$ is strictly convex if the equality holds only for $x_1 = x_2$. To prove the necessity of this, simply rearrange (B.1) as

$$f(x_1) - f(x_2) - \lambda^{-1}\{f(x_2 + \lambda(x_1 - x_2)) - f(x_2)\} \geq 0 \tag{B.3}$$

The limit $\lambda \to 0$ turns (B.3) into (B.2). To prove sufficiency, compute

$$\lambda f(x_1) + (1-\lambda)f(x_2) - f(\lambda x_1 + (1-\lambda)x_2)$$
$$= \lambda\{f(x_1) - f(\lambda x_1 + (1-\lambda)x_2)\} + (1-\lambda)\{f(x_2) - f(\lambda x_1 + (1-\lambda)x_2)\}$$
$$\geq \lambda(1-\lambda)(x_1 - x_2)f'(\lambda x_1 + (1-\lambda)x_2)$$
$$+ (1-\lambda)\lambda(x_2 - x_1)f'(\lambda x_1 + (1-\lambda)x_2)$$
$$= 0 \tag{B.4}$$

The inequality follows from (B.2).

For a twice-differentiable function, we have the theorem: If $f''(x)$ exists, for $f(x)$ to be convex it is sufficient that

$$f''(x) \geq 0 \quad \text{for } x \in I \tag{B.5}$$

This can be shown from the identity

$$\int_{x_1}^{x_2} dt \int_{x_1}^{t} f''(x)\,dx = \int_{x_1}^{x_2} [f'(t) - f'(x_1)]\,dt$$
$$= f(x_2) - f(x_1) - (x_2 - x_1)f'(x_1) \tag{B.6}$$

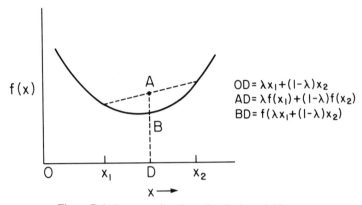

Figure B.1 A convex function of a single variable.

Inserting (B.5) on the left, (B.2) follows, and hence the convexity of $f(x)$. The sufficient criterion (B.5) often makes it easy to check the convexity of a function.

An important property of convex functions is *Jensen's Inequality*: For a convex function $f(x)$,

$$f(\langle x \rangle) \leq \langle f(x) \rangle \tag{B.7}$$

where $\langle \ \rangle$ denotes the average for some probability distribution λ_i,

$$\langle \phi(x) \rangle = \sum_i \lambda_i \phi(x_i) \tag{B.8}$$

where $\lambda_1 \geq 0$, and $\sum \lambda_i = 1$. Furthermore, if $f(x)$ is strictly convex, the equality shown in (B.7) holds only for the case $\lambda_i = 1$ for some i, $\lambda_j = 0$ for $j \neq i$. This theorem can be proved by induction, beginning from (B.1). The averages in (B.8) can also be extended to the continuous variable limit:

$$\langle \phi(x) \rangle = \int \lambda(x) \phi(x) \, dx \tag{B.9}$$

where $\lambda(x) \geq 0$, and $\int \lambda(x) \, dx = 1$.

Using Jensen's inequality, one can generate many other useful inequalities. For example, since e^x is convex,

$$\exp\left(\sum_i \lambda_i x_i\right) \leq \sum_i \lambda_i \exp(x_i) \tag{B.10}$$

Another example is the convex function $x \ln x$ defined for $x \geq 0$. Let $x_i = P_i/P_i'$, where $P_i \geq 0$, $\sum_i P_i = 1$, and $P_i' \geq 0$, $\sum_i P_i' = 1$. Then from

(B.7),
$$\left(\sum_i^N P'_i \cdot \frac{P_i}{P'_i}\right) \ln \left(\sum_j^N P'_j \frac{P_j}{P'_j}\right) \leq \sum_i^N P'_i \frac{P_i}{P'_i} \ln \frac{P_i}{P'_i}$$

which reduces to

$$\sum_i^N P_i \ln \frac{P_i}{P'_i} \geq 0 \tag{B.11}$$

This famous inequality is due to Gibbs (Gibbs 1931, Chapter 11, Hansen and McDonald 1976, p. 149).

If one lets $P'_i = 1/N$ in (B.11), there results

$$-\sum_i^N P_i \ln P_i \leq \ln N \tag{B.12}$$

which proves that if we define the entropy of a distribution as the left-hand side of this equality, the largest value it can have is $\ln N$, when all probabilites are equal.

Another property of convex functions is the theorem: If f is convex on I, then each $x_0 \in I$ for which $f'(x_0) = 0$ minimizes f on I; if f is strictly convex the minimum is unique. To prove this, simply let $x_2 = x_0$ (in B.2):

$$f(x_1) - f(x_0) \geq (x_1 - x_0) f'(x_0) = 0 \tag{B.13}$$

In other words, this theorem states that each and every extremum of a convex $f(x)$ is a minimum.

The foregoing discussion on convex functions can be easily generalized to convex functionals. A functional $F[f(x)]$ is said to be *convex* if

$$F[\lambda f_1(x) + (1 - \lambda) f_2(x)] \leq \lambda F[f_1(x)] + (1 - \lambda) F[f_2(x)] \tag{B.14}$$

where $0 \leq \lambda \leq 1$, and $f_1(x)$ and $f_2(x)$ are functions in the domain of $F[f(x)]$. $F[f(x)]$ is *strictly convex* if the equality holds only for $f_1(x) = f_2(x)$. If $F[f(x)]$ is differentiable (see Appendix A), then the convexity of $F[f(x)]$ is equivalent to

$$F[f_1(x)] - F[f_2(x)] - \int (f_1(x) - f_2(x)) \frac{\delta F[f(x)]}{\delta f(x)}\bigg|_{f=f_2} dx \geq 0 \tag{B.15}$$

Using (A.2), the proof of (B.15) follows the lines of the proof of (B.2).

The generalized Jensen inequality reads

$$F[\langle f(x) \rangle] \leq \langle F[f(x)] \rangle \tag{B.16}$$

where $\langle \ \rangle$ denotes the average in the sense of (B.8). Equation (B.16) can be proved by induction starting from (B.14). In parallel with the minimum principle for convex functions, we present the theorem: If

$F[f(x)]$ is convex, each $f_0(x)$ for which $\delta F[f(x)]/\delta f(x)|_{f=f_0} = 0$ minimizes $F[f]$ in the domain of $F[f]$. This theorem follows from (B.15).

Jensen's inequality may be extended to a convex function of a Hermitian operator. It becomes

$$f(\langle \hat{O} \rangle) \leq \langle f(\hat{O}) \rangle \tag{B.17}$$

For a convex function $f(x)$. The average "$\langle\ \rangle$" in this formula is the expectation value for some normalized state Ψ. To prove (B.17), express the operators in terms of their eigenvalues and eigenvectors:

$$\hat{O} = \sum_i O_i |O_i\rangle\langle O_i| \tag{B.18}$$

and

$$f(\hat{O}) = \sum_i f(O_i) |O_i\rangle\langle O_i| \tag{B.19}$$

Then

$$f(\langle \Psi| \hat{O} |\Psi\rangle) = f\left(\sum_i O_i |\langle O_i | \Psi\rangle|^2\right) = f\left(\sum_i O_i P_i\right) \tag{B.20}$$

and

$$\langle \Psi| f(\hat{O}) |\Psi\rangle = \sum_i f(O_i) |\langle O_i | \Psi\rangle|^2 = \sum_i f(O_i) P_i \tag{B.21}$$

where

$$P_i = |\langle O_i | \Psi\rangle|^2 \geq 0 \quad \text{and} \quad \sum_i P_i = \sum_i \langle \Psi | O_i\rangle\langle O_i | \Psi\rangle = \langle \Psi | \Psi\rangle = 1$$

Because $f(x)$ is a convex function, we thus have

$$f\left(\sum_i O_i P_i\right) \leq \sum_i f(O_i) P_i \tag{B.22}$$

proving (B.17).

APPENDIX C
SECOND QUANTIZATION FOR FERMIONS

This appendix deals with the so-called *second-quantization* techniques for many-fermion systems (Feynman 1972, Chapter 6, Blaizot and Ripka 1986, Chapter 1). Recall that in §2.1 we constructed a Hilbert space $\mathcal{H}_N$ of N particles, of which the symmetric subspace $\mathcal{H}_N^S$ describes boson systems and the antisymmetric subspace $\mathcal{H}_N^A$ describes fermion systems. It is often more convenient to work on an enlarged Hilbert space, the *Fock space* $\mathcal{F}$, which is the direct sum of the different N-particle Hilbert spaces for $N = 0, 1, 2, \ldots$. A general state in this Fock space is of the form

$$|\Psi\rangle = |\Psi_0\rangle + |\Psi_1\rangle + |\Psi_2\rangle + \cdots. \tag{C.1}$$

where $|\Psi_N\rangle$ (not necessarily normalized) is the component of $|\Psi\rangle$ in the N-particle space. In particular, $|\Psi_0\rangle$ is a state in the Hilbert space of no particles, the *vacuum*.

It is assumed that states of different numbers of particles are orthogonal to each other. The inner product of two states in $\mathcal{F}$ is then

$$\langle \Phi | \Psi \rangle = \langle \Phi_0 | \Psi_0 \rangle + \langle \Phi_1 | \Psi_1 \rangle + \langle \Phi_2 | \Psi_2 \rangle + \cdots. \tag{C.2}$$

Since we are concerned with fermion systems, only the antisymmetric subspace $\mathcal{F}^A$ of the Fock space is needed. Accordingly, each component $|\Psi_N\rangle$ of $|\Psi\rangle$ is an antisymmetric state in $\mathcal{H}_N^A$.

Within the Fock space, we are able to define operators that effect change of number of particles, the *creation operators* $\hat{a}_k^+$ and *destruction* (*annihilation*) operators $\hat{a}_k$. As will be shown, state vectors and other operators are expressible in terms of these. The creation operator $\hat{a}_k^+$ changes an N-particle state into an $(N + 1)$-particle state:

$$\hat{a}_k^+ |\alpha_1 \alpha_2 \cdots \alpha_N\rangle = |\alpha_k \alpha_1 \alpha_2 \cdots \alpha_N\rangle \tag{C.3}$$

where $|\alpha_1 \alpha_2 \cdots \alpha_N\rangle$ is an antisymmetric state describing N particles moving in N orbitals (see §2.1). If we let N be zero in (C.3), we create a single-particle state from the vacuum state $|0\rangle$,

$$\hat{a}_k^+ |0\rangle = |\alpha_k\rangle \tag{C.4}$$

In this way, any N-particle state can be built from $|0\rangle$,

$$|\alpha_1 \alpha_2 \cdots \alpha_N\rangle = \hat{a}_1^+ \hat{a}_2^+ \cdots \hat{a}_N^+ |0\rangle \tag{C.5}$$

Note that in this formula the creation operators are in a certain order.

In the bra space, the adjoint of an operator performs the corresponding transformation [see Equations (2.1.11) and (2.1.12)]; thus, with $\hat{a}_k$ as the adjoint of $\hat{a}_k^+$, we have

$$\langle \alpha_k | = \langle 0 | \hat{a}_k \tag{C.6}$$

and

$$\langle \alpha_1 \alpha_2 \cdots \alpha_N | = \langle 0 | \hat{a}_N \cdots \hat{a}_2 \hat{a}_1 \tag{C.7}$$

The requirement that the N-particle states in (C.5) and (C.7) be antisymmetric with respect to the permutation of any two particle indices imposes certain anticommutation relations on the creation and destruction operators; namely,

$$\{\hat{a}_\alpha^+, \hat{a}_\beta^+\} \equiv \hat{a}_\alpha^+ \hat{a}_\beta^+ + \hat{a}_\beta^+ \hat{a}_\alpha^+ = 0 \tag{C.8}$$

$$\{\hat{a}_\alpha, \hat{a}_\beta\} \equiv \hat{a}_\alpha \hat{a}_\beta + \hat{a}_\beta \hat{a}_\alpha = 0 \tag{C.9}$$

The special case $\beta = \alpha$ gives

$$\hat{a}_\alpha^+ \hat{a}_\alpha^+ = 0 \tag{C.10}$$

and

$$\hat{a}_\alpha \hat{a}_\alpha = 0 \tag{C.11}$$

which say that no orbital can be occupied by more than one fermion.

To ensure that the states in (C.5) and (C.7) are of the form (2.1.28), one further finds (For example, see Feynman 1972)

$$\hat{a}_\alpha |0\rangle = 0, \qquad \langle 0 | \hat{a}_\alpha^+ = 0 \tag{C.12}$$

$$\{\hat{a}_\alpha, \hat{a}_\beta^+\} \equiv \hat{a}_\alpha \hat{a}_\beta^+ + \hat{a}_\beta^+ \hat{a}_\alpha = \langle \alpha | \beta \rangle \tag{C.13}$$

The relations (C.8), (C.9), and (C.13) form the basis for the application of second-quantization techniques.

We can determine the effect of $\hat{a}_k$ on a general ket $|\alpha_1 \alpha_2 \cdots \alpha_N\rangle$ with the help of (C.12) and (C.13):

$$\hat{a}_k \hat{a}_1^+ \hat{a}_2^+ \cdots \hat{a}_N^+ |0\rangle = \{\hat{a}_k, \hat{a}_1^+\} \hat{a}_2^+ \cdots \hat{a}_N^+ |0\rangle - \hat{a}_1^+ \{\hat{a}_k, \hat{a}_2^+\} \hat{a}_3^+ \cdots \hat{a}_N^+ |0\rangle + \cdots$$
$$+ (-1)^{N-1} \hat{a}_1^+ \hat{a}_2^+ \cdots \{\hat{a}_k, \hat{a}_N^+\} |0\rangle$$
$$= \sum_{i=1}^{N} (-1)^{i-1} \langle \alpha_k | \alpha_i \rangle |\alpha_1 \cdots (\text{no } \alpha_i) \cdots \alpha_N\rangle \tag{C.14}$$

If the set $\{|\alpha_i\rangle\}$ is orthonormal, then this becomes

$$\hat{a}_k \hat{a}_1^+ \hat{a}_2^+ \cdots \hat{a}_N^+ |0\rangle = \begin{cases} (-1)^{k-1} |\alpha_1 \cdots (\text{no } \alpha_k) \cdots \alpha_N\rangle \\ \quad \text{if } \alpha_k \text{ is included in } |\alpha_1 \alpha_2 \cdots \alpha_N\rangle \\ 0, \quad \text{otherwise} \end{cases} \tag{C.15}$$

In these arguments there is no restriction to any one basis set. In particular, the creation and destruction operators associated with the coordinate representation are called *field operators*, denoted by $\hat{\psi}^+(\mathbf{x})$ and $\hat{\psi}(\mathbf{x})$, which create and destroy a one-particle state that is the eigenfunction of the coordinate operator $\hat{\mathbf{x}}$. In this representation, the commutation relations become

$$\{\hat{\psi}^+(\mathbf{x}), \hat{\psi}^+(\mathbf{x}')\} = 0, \qquad \{\hat{\psi}(\mathbf{x}), \hat{\psi}(\mathbf{x}')\} = 0 \tag{C.16}$$

$$\{\hat{\psi}(\mathbf{x}), \hat{\psi}^+(\mathbf{x}')\} = \langle \mathbf{x} | \mathbf{x}' \rangle = \delta(\mathbf{x} - \mathbf{x}') \tag{C.17}$$

The relation between creation and destruction operators in different representations can be obtained by noting, from (C.4) and (C.6), that $\hat{a}_k^+$ and $\hat{a}_k$ transform like $|\alpha_k\rangle$ and $\langle\alpha_k|$, respectively. For $|\alpha\rangle$ and $|\mathbf{x}\rangle$,

$$|\alpha\rangle = \int d\mathbf{x} |\mathbf{x}\rangle\langle\mathbf{x}|\alpha\rangle = \int d\mathbf{x}\, \psi_\alpha(\mathbf{x}) |\mathbf{x}\rangle \tag{C.18}$$

$$|\mathbf{x}\rangle = \sum_\alpha |\alpha\rangle\langle\alpha|\mathbf{x}\rangle = \sum_\alpha \psi_\alpha^*(\mathbf{x}) |\alpha\rangle \tag{C.19}$$

and so we have

$$\hat{a}_\alpha^+ = \int d\mathbf{x}\, \psi_\alpha(\mathbf{x}) \hat{\psi}^+(\mathbf{x}) \tag{C.20}$$

$$\hat{\psi}^+(\mathbf{x}) = \sum_\alpha \psi_\alpha^*(\mathbf{x}) \hat{a}_\alpha^+ \tag{C.21}$$

and their adjoints,

$$\hat{a}_\alpha = \int d\mathbf{x}\, \psi_\alpha^*(\mathbf{x}) \hat{\psi}(\mathbf{x}) \tag{C.22}$$

$$\hat{\psi}(\mathbf{x}) = \sum_\alpha \psi_\alpha(\mathbf{x}) \hat{a}_\alpha \tag{C.23}$$

In these formulas, $\psi_\alpha(\mathbf{x})$ is the orbital associated with the state $|\alpha\rangle$, in the coordinate representation.

We can express any symmetric operator in terms of creation and destruction operators. Let $\hat{A}(i)$ be an operator on the single-particle state,

$$\hat{A}(i) = \sum_{\alpha\beta} |\alpha\rangle\langle\alpha|\hat{A}|\beta\rangle\langle\beta| = \sum_{\alpha\beta} A_{\alpha\beta} |\alpha\rangle\langle\beta| \tag{C.24}$$

The simplest N-particle operator is of the form

$$\hat{A}_N = \hat{A}(1) + \hat{A}(2) + \cdots + \hat{A}(N) \tag{C.25}$$

where $\hat{A}(i)$ acts on the states of the ith particle:

$$\hat{A}(i)|\alpha_1\cdots\alpha_i\cdots\alpha_N\rangle = \sum_{\beta_i}\langle\beta_i|\hat{A}|\alpha_i\rangle|\alpha_1\cdots\beta_i\cdots\alpha_N\rangle$$

$$= \sum_{\beta_i} A_{\beta_i\alpha_i}|\alpha_1\cdots\beta_i\cdots\alpha_N\rangle \quad \text{(C.26)}$$

Thus

$$\hat{A}_N|\alpha_1\cdots\alpha_N\rangle = \sum_{i=1}^{N}\sum_{\beta_i} A_{\beta_i\alpha_i}|\alpha_1\cdots\beta_i\cdots\alpha_N\rangle \quad \text{(C.27)}$$

The operator $\hat{A}_N$ in the preceding formula can be written in terms of creation and destruction operators:

$$\hat{A} = \sum_{\alpha\beta} A_{\alpha\beta}\hat{a}_\alpha^+\hat{a}_\beta \quad \text{(C.28)}$$

To prove this, we first need the commutation relation

$$[\hat{A}, \hat{a}_\gamma^+] = \sum_{\alpha\beta} A_{\alpha\beta}[\hat{a}_\alpha^+\hat{a}_\beta, \hat{a}_\gamma^+]$$

$$= \sum_{\alpha\beta} A_{\alpha\beta}(\hat{a}_\alpha^+\{\hat{a}_\beta, \hat{a}_\gamma^+\} - \{\hat{a}_\alpha^+, \hat{a}_\gamma^+\}\hat{a}_\beta)$$

$$= \sum_{\alpha\beta} A_{\alpha\beta}\hat{a}^+\delta_{\beta\gamma} = \sum_{\alpha} A_{\alpha\gamma}\hat{a}_\alpha^+ \quad \text{(C.29)}$$

where we have used an operator identity $[\hat{A}\hat{B}, \hat{C}] = \hat{A}\{\hat{B}, \hat{C}\} - \{\hat{A}, \hat{C}\}\hat{B}$. Writing $|\alpha_i\cdots\alpha_N\rangle$ in terms of creation operators via (C.5), we have

$$\hat{A}\hat{a}_{\alpha_1}^+\cdots\hat{a}_{\alpha_N}^+|0\rangle = [\hat{A}, \hat{a}_{\alpha_1}^+]\hat{a}_{\alpha_2}^+\cdots\hat{a}_{\alpha_N}^+|0\rangle + \hat{a}_{\alpha_1}^+[\hat{A}, \hat{a}_{\alpha_2}^+]\hat{a}_{\alpha_3}^+\cdots\hat{a}_{\alpha_N}^+|0\rangle$$
$$+ \cdots + \cdots\hat{a}_{\alpha_1}^+\hat{a}_{\alpha_2}^+\cdots[\hat{A}, \hat{a}_{\alpha_N}^+]|0\rangle$$

$$= \sum_{i=1}^{N}\sum_{\beta_i} A_{\beta_i\alpha_i}\hat{a}_{\alpha_1}^+\cdots\hat{a}_{\beta_i}^+\cdots\hat{a}_{\alpha_N}^+|0\rangle \quad \text{(C.30)}$$

Thus the operator $\hat{A}$, in the *second-quantization form* (C.28), is equivalent to $\hat{A}_N$ when acting on an N-particle state. Note that the operator $\hat{A}$ does not have any N dependence. It is very convenient to have the same operator for a molecule and its ions.

Now we come to a symmetric operator $\hat{B}_n$ that is a sum of two-body operators,

$$\hat{B}_N = \sum_{i>j=1}^{N} \hat{B}(i, j) \quad \text{(C.31)}$$

where $\hat{B}(i, j)$ acts on the states of the ith and jth particles,

$$\hat{B}(i, j) |\alpha_1 \cdots \alpha_i \cdots \alpha_j \cdots \alpha_N\rangle$$
$$= \sum_{\beta_i \beta_j} (\beta_i \beta_j | \hat{B} | \alpha_i \alpha_j) |\alpha_1 \cdots \beta_i \cdots \beta_j \cdots \alpha_N\rangle \qquad (C.32)$$

Thus

$$\hat{B}_N |\alpha_1 \cdots \alpha_i \cdots \alpha_j \cdots \alpha_N\rangle$$
$$= \sum_{i>j=1}^{N} \sum_{\beta_i \beta_j} (\beta_i \beta_j | \hat{B} | \alpha_i \alpha_j) |\alpha_1 \cdots \beta_i \cdots \beta_j \cdots \alpha_N\rangle \qquad (C.33)$$

In a manner similar to the case of $\hat{A}_N$, the operator $\hat{B}_N$ can be written in terms of creation and destruction operators:

$$\hat{B} = \tfrac{1}{2} \sum_{\alpha\beta\gamma\delta} (\alpha\beta | \hat{B} | \gamma\delta) \hat{a}_\alpha^+ \hat{a}_\beta^+ \hat{a}_\delta \hat{a}_\gamma \qquad (C.34)$$

The proof of this statement is similar to that for the operator $\hat{A}_N$ and is omitted here (see, for example, Blaizot and Ripka 1986).

In case the one-particle states are continuous, the summations in (C.28) and (C.34) become integrations. Thus, in coordinate and spin representation,

$$\hat{A} = \int d\mathbf{x}_1 \, d\mathbf{x}_1' A(\mathbf{x}_1 \mathbf{x}_1') \hat{\psi}^+(\mathbf{x}_1) \hat{\psi}(\mathbf{x}_1') \qquad (C.35)$$

$$\hat{B} = \tfrac{1}{2} \int d\mathbf{x}_1 \, d\mathbf{x}_1' \, d\mathbf{x}_2 \, d\mathbf{x}_2' B(\mathbf{x}_1 \mathbf{x}_2, \mathbf{x}_1' \mathbf{x}_2') \hat{\psi}^+(\mathbf{x}_1) \hat{\psi}^+(\mathbf{x}_2) \hat{\psi}(\mathbf{x}_2') \hat{\psi}(\mathbf{x}_1') \qquad (C.36)$$

In particular, the kinetic-energy operator of (1.1.5) becomes

$$\hat{T} = \int d\mathbf{x}_1 \hat{\psi}^+(\mathbf{x}_1)(-\tfrac{1}{2}\nabla^2) \hat{\psi}(\mathbf{x}_1) \qquad (C.37)$$

the external potential-energy operator of (1.1.6) becomes

$$\hat{V} = \int d\mathbf{x}_1 \hat{\psi}^+(\mathbf{x}_1) \hat{\psi}(\mathbf{x}_1) v(\mathbf{x}_1) \qquad (C.38)$$

and the electron–electron interaction energy operator of (1.1.7) becomes

$$\hat{V}_{ee} = \frac{1}{2} \int d\mathbf{x}_1 \, d\mathbf{x}_2 \frac{1}{r_{12}} \hat{\psi}^+(\mathbf{x}_1) \hat{\psi}^+(\mathbf{x}_2) \hat{\psi}(\mathbf{x}_2) \hat{\psi}(\mathbf{x}_1) \qquad (C.39)$$

Finally, we derive the second-quantization expressions for the reduced density matrices. The expectation value of (C.35) in the state Ψ is

$$\langle \Psi | \hat{A} | \Psi \rangle = \int d\mathbf{x}_1 \, d\mathbf{x}_1' A(\mathbf{x}_1, \mathbf{x}_1') \langle \Psi | \hat{\psi}^+(\mathbf{x}_1) \hat{\psi}(\mathbf{x}_1') | \Psi \rangle \qquad (C.40)$$

which, in comparison with (2.3.22), gives

$$\gamma_1(\mathbf{x}_1', \mathbf{x}_1) = \langle \Psi | \hat{\psi}^+(\mathbf{x}_1) \hat{\psi}(\mathbf{x}_1') | \Psi \rangle \tag{C.41}$$

It follows that the particle (spin) density can be written as

$$\rho(\mathbf{x}_1) = \langle \Psi | \hat{\psi}^+(\mathbf{x}_1) \hat{\psi}(\mathbf{x}_1) | \Psi \rangle \tag{C.42}$$

Similarly, starting from (C.36), we obtain the second-order reduced density matrix for Ψ,

$$\gamma_2(\mathbf{x}_2'\mathbf{x}_1', \mathbf{x}_2\mathbf{x}_1) = \tfrac{1}{2} \langle \Psi | \hat{\psi}^+(\mathbf{x}_1) \hat{\psi}^+(\mathbf{x}_2) \hat{\psi}(\mathbf{x}_2') \hat{\psi}(\mathbf{x}_1') | \Psi \rangle \tag{C.43}$$

As a result of (C.42), the particle number operator as the integral of the diagonal of $\gamma_1(\mathbf{x}_1, \mathbf{x}_1')$ is

$$\hat{N} = \int \hat{\rho}(\mathbf{x}_1) \, d\mathbf{x}_1$$

$$= \int d\mathbf{x}_1 \hat{\psi}^+(\mathbf{x}_1) \hat{\psi}(\mathbf{x}_1) \tag{C.44}$$

If the basis set is discrete,

$$\hat{N} = \sum_i \hat{a}_i^+ \hat{a}_i = \sum_i \hat{N}_i \tag{C.45}$$

where

$$\hat{N}_i = \hat{a}_i^+ \hat{a}_i \tag{C.46}$$

is the occupation operator for the state $|\alpha_i\rangle$.

APPENDIX D
THE WIGNER DISTRIBUTION FUNCTION AND THE $\hbar$ SEMICLASSICAL EXPANSION

In contrast to the classical description of a state by a distribution function in phase space $(\mathbf{r}, \mathbf{p})$, a quantum state is normally described by a density operator $\hat{\Gamma}$. (The properties of $\hat{\Gamma}$ are discussed in Chapter 2 of the text.) A precise probability distribution in phase space is impossible for a quantum state because of the uncertainty principle; that is, one cannot talk about the probability of an electron being at a certain point in phase space, knowing that an electron cannot simultaneously have definite values of $\mathbf{r}$ and $\mathbf{p}$.

However, it is possible to represent $\hat{\Gamma}$ in the phase space $(\mathbf{r}, \mathbf{p})$ by some function, a quasiprobability distribution function, resembling the classical distribution function, such that the expectation value of a quantum operator $\hat{A}$ in the state described by $\hat{\Gamma}$ can be expressed in a form similar to that for a classical average. There are many such quantum distribution functions; for recent reviews see Hillery, O'Connell, Scully, and Wigner (1984) and Balazs and Jennings (1984). In this appendix, we focus on the *Wigner distribution function* (Wigner 1932). We derive the $\hbar$ expansion of the Wigner distribution function associated with the density operator $e^{-\beta \hat{H}}$, which has many applications in statistical mechanics and in density-functional theory.

The Wigner distribution associated with a density operator $\hat{\Gamma}$ is defined as the Fourier transform

$$f(\mathbf{r}, \mathbf{p}) = \frac{1}{(2\pi\hbar)^3} \int d\mathbf{s} \left\langle \mathbf{r} - \frac{\mathbf{s}}{2} \middle| \hat{\Gamma} \middle| \mathbf{r} + \frac{\mathbf{s}}{2} \right\rangle e^{i\mathbf{p}\cdot\mathbf{s}/\hbar} \tag{D.1}$$

where $\langle \mathbf{r} - (\mathbf{s}/2) | \hat{\Gamma} | \mathbf{r} + (\mathbf{s}/2) \rangle$ is the matrix representation of $\hat{\Gamma}$ in coordinate space. We restrict our discussion to one particle in three dimensions, but the generalization to many particles is straightforward (Hillery, O'Connell, Scully, and Wigner 1984). The Wigner distribution function can also be defined for reduced density matrices by putting into (D.1), in place of $\hat{\Gamma}$, either $\hat{\gamma}_1$ of (2.3.3) or $\hat{\Gamma}_1$ of (2.3.20). For a pure state, $\hat{\Gamma} = |\psi\rangle\langle\psi|$ and

$$f(\mathbf{r}, \mathbf{p}) = \frac{1}{(2\pi\hbar)^3} \int d\mathbf{s}\, \psi\left(\mathbf{r} - \frac{\mathbf{s}}{2}\right) \psi^*\left(\mathbf{r} + \frac{\mathbf{s}}{2}\right) e^{i\mathbf{p}\cdot\mathbf{s}/\hbar} \tag{D.2}$$

The inverse of (D.1) is

$$\langle \mathbf{r}| \hat{\Gamma} |\mathbf{r}'\rangle = \int d\mathbf{p}\, e^{-i\mathbf{p}\cdot(\mathbf{r}'-\mathbf{r})/\hbar} f(\tfrac{1}{2}(\mathbf{r}+\mathbf{r}'), \mathbf{p}) \tag{D.3}$$

The function $f(\mathbf{r}, \mathbf{p})$ cannot be taken as the probability of the system being found at $(\mathbf{r}, \mathbf{p})$ in phase space. [Note, for example, that f is not even everywhere positive (Cohen and Zaparovanny 1980).] Nevertheless, it has many of properties of a classical probability distribution. Among them we have

$$\int f(\mathbf{r}, \mathbf{p})\, d\mathbf{p} = \langle \mathbf{r}| \hat{\Gamma} |\mathbf{r}\rangle = \rho(\mathbf{r}) \tag{D.4}$$

$$\int f(\mathbf{r}, \mathbf{p})\, d\mathbf{r} = \langle \mathbf{p}| \hat{\Gamma} |\mathbf{p}\rangle = \rho_p(\mathbf{p}) \tag{D.5}$$

and

$$\int f(\mathbf{r}, \mathbf{p})\, d\mathbf{r}\, d\mathbf{p} = \operatorname{tr} \hat{\Gamma} \tag{D.6}$$

where $\rho(\mathbf{r})$ and $\rho_p(p)$ are the coordinate and momentum densities respectively. To prove (D.4), use the identity

$$\int d\mathbf{p}\, e^{i\mathbf{p}\cdot\mathbf{s}/\hbar} = (2\pi\hbar)^3 \delta(\mathbf{s}) \tag{D.7}$$

where $\delta(\mathbf{s})$ is the Dirac delta function. For (D.5), one needs the coordinate transformation from $\mathbf{s}$ and $\mathbf{r}$ to $\mathbf{r}+(\mathbf{s}/2)$ and $r-(\mathbf{s}/2)$, the Jacobian of which is 1. Equation (D.6) follows from either (D.4) or (D.5). According to (D.4) and (D.5), probability in coordinate space or momentum space can be obtained by integrating $f(\mathbf{r}, \mathbf{p})$ over $\mathbf{p}$ or $\mathbf{r}$.

Another property of $f(\mathbf{r}, \mathbf{p})$ is that the expectation value of an operator [see (1.1.12) in the text for a pure state] can be expressed in terms of $f(\mathbf{r}, \mathbf{p})$ in a manner that is essentially classical:

$$\operatorname{Tr}(\hat{\Gamma}\hat{A}) = \int f(\mathbf{r}, \mathbf{p}) A(\mathbf{r}, \mathbf{p})\, d\mathbf{r}\, d\mathbf{p} \tag{D.8}$$

where $A(\mathbf{r}, \mathbf{p})$ is the *Weyl classical function* corresponding to the operator $\hat{A}$,

$$A(\mathbf{r}, \mathbf{p}) = \int d\mathbf{s} \left\langle \mathbf{r} - \frac{\mathbf{s}}{2} \right| \hat{A} \left| \mathbf{r} + \frac{\mathbf{s}}{2} \right\rangle e^{i\mathbf{p}\cdot\mathbf{s}/\hbar} \tag{D.9}$$

the inverse of which is [compare (D.3)]

$$\langle \mathbf{r}| \hat{A} |\mathbf{r}'\rangle = \frac{1}{(2\pi\hbar)^3} \int d\mathbf{p}\, e^{-i\mathbf{p}\cdot(\mathbf{r}'-\mathbf{r})/\hbar} A(\tfrac{1}{2}(\mathbf{r}+\mathbf{r}'), \mathbf{p}) \tag{D.10}$$

For operators of the type $\hat{C} = A(\hat{\mathbf{r}}) + B(\hat{\mathbf{p}})$, it can readily be shown that the corresponding classical function is

$$C(\mathbf{r}, \mathbf{p}) = A(\mathbf{r}) + B(\mathbf{p}) \tag{D.11}$$

Thus, corresponding to the Hamiltonian $H = \hat{\mathbf{p}}^2/2m + v(\hat{\mathbf{r}})$,

$$H(\mathbf{r}, \mathbf{p}) = \frac{\mathbf{p}^2}{2m} + v(\mathbf{r}) \tag{D.12}$$

For operators involving products of $\hat{\mathbf{r}}$ and $\hat{\mathbf{p}}$, the corresponding classical functions are more complicated (see Hillery, O'Connell, Scully, and Wigner 1984), but can be derived following (D.9).

The difference between (D.1) and (D.9) indicates that the Wigner distribution function is $(2\pi\hbar)^{-3}$ times the Weyl classical function corresponding to the density operator $\hat{\Gamma}$.

We now prove (D.8). For this we need the phase-space representation of an operator $\hat{C} = \hat{A}\hat{B}$ in terms of the representations of $\hat{A}$ and $\hat{B}$. From (D.9), we have

$$C(\mathbf{r}, \mathbf{p}) = \int d\mathbf{s}\, e^{i\mathbf{p}\cdot\mathbf{s}/\hbar} \left\langle \mathbf{r} - \frac{\mathbf{s}}{2} \middle| \hat{A}\hat{B} \middle| \mathbf{r} + \frac{\mathbf{s}}{2} \right\rangle$$

$$= \iint d\mathbf{s}\, d\mathbf{r}'\, e^{i\mathbf{p}\cdot\mathbf{s}/\hbar} \left\langle \mathbf{r} - \frac{\mathbf{s}}{2} \middle| \hat{A} \middle| \mathbf{r}' \right\rangle \left\langle \mathbf{r}' \middle| \hat{B} \middle| \mathbf{r} + \frac{\mathbf{s}}{2} \right\rangle \tag{D.13}$$

where we have inserted an identity operator $\int d\mathbf{r}' |\mathbf{r}'\rangle\langle\mathbf{r}'| = \hat{I}$. The matrices of $\hat{A}$ and $\hat{B}$ can now be expressed in terms of their classical functions; namely, following (D.10),

$$\left\langle \mathbf{r} - \frac{\mathbf{s}}{2} \middle| \hat{A} \middle| \mathbf{r}' \right\rangle = \frac{1}{(2\pi\hbar)^3} \int d\mathbf{p}_1\, e^{+i\mathbf{p}_1 \cdot [\mathbf{r}-(\mathbf{s}/2)-\mathbf{r}']/\hbar} A\left(\frac{1}{2}\left(\mathbf{r} - \frac{\mathbf{s}}{2} + \mathbf{r}'\right), \mathbf{p}_1\right) \tag{D.14}$$

and

$$\left\langle \mathbf{r}' \middle| \hat{B} \middle| \mathbf{r} + \frac{\mathbf{s}}{2} \right\rangle = \frac{1}{(2\pi\hbar)^3} \int d\mathbf{p}_2\, e^{+i\mathbf{p}_2 \cdot [\mathbf{r}'-\mathbf{r}-(\mathbf{s}/2)]/\hbar} B\left(\frac{1}{2}\left(\mathbf{r}' + \mathbf{r} + \frac{\mathbf{s}}{2}\right), \mathbf{p}_2\right) \tag{D.15}$$

Inserting (D.14) and (D.15) into (D.13), we find

$$C(\mathbf{r}, \mathbf{p}) = \iiiint d\mathbf{s}\, d\mathbf{r}'\, d\mathbf{p}_1\, d\mathbf{p}_2 \left(\frac{1}{2\pi\hbar}\right)^6 A\left(\frac{1}{2}\left(\mathbf{r} - \frac{\mathbf{s}}{2} + \mathbf{r}'\right), \mathbf{p}_1\right)$$
$$\times B\left(\frac{1}{2}\left(\mathbf{r}' + \mathbf{r} + \frac{\mathbf{s}}{2}\right), \mathbf{p}_2\right) \exp\frac{i}{\hbar}\left[\mathbf{p}\cdot\mathbf{s} + \mathbf{p}_1 \cdot \left(\mathbf{r} - \frac{\mathbf{s}}{2} - \mathbf{r}'\right)\right.$$
$$\left. + \mathbf{p}_2 \cdot \left(\mathbf{r}' - \mathbf{r} - \frac{\mathbf{s}}{2}\right)\right] \tag{D.16}$$

Thus

$$\iint d\mathbf{r}\, d\mathbf{p}\, C(\mathbf{r}, \mathbf{p}) = \iiiint d\mathbf{r}\, d\mathbf{r}'\, d\mathbf{p}_1\, d\mathbf{p}_2 \left(\frac{1}{2\pi\hbar}\right)^3$$
$$\times A\left(\frac{1}{2}(\mathbf{r}+\mathbf{r}'), \mathbf{p}_1\right) B\left(\frac{1}{2}(\mathbf{r}'+\mathbf{r}), \mathbf{p}_2\right)$$
$$\times \exp\left\{+\frac{i}{\hbar}(\mathbf{p}_1-\mathbf{p}_2)\cdot(\mathbf{r}-\mathbf{r}')\right\}$$
$$= \iint d\mathbf{r}_1\, d\mathbf{p}_1\, A(\mathbf{r}_1, \mathbf{p}_1) B(\mathbf{r}_1, \mathbf{p}_1) \quad (D.17)$$

where the first equality follows the integration over $\mathbf{p}$ and $\mathbf{s}$, and the second equality follows the transformation from $\mathbf{r}$ and $\mathbf{r}'$ to $1/2(\mathbf{r}+\mathbf{r}')$ and $\mathbf{r}-\mathbf{r}'$ and integration over $\mathbf{r}-\mathbf{r}'$, using (D.7). Using (D.1), and noting the difference in the factor $(1/2\pi\hbar)^3$, we obtain

$$\text{Tr}\, \hat{C} = (2\pi\hbar)^{-3} \iint d\mathbf{r}\, d\mathbf{p}\, C(\mathbf{r}, \mathbf{p})$$
$$= (2\pi\hbar)^{-3} \iint d\mathbf{r}\, d\mathbf{p}\, A(\mathbf{r}, \mathbf{p}) B(\mathbf{r}, \mathbf{p}) \quad (D.18)$$

proving (D.8).

As shown in (D.16), the classical function of the product of two operators is an involved nonlocal functional of the corresponding classical functions. To further simplify (D.16), assume that $A(\mathbf{r}, \mathbf{p})$ and $B(\mathbf{r}, \mathbf{p})$ have convergent Taylor expansions anywhere in phase space, and use the identity for convergent Taylor expansions,

$$f(x+y) = \sum_{n=0}^{\infty} \frac{1}{n!} y^n \left(\frac{\partial}{\partial x}\right)^n f(x) \equiv e^{y(\partial/\partial x)} f(x) \quad (D.19)$$

Then

$$A\left(\frac{1}{2}\left(\mathbf{r}-\frac{\mathbf{s}}{2}+\mathbf{r}'\right), \mathbf{p}_1\right) = \exp\left[-\frac{1}{2}\left(\mathbf{r}-\mathbf{r}'+\frac{\mathbf{s}}{2}\right)\cdot\nabla_\mathbf{r}^A\right] \exp\left[(\mathbf{p}_1-\mathbf{p})\cdot\nabla_\mathbf{p}^A\right] A(\mathbf{r}, \mathbf{p})$$
$$(D.20)$$

and

$$B\left(\frac{1}{2}\left(\mathbf{r}'+\mathbf{r}+\frac{\mathbf{s}}{2}\right), \mathbf{p}_2\right) = \exp\left[-\frac{1}{2}\left(\mathbf{r}-\mathbf{r}'-\frac{\mathbf{s}}{2}\right)\cdot\nabla_\mathbf{r}^B\right] \exp\left[(\mathbf{p}_2-\mathbf{p})\cdot\nabla_\mathbf{p}^B\right] B(\mathbf{r}, \mathbf{p})$$
$$(D.21)$$

where, for example, $\nabla_\mathbf{r}^A$ is the gradient operator acting on the coordinates of $A(\mathbf{r}, \mathbf{p})$. Inserting (D.20) and (D.21) into (D.16), one finds

$$C(\mathbf{r}, \mathbf{p}) = \iiiint d\mathbf{s}\, d\mathbf{r}'\, d\mathbf{p}_1\, d\mathbf{p}_2 \left(\frac{1}{2\pi\hbar}\right)^6$$

$$\times \exp\frac{i}{\hbar}\left[\mathbf{p}\cdot\left(\mathbf{s} - \frac{\hbar}{i}\nabla_\mathbf{p}^A - \frac{\hbar}{i}\nabla_\mathbf{p}^B\right) + \mathbf{p}_1\cdot\left(\mathbf{r} - \frac{\mathbf{s}}{2} - \mathbf{r}' + \frac{\hbar}{i}\nabla_\mathbf{p}^A\right)\right.$$

$$+ \mathbf{p}_2\cdot\left(\mathbf{r}' - \mathbf{r} - \frac{\mathbf{s}}{2} + \frac{\hbar}{i}\nabla_\mathbf{p}^B\right) - \frac{1}{2}\left(\mathbf{r} - \mathbf{r}' + \frac{\mathbf{s}}{2}\right)\cdot\frac{\hbar}{i}\nabla_\mathbf{r}^A$$

$$\left. - \frac{1}{2}\left(\mathbf{r} - \mathbf{r}' - \frac{\mathbf{s}}{2}\right)\cdot\frac{\hbar}{i}\nabla_\mathbf{r}^B\right] A(\mathbf{r}, \mathbf{p}) B(\mathbf{r}, \mathbf{p}) \quad \text{(D.22)}$$

Each of the four integrations now can be carried out using (D.7). Finally,

$$C(\mathbf{r}, \mathbf{p}) = \exp\left[\frac{i\hbar}{2}(\nabla_\mathbf{p}^B\cdot\nabla_\mathbf{r}^A - \nabla_\mathbf{r}^B\cdot\nabla_\mathbf{p}^A)\right] A(\mathbf{r}, \mathbf{p}) B(\mathbf{r}, \mathbf{p}) \quad \text{(D.23)}$$

This is the fundamental relation giving the classical function for the product of two operators. If one sets $\hbar = 0$ in (D.23), then $C(\mathbf{r}, \mathbf{p}) = A(\mathbf{r}, \mathbf{p})B(\mathbf{r}, \mathbf{p})$ in agreement with classical mechanics. The power of (D.22) lies in the possibility it affords to expand the exponential in power of $\hbar$ to achieve quantum corrections to classical mechanics. This is demonstrated below for the Boltzmann density matrix.

The equilibrium state of a quantum system in the canonical ensemble is completely described by the Boltzmann density matrix (unnormalized)

$$\hat{G} = e^{-\beta\hat{H}} \quad \text{(D.24)}$$

where $\beta = 1/k\theta$, k is the Boltzmann constant, and θ is the temperature. The quantum partition function is just the trace of $\hat{\rho}$. It is easily shown by simple differentiation that $\hat{G}$ satisfies the Bloch equation,

$$\frac{\partial\hat{G}}{\partial\beta} = -\hat{H}\hat{G} = -\hat{G}\hat{H} \quad \text{(D.25)}$$

subject to the initial condition $\hat{G}(\beta = 0) = \hat{I}$, where $\hat{I}$ is the identity operator. Using the product rule (D.23), we can represent the Bloch equation in phase space:

$$\frac{\partial G(\mathbf{r}, \mathbf{p})}{\partial\beta} = -\exp\left[\frac{i\hbar}{2}(\nabla_\mathbf{p}^G\cdot\nabla_\mathbf{r}^H - \nabla_\mathbf{r}^G\cdot\nabla_\mathbf{p}^H)\right] H(\mathbf{r}, \mathbf{p}) G(\mathbf{r}, \mathbf{p})$$

$$= -\exp\left[\frac{i\hbar}{2}(\nabla_\mathbf{p}^H\cdot\nabla_\mathbf{r}^G - \nabla_\mathbf{r}^H\cdot\nabla_\mathbf{p}^G)\right] G(\mathbf{r}, \mathbf{p}) H(\mathbf{r}, \mathbf{p}) \quad \text{(D.26)}$$

Taking the average of these two expressions, we obtain

$$\frac{\partial G(\mathbf{r},\mathbf{p})}{\partial \beta} = -\cos\left[\frac{\hbar}{2}(\nabla^G_\mathbf{p} \cdot \nabla^H_\mathbf{r} - \nabla^G_\mathbf{r} \cdot \nabla^H_\mathbf{p})\right] H(\mathbf{r},\mathbf{p}) G(\mathbf{r},\mathbf{p}) \quad (D.27)$$

This is called the Wigner translation of the Bloch equation (Hillery, O'Connell, Scully, and Wigner 1984). To obtain the $\hbar$ semiclassical expansion, expand the cosine function

$$\cos x = 1 - \frac{x^2}{2!} + \frac{x^4}{4!} - \cdots \quad (D.28)$$

and use (D.12) for $H(\mathbf{r},\mathbf{p})$. To the order of $\hbar^2$, there results

$$\frac{\partial G(\mathbf{r},\mathbf{p})}{\partial \beta} = -\left[1 - \frac{\hbar^2}{8}(\nabla^G_\mathbf{p} \cdot \nabla^H_\mathbf{r} - \nabla^G_\mathbf{r} \cdot \nabla^H_\mathbf{p})^2\right] H(\mathbf{r},\mathbf{p}) G(\mathbf{r},\mathbf{p})$$

$$= \left[H(\mathbf{r},\mathbf{p}) - \frac{\hbar^2}{8}\sum_{ij}\frac{\partial^2}{\partial r_i \partial r_j}v(\mathbf{r})\frac{\partial^2}{\partial p_i \partial p_j} - \frac{\hbar^2}{8m}\sum_i \frac{\partial^2}{\partial r_i^2}\right] G(\mathbf{r},\mathbf{p})$$

$$(D.29)$$

Let

$$G(\mathbf{r},\mathbf{p}) = G_0(\mathbf{r},\mathbf{p})(1 + \hbar X_1(\mathbf{r},\mathbf{p}) + \hbar^2 X_2(\mathbf{r},\mathbf{p}))$$

in (D.29) and equate the two sides of the equation for different order of $\hbar$. Using the initial condition $G(\mathbf{r},\mathbf{p}) = 1$ for $\beta = 0$, we get

$$G(\mathbf{r},\mathbf{p}) = e^{-\beta H(\mathbf{r},\mathbf{p})}\left[1 + \hbar^2\left(-\frac{\beta^2}{8m}\sum_i \frac{\partial^2}{\partial r_i^2}v(\mathbf{r}) + \frac{\beta^3}{24m}\sum_i \left(\frac{\partial}{\partial r_i}v(\mathbf{r})\right)^2\right.\right.$$

$$\left.\left. + \frac{\beta^3}{24m^2}\sum_{ij}\frac{\partial^2}{\partial r_i \partial r_j}v(\mathbf{r}) p_i p_j\right)\right] \quad (D.30)$$

This is the famous Wigner expansion (Wigner 1932). The higher orders of the expansion can be derived similarly. This formula is useful for finding the quantum corrections to classical mechanics. In this book, it is applied to calculation of the gradient corrections to the Thomas–Fermi local kinetic energy functional (see §6.7).

For discussion of other possible phase-space distributions, see Cohen (1966) and Cohen and Zaparovanny (1980).

APPENDIX E
THE UNIFORM ELECTRON GAS

Consider a system comprising a large number N of electrons moving in a cubical box of volume $V = l^3$, throughout which there is uniformly spread out a positive charge sufficient to make the system neutral. A *uniform electron gas* is such a system in the limit $N \to \infty$, $V \to \infty$, with the density, $\rho = N/V$, remaining finite.

We are interested in the total ground-state energy of a uniform electron gas including the positive background; namely,

$$E[\rho] = T_s[\rho] + \int \rho(\mathbf{r}) v(\mathbf{r}) \, d\mathbf{r} + J[\rho] + E_{xc}[\rho] + E_b \tag{E.1}$$

where the external potential, due to the positive charge density $n(\mathbf{r})$, is

$$v(\mathbf{r}) = -\int \frac{n(\mathbf{r})}{|\mathbf{r} - \mathbf{r}'|} \, d\mathbf{r}' \tag{E.2}$$

and E_b is the electrostatic energy of the positive background,

$$E_b = \frac{1}{2} \iint \frac{n(\mathbf{r}) n(\mathbf{r}')}{|\mathbf{r} - \mathbf{r}'|} \, d\mathbf{r} \, d\mathbf{r}' \tag{E.3}$$

Since the system is neutral everywhere, $\rho(\mathbf{r}) = n(\mathbf{r})$, we have

$$\int \rho(\mathbf{r}) v(\mathbf{r}) \, d\mathbf{r} + J[\rho] + E_b = 0 \tag{E.4}$$

Therefore

$$\begin{aligned} E[\rho] &= T_s[\rho] + E_{xc}[\rho] \\ &= T_s[\rho] + E_x[\rho] + E_c[\rho] \\ &= \int \rho \varepsilon_t[\rho] \, d\mathbf{r} + \int \rho \varepsilon_x[\rho] \, d\mathbf{r} + \int \rho \varepsilon_c[\rho] \, d\mathbf{r} \end{aligned} \tag{E.5}$$

The Kohn–Sham equations (7.2.7)–(7.2.9), under the condition that $\rho(\mathbf{r}) = N/V$, a constant, are satisfied by the plane waves

$$\psi_k(\mathbf{r}) = \frac{1}{V^{1/2}} e^{i\mathbf{k} \cdot \mathbf{r}} \tag{E.6}$$

where the periodic boundary conditions require

$$k_x = \frac{2\pi}{l} n_x, \quad k_y = \frac{2\pi}{l} n_y, \quad k_z = \frac{2\pi}{l} n_z \quad n_x, n_y, n_z = 0, \pm 1, \pm 2, \ldots$$
(E.7)

In §6.1 we derived the kinetic energy and exchange energy of a system described by such a set of orbitals, each doubly occupied. The results are (6.1.19) and (6.1.20);

$$T_s^0[\rho] = C_F \int \rho^{5/3} \, d\mathbf{r}$$
(E.8)

or

$$\varepsilon_t^0(\rho) = C_F \rho^{2/3} = \frac{1.1049}{r_s^2}$$
(E.9)

and

$$E_x^0[\rho] = -C_x \int \rho^{4/3} \, d\mathbf{r}$$
(E.10)

or

$$\varepsilon_x^0[\rho] = -C_x \rho^{1/3} = -\frac{0.4582}{r_s}$$
(E.11)

where the superscript 0 denotes that all electrons are paired (zero spin polarization). The quantity r_s is the radius of a sphere whose volume is the effective volume of an electron,

$$\tfrac{4}{3}\pi r_s^3 = \frac{1}{\rho}$$
(E.12)

Therefore, the total energy for the spin-compensated case is

$$E^0[\rho] = \int \rho \varepsilon^0(\rho) \, d\mathbf{r}$$
(E.13)

where

$$\varepsilon^0(\rho) = \frac{1.1049}{r_s^2} - \frac{0.4582}{r_s} + \varepsilon_c^0(r_s)$$
(E.14)

We discuss the correlation contribution ε_c^0 later.

We note that large r_s means low density, small r_s means high density: r_s at the nucleus of an hydrogen-like atom of nuclear charge Z is $0.91/Z$; at the first Bohr radius it is $1.77/Z$. Values of r_s associated with the interstitial charge densities in metals range from a shade less than 2 to about 5 (see p. 26 of Moruzzi, Janak, and Williams 1978). Figure E.1

E THE UNIFORM ELECTRON GAS

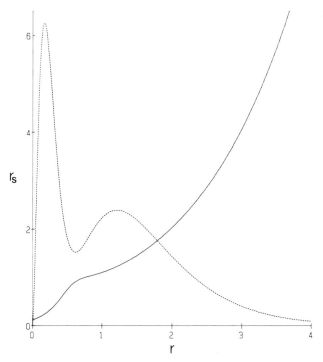

Figure E.1 The radius r_s for the ground state of the carbon atom as a function of distance from the nucleus. Shown as a dotted line is the radial distribution function $D(r) = 4\pi r^2 \rho(r)$.

shows the radius r_s for the carbon atom as a function of distance from the nucleus.

In the case of a spin-polarized system where there is nonvanishing spin polarization,

$$\zeta = \frac{\rho^\alpha - \rho^\beta}{\rho^\alpha + \rho^\beta} \neq 0$$

the kinetic and exchange energy are also known: (8.2.9) and (8.2.18). From (8.2.19), the exchange energy is

$$E_x[\rho^\alpha, \rho^\beta] = \int \rho \varepsilon_x(\rho, \xi) \, d\mathbf{r} \tag{E.15}$$

$$\varepsilon_x(\rho, \xi) = \varepsilon_x^0(\rho) + [\varepsilon_x^1(\rho) - \varepsilon_x^0(\rho)] f(\zeta) \tag{E.16}$$

and

$$f(\zeta) = \tfrac{1}{2}(2^{1/3} - 1)^{-1}\{(1 + \zeta)^{4/3} + (1 - \zeta)^{4/3} - 2\} \qquad (E.17)$$

Here the superscript 1 indicates the completely spin-polarized "ferromagnetic" case, $\varepsilon_x^1(\rho) = \varepsilon_x(\rho, 1)$. The corresponding correlation energy formula is

$$\varepsilon_c[\rho^\alpha, \rho^\beta] = \int \rho \varepsilon_c(\rho, \zeta) \, d\mathbf{r} \qquad (E.18)$$

but there is no simple exact formula like (E.16) relating ε_c to the limiting spin-compensated and spin-polarized correlation energies per particle, ε_c^0 and ε_c^1.

Even ε_c^0 is a very difficult problem; only limiting cases are known in analytic form. The high-density limit is (Gell-Mann and Brueckner 1957; Carr and Maradudin 1964)

$$\varepsilon_c^0 = 0.0311 \ln r_s - 0.048 + r_s(A^0 \ln r_s + C^0), \qquad r_s \ll 1 \qquad (E.19)$$

and the low-density limit has the form (Carr 1961; Nozieres and Pines 1966)

$$\varepsilon_c^0 = \frac{1}{2} \left(\frac{g_0}{r_s} + \frac{g_1}{r_s^{3/2}} + \frac{g_2}{r_s^2} + \cdots \right), \qquad r_s \gg 1 \qquad (E.20)$$

Similar formulae are available for ε_c^1.

From a random-phase-type analysis, von Barth and Hedin (1972) proposed

$$\varepsilon_c^{BH}(\rho, \zeta) = \varepsilon_c^0(r_s) + [\varepsilon_c^1(r_s) - \varepsilon_c^0(r_s)] f(\zeta) \qquad (E.21)$$

which has precisely the same form as the exchange energy ε_x in (E.16). They further suggested use of analytic expressions for ε_c^0 and ε_c^1 due to Hedin and Lundqvist (1971),

$$\varepsilon_c^{0,HL} = -\frac{1}{2} c_0 F\left(\frac{r_s}{r_0}\right), \qquad \varepsilon_c^{1,HL} = -\frac{1}{2} c_1 F\left(\frac{r_s}{r_1}\right) \qquad (E.22)$$

where

$$F(Z) = (1 + Z^3) \ln\left(1 + \frac{1}{Z}\right) + \frac{Z}{2} - Z^2 - \frac{1}{3} \qquad (E.23)$$

and

$$c_0 = 0.0504, \qquad c_1 = 0.0254$$
$$r_0 = 30, \qquad r_1 = 75 \qquad (E.24)$$

The factor $\tfrac{1}{2}$ appears in (E.20) because atomic units are used here;

Rydbergs were used as energy unit in the original paper. Unfortunately, (E.21) and (E.22) do not reproduce the known high- and low-density limits of the uniform-gas correlation energy, that is (E.19) and (E.20) (Williams and von Barth 1983). Nevertheless, ε_c^{BH} has found extensive applications (for example, Moruzzi, Janak, and Williams 1978). Gunnarsson and Lundqvist (1976) suggested calculating ε_c through (E.21)–(E.23), but with coefficients different from those in (E.24).

Vosko, Wilk, and Nusair (1980) made a careful analysis of the random phase approximation and proposed the more accurate description

$$\varepsilon_c^{VWN}(r_s, \zeta) = \varepsilon_c^0(r_s) + \alpha(r_s)\left[\frac{f(\zeta)}{f''(0)}\right][1 + \beta(r_s)\zeta^4] \quad \text{(E.25)}$$

where $\alpha(r_s)$ is the spin stiffness, which is related to spin susceptibilities, and $\beta(r_s)$ is chosen to satisfy $\varepsilon_c(r_s, 1) = \varepsilon_c^1(r_s)$, namely,

$$1 + \beta(r_s) = f''(0)\frac{\varepsilon_c^1(r_s) - \varepsilon_c^0(r_s)}{\alpha(r_s)}. \quad \text{(E.26)}$$

Tables of $\alpha(r_s)$ and $\beta(r_s)$ may be found in the original paper. Another advance in the work of Vosko, Wilk, and Nusair is that Padé-approximant interpolations were made of accurate numerical calculations of uniform electron gas ε_c^0 and ε_c^1 that had been made by Ceperley and Alder. Ceperley and Alder (1980) had calculated the total energy for the uniform electron gas in spin-compensated and ferromagnetic states for several different values of r_s, using the quantum Monte Carlo method. The correlation energy was obtained by subtracting the corresponding kinetic and exchange energies from the total energy (E.5). The VWN interpolations for ε_c^0 and ε_c^1 also incorporate the high- and low-density limits. The final form for both ε_c^0 and ε_c^1 is

$$\varepsilon_c(r_s) = \frac{A}{2}\left\{\ln\frac{x}{X(x)} + \frac{2b}{Q}\tan^{-1}\frac{Q}{2x+b}\right.$$
$$\left. - \frac{bx_0}{X(x_0)}\left[\ln\frac{(x-x_0)^2}{X(x)} + \frac{2(b+2x_0)}{Q}\tan^{-1}\frac{Q}{2x+b}\right]\right\} \quad \text{(E.27)}$$

where $x = r_s^{1/2}$, $X(x) = x^2 + bx + c$, and $Q = (4c - b^2)^{1/2}$. For $\varepsilon_c^0(r_s)$, $A = 0.0621814$, $x_0 = -0.409286$, $b = 13.0720$, and $c = 42.7198$; for $\varepsilon_c^1(r_s)$, $A = \frac{1}{2}(0.0621814)$, $x_0 = -0.743294$, $b = 20.1231$, and $c = 101.578$. These formulas are generally accepted as the most accurate available for the uniform-gas correlation energy per particle.

Inserting (E.27) into (E.18) and adding (E.15), one gets the exchange-correlation energy for the uniform gas as a functional of the density; when applied to a nonuniform system, this defines the local-spin-density approximation (LSD) of §8.2 of the text.

APPENDIX F

TABLES OF VALUES OF ELECTRONEGATIVITIES AND HARDNESSES

We give here tables of values of absolute electronegativities and absolute hardnesses of many species, as compiled by Pearson (1988).

Electronegativity and hardness are calculated from ionization potentials and electron affinities, the values of these being taken from experimental data where possible; sources may be found in Pearson (1988). The defining formulas are

$$\chi = \frac{I+A}{2}, \qquad \eta = \frac{I-A}{2} \qquad (F.1)$$

where I and A are, respectively, ground-state ionization potential and electron affinity. It is important to note, as has been emphasized in the present text, that the values of these "absolute" quantities have no necessary relation to values of electronegativity or hardness defined in other ways, though there is a good overall general correlation between these electronegativity values and many other electronegativity scales in the literature.

To recover the original I and A values from these tables, one of course uses

$$I = \chi + \eta, \qquad A = \chi - \eta \qquad (F.2)$$

In a number of cases—all of those for which $\eta > \chi$, negative values of A are found. Strictly speaking, this departs from the "ground state" requirement, since any species S with negative electron affinity is unstable relative to dissociation into S^+ and an electron at affinity. Nevertheless, such species are important to chemistry, and the unstable states they represent can be thought of as ground states for some purposes.

The chemical potential is the negative of the electronegativity,

$$\mu = -\chi = -\frac{I+A}{2} \qquad (F.3)$$

This is the key density-functional quantity discussed at great length in the text: the escaping tendency of the electron cloud.

For a recent review of the many extant electronegativity scales, see Mullay (1987). For alternative scales of hardness values, see Orsky and

Whitehead (1987), and Vinayagam and Sen (1988). Lackner and Zweig (1983) give both electronegativities and hardnesses for the whole periodic table.

Table F.1 contains χ and η values for atoms. Table F.2 has values for monoatomic cations, Table F.3 for radicals, and Table F.4 for molecules. Pearson (1988) argues that the values in Table F.3 are also appropriate for corresponding negative ions.

Table F.1 Absolute Electronegativities and Absolute Hardness for Atoms[a]

Atom	χ (eV)	η (eV)	Atom	χ (eV)	η (eV)
H	7.18	6.43	Rb	2.34	1.85
Li	3.01	2.39	Sr	2.0	3.7
Be	4.9	4.5	Y	3.19	3.19
B	4.29	4.01	Zr	3.64	3.21
C	6.27	5.00	Nb	4.0	3.0
N	7.30	7.23	Mo	3.9	3.1
O	7.54	6.08	Ru	4.5	3.0
F	10.41	7.01	Rh	4.30	3.16
Na	2.85	2.30	Pd	4.45	3.89
Mg	3.75	3.90	Ag	4.44	3.14
Al	3.23	2.77	Cd	4.33	4.66
Si	4.77	3.38	In	3.1	2.8
P	5.62	4.88	Sn	4.30	3.05
S	6.22	4.14	Sb	4.85	3.80
Cl	8.30	4.68	Te	5.49	3.52
K	2.42	1.92	I	6.76	3.69
Ca	2.2	4.0	Cs	2.18	1.71
Sc	3.34	3.20	Ba	2.4	2.9
Ti	3.45	3.37	La	3.1	2.6
V	3.6	3.1	Hf	3.8	3.0
Cr	3.72	3.06	Ta	4.11	3.79
Mn	3.72	3.72	W	4.40	3.58
Fe	4.06	3.81	Re	4.02	3.87
Co	4.3	3.6	Os	4.9	3.8
Ni	4.40	3.25	Ir	5.4	3.8
Cu	4.48	3.25	Pt	5.6	3.5
Zn	4.45	4.94	Au	5.77	3.46
Ga	3.2	2.9	Hg	4.91	5.54
Ge	4.6	3.4	Tl	3.2	2.9
As	5.3	4.5	Pb	3.90	5.50
Se	5.89	3.87	Bi	4.69	3.74
Br	7.59	4.22			

[a] From Pearson (1988).

Table F.2 Absolute Electronegativities and Absolute Hardnesses for Monoatomic Cations[a]

Ion	χ (eV)	η (eV)	Ion	χ (eV)	η (eV)
Li^+	40.52	35.12	Ag^{2+}	28.2	6.7
Na^+	26.21	21.08	Cd^{2+}	27.20	10.29
K^+	17.99	13.64	Sn^{2+}	22.57	7.49
Rb^+	15.77	11.55	Hf^{2+}	19.1	4.2
Cs^+	14.5	10.6	W^{2+}	20.9	4.5
Cu^+	14.01	6.28	Os^{2+}	22.3	5.7
Ag^+	14.53	6.96	Pt^{2+}	27.2	8.0
Au^+	14.9	5.6	Hg^{2+}	26.5	7.7
Tl^+	13.27	7.16	Pb^{2+}	23.49	8.46
Co^+	12.46	4.60	B^{3+}	148.65	110.72
Rh^+	12.77	5.31	Al^{3+}	74.22	45.77
Ir^+	13.0	3.9	Ga^{3+}	47	17
Cl^+	18.39	5.42	In^{3+}	41	13
Br^+	16.8	5.0	Tl^{3+}	40.3	10.4
I^+	14.79	4.34	Sc^{3+}	49.11	24.36
Be^{2+}	86.05	67.84	Y^{3+}	41.2	20.6
Mg^{2+}	47.59	32.55	La^{3+}	34.57	15.39
Ca^{2+}	31.39	19.52	Ce^{3+}	28.48	8.28
Sr^{2+}	27.3	16.3	Lu^{3+}	33.08	12.12
Sc^{2+}	18.78	5.98	Ti^{3+}	35.38	7.89
Ti^{2+}	20.54	6.96	V^{3+}	38.01	8.70
V^{2+}	21.98	7.33	Cr^{3+}	40.0	9.1
Cr^{2+}	23.73	7.23	Mn^{3+}	42.4	8.8
Mn^{2+}	24.66	9.02	Fe^{3+}	42.73	12.08
Fe^{2+}	23.42	7.24	Co^{3+}	42.4	8.9
Co^{2+}	25.28	8.22	Ni^{3+}	45.0	9.9
Ni^{2+}	26.67	8.50	Zr^{3+}	28.65	5.68
Cu^{2+}	28.56	8.27	Nb^{3+}	31.7	6.6
Zn^{2+}	28.84	10.88	Mo^{3+}	36.8	9.6
Ge^{2+}	25.08	9.15	Ru^{3+}	39.2	10.7
Y^2	16.38	4.14	Rh^{3+}	42.2	11.2
Zr^{2+}	18.06	4.93	Hf^{3+}	28.3	5.0
Nb^{2+}	19.68	5.36	W^{3+}	32.4	7.0
Mo^{2+}	21.60	5.51	Re^{3+}	33.7	7.8
Ru^{2+}	22.62	5.86	Os^{3+}	35.2	7.5
Rh^{2+}	24.57	6.49	Ir^{3+}	37.4	7.9
Pd^{2+}	26.18	6.75	Au^{3+}	45.8	8.4

[a] From Pearson (1988).

Table F.3 Absolute Electronegativities and Absolute Hardnesses for Radicals[a,b]

Species	χ	η	Species	χ	η
F	10.41	7.01	HO_2	6.36	5.17
OH	7.50	5.67	NO_2	>6.2	>3.9
NH_2	6.07	5.33	CN	8.92	5.10
CH_3	4.96	4.87	CH_3S	5.0	3.1
Cl	8.31	4.70	C_6H_5S	5.50	3.08
SH	6.4	4.1	C_6H_5O	5.60	3.25
PH_2	5.54	4.29	C_2H_5	4.00	4.39
SiH_3	4.78	3.37	$i\text{-}C_3H_7$	3.55	4.03
Br	7.60	4.24	$t\text{-}C_4H_9$	3.31	3.61
SeH	6.0	2.8	C_6H_5	5.2	4.1
I	6.76	3.70	C_2H_3	4.85	4.10
H	7.17	6.42			

[a] From Pearson (1988).

[b] These values are recommended by Pearson (1988) as also appropriate for the corresponding negative ions.

Table F.4 Absolute Electronegativities and Absolute Hardnesses for Molecules[a]

Species	χ (eV)	η (eV)	Species	χ (eV)	η (eV)
SF_6	8.0	7.4	CH_3CHO	4.5	5.7
BF_3	6.2	9.7	C_2H_4	4.4	6.2
SO_3	7.2	5.5	C_5H_5N	4.4	5.0
Cl_2	7.0	4.6	Butadiene	4.3	4.9
H_2	6.7	8.7	H_2S	4.2	6.2
SO_2	6.7	5.6	C_2H_2	4.4	7.0
N_2	6.70	8.9	$HCONH_2$	4.2	6.2
Br_2	6.6	4.0	Styrene	4.11	4.36
O_2	6.3	5.9	CH_3COCH_3	4.1	5.6
CO	6.1	7.9	PH_3	4.1	6.0
I_2	6.0	3.4	C_6H_6	4.1	5.3
BCl_3	5.97	5.64	Toluene	3.9	5.0
CS	5.96	5.76	Propylene	3.9	5.9
HNO_3	5.80	5.23	C_6H_5OH	3.8	4.8
CH_3NO_2	5.79	5.34	C_6H_5SH	3.8	4.6
PF_3	5.7	6.7	CH_3Cl	3.8	7.5
HCN	5.7	8.0	p-Xylene	3.7	4.8
BBr_3	5.67	4.85	1,2,5-Trimethylbenzene	3.69	4.72
PBr_3	5.6	4.2	Cyclohexene	3.4	5.5
S_2	5.51	3.85	DMF	3.4	5.8
$C_6H_5NO_2$	5.5	4.4	$C_6H_5NH_2$	3.3	4.4
PCl_3	5.5	4.7	$CH_3CH\!=\!C(CH_3)_2$	3.3	5.5
NO_2	5.4	7.6	CH_3F	3.2	9.4
Acrylonitrile	5.35	5.56	H_2O	3.1	9.5
CS_2	5.35	4.73	$(CH_3)_3As$	3.0	5.7
HI	5.3	5.3	$(CH_3)_3P$	2.8	5.9
CO_2	5.0	8.8	$(CH_3)_2S$	2.7	6.0
HF	5.0	11.0	NH_3	2.6	8.2
CH_3I	4.9	4.7	CH_4	2.5	10.3
HCl	4.7	8.0	$C(CH_3)_4$	2.2	8.3
CH_3CN	4.7	7.5	$(CH_3)_2O$	2.0	8.0
CH_2O	4.7	6.2	$(CH_3)_3N$	1.5	6.3
HCO_2CH_3	4.6	6.4			

[a] From Pearson (1988).

APPENDIX G
THE REVIEW LITERATURE OF DENSITY-FUNCTIONAL THEORY

Density-functional theory is a relatively new and fast-developing part of the theory of matter. In the text of this book we have referred to many if not most of the important specific original research papers in the subject; these are listed alphabetically by author, in the Bibliography. Here we list the main reviews that have been written, with a remark or two about each.

Bamzai, A.S. and Deb, B.M. (1981). The role of single-particle density in chemistry. *Rev. Mod. Phys.* 53: 95–126, 593. More than 400 references. Some discussion of density-functional theory.

Callaway, J. and March, N.H. (1984). Density functional methods: Theory and applications. *Solid State Phys.* 38: 135–221. Comprehensive review with more than 150 references.

Connolly, J.W.D. (1977). The $X\alpha$ method. In *Semiempirical Methods of Electronic Structure Calculation, Part A: Techniques*. Segal, G.A. (ed.). New York: Plenum, pp. 105–132. A good review of the $X\alpha$ method and its applications, including discussion of its relations with density-functional and density-matrix-functional theories; 135 references.

Dahl, J.P. and Avery, J. (eds.) (1984). *Local Density Approximations in Quantum Chemistry and Solid State Physics*. New York and London: Plenum, 850 pp., including 18-page subject index. Comprises 44 contributions from a conference in Copenhagen in June of 1982. A broad range of papers, mostly theoretical, focusing on the LDA or $X\alpha$ method and its chemical applications. Two papers by Davidson are of particular originality, one on v-representability and one on the chemical potential. Especially recommended are the papers by Perdew on self-interaction corrections and the paper by Gross and Dreizler on implementation of and approximations in relativistic density-functional theory.

Davidson, E.R. (1976). *Reduced Density Matrices in Quantum Chemistry*. New York: Academic Press, 135 pp., with alphabetical references and subject index. Terse but definitive, with only very brief discussion of density-functional theory.

Deb, B.M. (ed.) (1981). *The Force Concept in Chemistry*. New York:

Van Nostrand Reinhold, 502 pp., with author and subject indices. Beautifully organized and produced treatise on the theme of the title, with multitudinous references. No article on density-functional theory *per se,* but mention of it in several places. Authors are Bader, Coleman, Deb, Epstein, Goodisman, Nakatsuji and Koga, Politzer and Daiker, and Pulay.

Dreizler, R.M. and Providencia, J. de (1985). *Density Functional Methods in Physics.* New York and London: Plenum Press, 533 pp., including 4-page subject index. Comprises 18 major contributions plus abstracts of a poster session, from a NATO Summer School held in Alcabideche, Portugal, in September of 1983. Topics covered include the constrained search formulation (Levy and Perdew, 51 references), rigorous mathematical justification for the theory (Lieb, 31 references), time-dependent and relativistic systems (Gross and Dreizler, 89 references), correlation energies (Stoll and Savin, 77 references), excitation energies (Almladh and von Barth, 30 references), the meaning of the Kohn–Sham orbital energies (Perdew, 81 references), and application to nuclear bulk properties in nuclear physics (Brack, 93 references).

Erdahl, R. and Smith, V.H. Jr. (eds.) (1987). *Density Matrices and Density Functionals.* Dordrecht: Reidel Publishing Company, 722 pp., with 6-page subject index. Comprises 38 contributions from a symposium in honor of John Coleman held August, 1985, in Kingston, Ontario. A remarkable collection of 39 timely contributions that update the subject of density-matrix theory. Of special relevance for density-functional theory are the discussions of Levy (24 references) of the correlation energy functionals of one matrices, Lindholm and Lundquist (40 references) on derivation of semiempirical MO methods from density-functional theory, Hunter (27 references) on the differential equation for the electron density, Vosko and Lagowski (19 references) on understanding energy differences in density-functional theory, and Becke (28 references) on density-functional calculation of molecular dissociation energies.

Ghosh, S.K. and Deb, B.M. (1982). Densities, density-functionals and electron fluids. *Phys. Rep.* 92: 1–44. Nice review of density-functional theory as it relates to quantum fluid dynamics; 380 references.

Gombas, P. (1949). *Die Statistische Theorie des Atoms und Ihre Anwendungen.* Wein: Springer-Verlag, 406 pp., with author and subject indices. Definitive summary of the first 20 years of work on Thomas–Fermi and derivative theories and their applications.

Jones, R.O. (1987). Molecular calculations with the density functional formalism. Pages 413–437 in Part I of Lawley (1987).

Keller, J. and Gázquez, J.L. (eds.) (1983). *Density Functional Theory.* Berlin: Springer-Verlag, 301 pp. Contributions from a workshop held in Mexico City in October 1980. Excellent reviews by Levy, Harriman, Donnelly, Perdew, Zaremba, Gázquez, Das, Keller and Amador, and Robledo and Varea.

Langreth, D. and Suhl, H. (eds.) (1984). *Many-Body Phenomena at Surfaces.* Orlando: Academic Press, 578 pp. Comprises 30 conributions from a workshop in Santa Barbara, California, held in July of 1983, surveying both experimental and theoretical work, and encompassing both density-functional and other theoretical methods. Of special density-functional interest are the papers by von Barth (overview of density-functional theory, 74 references), Langreth (critical discussion of the local-density approximation and its improvements, 20 references), Perdew and Levy (density-functional theory for open systems, 45 references), and Jones (LDA calculations on molecules, 29 references).

Lundqvist, S. and March, N.H. (eds.) (1983). *Theory of the Inhomogeneous Electron Gas.* New York and London: Plenum Press, 395 pp., including 5-page subject index. Integrated five-chapter account. Chapter 1, by March (109 references), covers Thomas–Fermi theory; Chapter 2, by Kohn and Vashishta (96 references), general density-functional theory; Chapter 3, by Lundquist (50 references), dynamic aspects; Chapter 4, by Williams and von Barth (223 references), applications to atoms, molecules and solids; Chapter 5, by Lang (235 references), applications to metal surfaces and metal adsorbate systems. The chapter by Kohn and Vashishta is particularly recommended, constituting as it does a lucid review written almost 20 years after the Hohenberg–Kohn paper.

March, N.H. (1957). The Thomas–Fermi approximation in quantum mechanics. *Adv. Phys.* 6: 1–101. Superb, detailed review with about 100 references.

March, N.H. (1975). *Self-Consistent Fields in Atoms.* Oxford: Pergamon, 233 pp., including 2-page subject index and reprints of the classic papers of Thomas (1927), Fermi (1928), and Slater (1951). A beautiful summary of Thomas–Fermi–Dirac theory.

March, N.H. and Deb, B.M (eds.) (1987). *The Single-Particle Density in Physics and Chemistry.* New York: Academic Press, 385 pp., including 7-page subject index. Chapters by Bartolotti, Deb and Ghosh, Geldart and Rasolt, Jones, Levy, March (three), Politzer, and Theophilou.

Parr, R.G. (1983). Density functional theory. *Ann. Rev. Phys. Chem.* 34: 631–656. Includes 263 references.

Rajagopal, A.K. (1980). Theory of inhomogeneous electron systems:

spin-density-functional formalism. *Adv. Chem. Phys.* 41: 59–192. Fine critical review with 175 references.

Salahub, D.R. (1987). Transition-metal atoms and dimers. Pages 447–520 in part II of Lawley (1987). 389 references.

Sen, K.D. (ed.) (1987). *Electronegativity*. Berlin: Springer-Verlag, 198 pp. Seven up-to-date reviews on this topic. The last paper has an erroneous appendix.

Slater, J.C. (1974). *The Self-Consistent Field for Molecules and Solids*. New York: McGraw-Hill, 583 pp., including 215 pages of references and a 3-page subject index. Description of the $X\alpha$ method, unfortunately ignoring its status as a density-functional method.

The following books are in prospect:

Kryachko, E.S. and Ludena, E.V. (1990). *Density Functional Theory of Many-Electron Systems*. Dordrecht: Kluwer Press, 850 pp.

March, N.H. (1991). *Electron Density Theory of Atoms and Molecules*. New York: Academic Press.

Sham, L.J. and Schluter, M. (eds.) (1991). *Principles and Applications of Density Functional Theory*. Teaneck: World Scientific. Notes and reprints; about 500 pages.

BIBLIOGRAPHY

Acharya, P.K. (1983). Comment on the derivation of atomic kinetic energy functionals with full Weizsacker correction. *J. Chem. Phys.* **78:** 2101–2102.

Acharya, P.K., Bartolotti, L.J., Sears, S.B., and Parr, R.G. (1980). An atomic kinetic energy functional with full Weizsacker correction. *Proc. Natl. Acad. Sci. USA* **77:** 6978–6982.

Adams, W.H. and Clayton, M.M. (1986). The hydrogen atom in the hydrogen molecule. *Int. J. Quantum Chem. Symp.* **19:** 333–348.

Allan, N.L., West, C.G., Cooper, D.L., Grout, P.J., and March, N.H. (1985). The gradient expansions of the kinetic energy and the mean momentum for light diatomic molecules. *J. Chem. Phys.* **83:** 4562–4564.

Almbladh, C.-O. and Pedroza, A.C. (1984). Density functional exchange-correlation potentials and orbital eigenvalues for light atoms. *Phys. Rev. A* **29:** 2322–2330.

Alonso, J.A. and Girifalco, L.A. (1977). A non-local approximation to the exchange energy of the non-homogeneous electron gas. *Solid State Commun.* **24:** 135–138.

Alonso, J.A. and Girifalco, L.A. (1978a). Gradient corrections in the energy density functional. *Chem. Phys. Lett.* **53:** 190–191.

Alsonso, J.A. and Girifalco, L.A. (1978b). Nonlocal approximation to the exchange potential and kinetic energy of an inhomogeneous electron gas. *Phys. Rev. B* **17:** 3735–3743.

Alvarellos, J.E., Tarazona, P., and Chacón, E. (1986). Nonlocal density functional for exchange and correlation energy of electrons. *Phys. Rev. B* **33:** 6579–6587.

Amaral, O.A.V. (1985). Simple molecular wavefunctions with correlation corrections. *Theor. Chim. Acta* **67:** 193–197.

Amaral, O.A.V. and McWeeny, R. (1983). Simple molecular wavefunctions with correlation corrections. *Theor. Chim. Acta* **64:** 171–180.

Arfken, G. (1980). *Mathematical Methods for Physicists.* New York: Academic Press.

Aryasetiawan, F. and Stott, M.J. (1988). Effective potentials in desity-functional theory. *Phys. Rev. B* **38:** 2974–2987.

Ashby, N. and Holzman, M.A. (1970). Semistatistical model for electrons in atoms. *Phys. Rev. A* **1:** 764–777.

Ashcroft, N.W. and Mermin, N.D. (1976). *Solid State Physics.* Philadelphia: Saunders College.

Averill, F.W. and Painter, G.S. (1981). Virial theorem in the density-functional formalism: Forces in H_2. *Phys. Rev. B* **24:** 6795–6800.

Bader, R.F.W. (1970). *An Introduction to the Electronic Structure of Atoms and Molecules.* Toronto: Clarke, Irwin.

Bader, R.F.W. (1986). Reply to comments of Li and Parr on Bader's definition of an atom. *J. Chem. Phys.* **85**: 3133–3134.

Bader, R.F.W. and Nguyen-Dang, T.T. (1981). Quantum theory of atoms in molecules—Dalton revisited. *Adv. Quantum Chem.* **14**: 63–124.

Baird, N.C., Sichel, J.M., and Whitehead, M.A. (1968). A molecular orbital approach to electronegativity equalization. *Theor. Chim. Acta* **11**: 38–50.

Balazs, N.L. (1967). Formation of stable molecules within the statistical theory of atoms. *Phys. Rev.* **156**: 42–47.

Balazs, N.L. and Jennings, B.K. (1984). Wigner's function and other distribution functions in mock spaces. *Phys. Rep.* **104**: 347–391.

Balbás, L.C., Alonso, J.A., and Vega, L.A. (1986). Density functional theory of the chemical potential of atoms and its relations to electrostatic potentials and bonding distances. *Z. Phys. D.* **1**: 215–221.

Baltin, R. (1972). The energy-density functional of an electron gas in locally linear approximation of the one-body potential. *Z. Naturforsch. A* **27**: 1176–1186.

Baltin, R. (1987). The three-dimensional kinetic energy functional compatible with the exact differential equation for its associated tensor. *J. Chem. Phys.* **86**: 947–952.

Baltin, R. (1988a). Differential force law and related integral theorems for a system of N identical interacting particles. I. General geometries. *J. Chem. Phys.* **88**: 6962–6969.

Baltin, R. (1988b). Differential force law and related integral theorems for a system of N identical interacting particles. II. Spherical symmetry. *J. Chem. Phys.* **88**: 6970–6976.

Baroni, S. (1984). Exact-exchange extension of the local spin-density approximation in atoms. II—The iron series. *J. Chem. Phys.* **80**: 5703–5708.

Baroni, S., Giannozzi, P., and Testa, A. (1987). Green's function approach to linear response in solids. *Phys. Rev. Lett.* **58**: 1861–1864.

Baroni, S. and Turcel, E. (1983). Exact-exchange extension of the local-spin-density approximation in atoms: Calculation of total energies and electron affinities. *J. Chem. Phys.* **79**: 6140–6144.

Bartolotti, L. (1980). Atomic properties from C. Fischer's numerical Hartree–Fock program. Calculations done at UNC-Chapel Hill.

Bartolotti, L.J. (1981). Time-dependent extension of the Hohenberg–Kohn–Levy energy-density functional. *Phys. Rev. A* **24**: 1661–1667.

Bartolotti, L. (1982a). A new gradient expansion of the exchange energy to be used in density functional calculations on atoms. *J. Chem. Phys.* **76**: 6057–6059.

Bartolotti, L.J. (1982b). Time-dependent Kohn–Sham density-functional theory. *Phys. Rev. A* **26**: 2243–2244. Erratum: *Phys. Rev. A* **27**: 2248 (1983).

Bartolotti, L.J. (1984). Variational-perturbation theory within a time-dependent Kohn–Sham formalism: An application to the determination of multiple polarizibilities, spectral sums, and dispersion coefficients. *J. Chem. Phys.* **80**: 5687–5695.

Bartolotti, L.J. (1986). The hydrodynamic formalism of time-dependent Kohn–Sham orbital density functional theory. *J. Phys. Chem.* **90:** 5518–5523.

Bartolotti, L.J. and Acharya, P.K. (1982). On the functional derivative of the kinetic energy density functional. *J. Chem. Phys.* **77:** 4576–4585.

Bartolotti, L.J., Gadre, S.R., and Parr, R.G. (1980). Electronegativities of the elements from simple $X\alpha$ theory. *J. Am. Chem. Soc.* **102:** 2945–2948.

Bartolotti, L.J. and Parr, R.G. (1980). The concept of pressure in density functional theory. *J. Chem. Phys.* **72:** 1593–1596.

Baskes, M.I. (1987). Application of the embedded-atom method to covalent materials: a semiempirical potential for silicon. *Phys. Rev. Lett.* **59:** 2666–2669.

Becke, A.D. (1986a). Completely numerical calculations on diatomic molecules in the local-density approximation. *Phys. Rev. A* **33:** 2786–2788.

Becke, A.D. (1986b). Density functional calculations of molecular bond energies. *J. Chem. Phys.* **84:** 4524–4529.

Becke, A.D. (1988a). Correlation energy of an inhomogeneous electron gas: a coordinate-space model. *J. Chem. Phys.* **88:** 1053–1062.

Becke, A.D. (1988b). Density-functional exchange-energy approximation with correct asymptotic behavior. *Phys. Rev. A* **38:** 3098–3100.

Berk, A. (1983). Lower bounds energy functionals and their application to diatomic systems. *Phys. Rev. A* **28:** 1908–1923.

Berkowitz, M. (1986). Exponential approximation for the density matrix and the Wigner's distribution. *Chem. Phys. Lett.* **129:** 486–488.

Berkowitz, M. (1987). Density functional approach to frontier controlled reactions. *J. Am. Chem. Soc.* **109:** 4823–4825.

Berkowitz, M., Ghosh, S.K., and Parr, R.G. (1985). On the concept of local hardness in chemistry. *J. Am. Chem. Soc.* **107:** 6811–6814.

Berkowitz, M. and Parr, R.G. (1988). Molecular hardness and softness, local hardness and softness, hardness and softness kernels, and relations among these quantities. *J. Chem. Phys.* **88:** 2554–2557.

Bernholc, J. and Holzwarth, N.A.W. (1983). Local spin-density descriptions of multiple metal-metal bonding: Mo_2 and Cr_2 *Phys. Rev. Lett.* **50:** 1451–1454.

Blaizot, J.-P. and Ripka, G. (1986). *Quantum Theory of Finite Systems.* Cambridge, Mass.: MIT Press.

Borkman, R. F. (1969). Simple bond-charge model for symmetric stretching vibrations of XY_n molecules. *J. Chem. Phys.* **51:** 5596–5601.

Borkman, R.F. and Parr, R.G. (1968). Toward an understanding of potential-energy curves for diatomic molecules. *J. Chem. Phys.* **48:** 1116–1126.

Borkman, R.F. and Settle, F.A., Jr. (1971). Bond orders and electric quadrupole moments for ethane, ethylene, and acetylene from a point-charge model. *J. Am. Chem Soc.* **93:** 5640–5644.

Borkman, R.F., Simons, G. and Parr, R.G. (1969). Simple bond-charge model for potential-energy curves of heteronuclear diatomic molecules. *J. Chem. Phys.* **50:** 58–65.

Boyd, R.J. and Markus, G.E. (1981). Electronegativities of the elements from a nonempirical electrostatic model. *J. Chem. Phys.* **75:** 5385–5388.

Brack, M. (1985). Semiclassical description of nuclear bulk properties. In *Density*

Functional Methods in Physics. Dreizler, R.M. and da Providencia, J. (eds.). New York: Plenum, pp. 331–379.

Brown, J.E. and Parr, R.G. (1971). Vibrational potential functions for CO_2, OCS, HCN and N_2O. *J. Chem. Phys.* **54**: 3429–3438.

Brual, G., Jr. and Rothstein, S.M. (1978). Rare gas interactions using an improved statistical method. *J. Chem. Phys.* **69**: 1177–1183.

Calder, G.V. and Ruedenberg, K. (1968). Quantitative correlations between rotational and vibrational spectroscopic constants in diatomic molecules. *J. Chem. Phys.* **49**: 5399–5415.

Callaway, J. and March, N.H. (1984). Density functional methods: Theory and applications. *Solid State Phys.* **38**: 135–221.

Capitani, J.F., Nalewajski, R.F., and Parr, R.G. (1982). Non Born–Oppenheimer density functional theory of molecular systems. *J. Chem. Phys.* **76**: 568–573.

Car, R. and Parrinello, M. (1985). Unified approach for molecular dynamics and density-functional theory. *Phys. Rev. Lett.* **55**: 2471–2474.

Carr, W.J., Jr. (1961). Energy, specific heat, and magnetic properties of the low-density electron gas. *Phys. Rev.* **122**: 1437–1446.

Carr, W.J. and Maradudin, A.A. (1964). Ground-state energy of a high-density electron gas. *Phys. Rev. A* **133**: 371–374.

Carroll, M.T., Bader, R.F.W., and Vosko, S.H. (1987). Local and non-local spin density functional calculations of the correlation energy of atoms and molecules. *J. Phys. B* **20**: 3599–3629.

Cedillo, A., Robles, J., and Gázquez, J.L. (1988). New nonlocal exchange-energy functional from a kinetic-energy-density Padé-approximant model. *Phys. Rev. A* **38**: 1697–1701.

Ceperley, D.M. and Alder, B.J. (1980). Ground state of the electron gas by a stochastic method. *Phys. Rev. Lett.* **45**: 566–569.

Chandler, D., McCoy, J.D., and Singer, S.J. (1986a). Density functional theory of nonuniform polyatomic systems. I. General formulation. *J. Chem. Phys.* **85**: 5971–5976.

Chandler, D., McCoy, J.D., and Singer, S.J. (1986b). Density functional theory of nonuniform polyatomic systems. II. Rational closures for integral equations. *J. Chem. Phys.* **85**: 5977–5982.

Chattaraj, P.K. and Deb, B.M. (1984). Role of kinetic energy in density functional theory. *J. Scient. Ind. Res.* **43**: 238–249.

Chayes, J.T., Chayes, L., and Lieb, E.H. (1984). The inverse problem in classical statistical mechanics. *Commun. Math. Phys.* **93**: 57–121.

Chayes, J.T., Chayes, L., and Ruskai, M.B. (1985). Density functional approach to quantum lattice systems. *J. Statist. Phys.* **38**: 493–518.

Chen, Z. and Spruch, L. (1987). Relative contribution of the electron–electron and electron–nuclear interactions to the ground-state energy of a neutral atom. *Phys. Rev. A* **35**: 4035–4043.

Choquet-Bruhat, Y., Dewitt-Morette, C., and Dillard-Bleick, M. (1982). *Analysis, Manofolds and Physics,* Revised Edition. Amsterdam: North-Holland.

Cioslowski, J. (1988a). Density functional for the energy of electronic systems—explicit variational construction. *Phys. Rev. Lett.* **60:** 2141–2143.

Cioslowski, J. (1988b). Note on the relation between total electronic energy and the sum of orbital energies. *J. Chem. Phys.* **88:** 2089–2090.

Cioslowski, J. and Yang, W. (1989). Integral formulation of density functional theory: Fermions in a one-dimensional quadratic potential. Unpublished.

Clementi, E. and Roetti, E. (1974). Roothan–Hartree–Fock atomic wave functions. *Atomic Data and Nuclear Data Tables* **14:** 177–478.

Clugston, M.J. and Gordon, R.G. (1977a). Electron-gas model for open shell–closed shell interactions. I. Application to the emission spectra of the diatomic noble-gas halides. *J. Chem. Phys.* **66:** 239–243.

Clugston, M.J. and Gordon, R.J. (1977b). An electron-gas study of the bonding, structure and octahedral ligand-field splitting in transition-metal halides. *J. Chem. Phys.* **67:** 3965–3969.

Cohen, J.S. and Pack, R.T. (1974). Modified statistical method for intermolecular potentials. Combining rules for higher Van der Waals coefficients. *J. Chem. Phys.* **61:** 2372–2382.

Cohen, L. (1966). Generalized phase-space distribution functions. *J. Math. Phys.* **7:** 781–786.

Cohen, L. and Frishberg, C. (1976). Hierarchy equations for reduced density matrices. *Phys. Rev. A* **13:** 927–930.

Cohen, L., Frishberg, C., Lee, C., and Massa, L.J. (1986). Correlation energy for a Slater determinant fitted to the electron density. *Int. J. Quantum Chem. Symp.* **19:** 525–533.

Cohen, L., Santhanan, P., and Frishberg, C. (1980). Reduced density matrix equations and functionals for correlation energy. *Int. J. Quantum Chem. Symp.* **14:** 143–154.

Cohen, L. and Zaparovanny, Y.I. (1980). Positive quantum joint distributions. *J. Math. Phys.* **21:** 794–796.

Coleman, J. (1963). Structure of fermion density matrices. *Rev. Mod. Phys.* **35:** 668–687.

Coleman, J. (1981). Calculation of first- and second-order density matrices. In *The Force Concept in Chemistry*. Deb, B.M. (ed.). New York: Van Nostrand Reinhold, pp. 418–448.

Colle, R. and Salvetti, O. (1975). Approximate calculation of the correlation energy for the closed shells. *Theor. Chim. Acta* **37:** 329–334.

Conolly, J.W.D. (1977). The $X\alpha$ method. In *Semiempirical Methods of Electronic Structure Calculations Part A: Techniques*. Segal, G.A. (ed.). New York: Plenum, pp. 105–132.

Cook, K. and Karplus, M. (1987). Electron correlation and density functional methods. *J. Phys. Chem.* **91:** 31–37.

Cortona, P. (1988). Self-interaction correction: the transition-metal atoms. *Phys. Rev. A* **38:** 3850–3856.

Coulson, C.A., O'Leary, B., and Mallion, R.B. (1978). *Hückel Theory for Organic Chemists*. London: Academic Press.

Courant, R. and Hilbert, D. (1953). *Methods of Mathematical Physics*, Volume I. New York: Interscience.

Csavinsky, P. (1968). Approximate variational solution of the TF equation for atoms. *Phys. Rev.* **166**: 53–56.

Cummins, P.L. and Nordholm, S. (1984). Inhomogeneity corrected Thomas–Fermi theory for atoms. *J. Chem. Phys.* **80**: 272–278.

Curtin, W.A. (1987). Density-functional theory of the solid–liquid interface. *Phys. Rev. Lett.* **59**: 1228–1231.

Curtin, W.A. and Ashcroft, N.W. (1986). Density-functional theory of freezing of simple liquids. *Phys. Rev. Lett.* **56**: 2775–2778.

Curtin, W.A. and Runge, K. (1987). Weighted-density-functional and simulation studies of the bcc hard-sphere solid. *Phys. Rev. A* **35**: 4755–4762.

Dahl, J.P. and Avery, J. (eds.) (1984). *Local Density Approximations in Quantum Chemistry and Solid State Physics*. New York and London: Plenum.

Dalgarno, A. (1962). Atomic polarizabilities and shielding factors. *Adv. Phys.* **11**: 281–315.

Das, G.P., Ghosh, S.K., and Sahni, V.C. (1988). On the correlation energy density functional in momentum space. *Solid State Comm.* **65**: 719–721.

Datta, D., (1986). Geometric mean principle for hardness equalization: a corollary of Sanderson's geometric mean principle of electronegativity equalization. *J. Phys. Chem.* **90**: 4216–4217.

Davidson, E.R. (1976). *Reduced Density Matrices in Quantum Chemistry*. New York: Academic Press.

Deb, B.M. (ed.) (1981). *The Force Concept in Chemistry*. New York: Van Nostrand Reinhold.

Deb, B.M. and Chattaraj, P.K. (1988). New quadratic nondifferential Thomas–Fermi–Dirac-type equation for atoms. *Phys. Rev. A* **37**: 4030–4033.

Deb, B.M. and Ghosh, S.K. (1982). Schrödinger fluid dynamics of many-electron systems in a time-dependent-density-functional framework. *J. Chem. Phys.* **77**: 342–348.

Deb, B.M. and Ghosh, S.K. (1983). New method for the direct calculation of electron density in many-electron systems. I. Application to closed-shell atoms. *Int. J. Quantum Chem.* **23**: 1–26.

Deb, B.M. and Ghosh, S.K. (1987). Quantum fluid dynamical significance of the single-particle density. In *Single-Particle Density in Physics and Chemistry*. March, N.H. and Delo, B.M. (eds.). London: Academic Press, pp. 219–284.

Dederichs, P.H., Blügel, S., Zeller, R., and Akai, H. (1984). Ground states of constrained systems: application to cerium impurities. *Phys. Rev. Lett.* **53**: 2512–2515.

Delley, G., Freeman, A.J., and Ellis, D.E. (1983). Metal-metal bonding in Cr–Cr and Mo–Mo dimers: Another success of local spin-density theory. *Phys. Rev. Lett.* **50**: 488–491.

DePristo, A.E. and Kress, J.D. (1986). Kinetic energy functionals via Padé approximations. *Phys. Rev. A* **35**: 438–441.

DePristo, A.E. and Kress, J.D. (1987). Rational function representation for accurate exchange energy functionals. *J. Chem. Phys.* **86**: 1425–1428.

Dhar, S. and Kestner, N.R. (1988). Electronic structure of the Fe_2 molecule in the local-spin-density approximation. *Phys. Rev. A* **38**: 1111–1119.

Dhara, A.K. and Ghosh, S.K. (1987). Density-functional theory for time-dependent systems. *Phys. Rev. A* **35**: 442–444.

Dharma-wardana, M.W. (1987). Density functional methods in hot dense plasmas. In *Strongly Coupled Plasma Physics*. Rogers, F.J. and Dewitt, H.E. (eds.). New York: Plenum, pp. 275–292.

Ding, K., Chandler, D., Smithline, S.J., and Haymet, A.D.J. (1987). Density-functional theory for the freezing of water. *Phys. Rev. Lett.* **59**: 1698–1701.

Dirac, P.A.M. (1930). Note on exchange phenomena in the Thomas atom. *Proc. Cambridge Phil. Soc.* **26**: 376–385.

Dirac, P.A.M. (1947). *The Principles of Quantum Mechanics*, 3d edn. Oxford: Oxford University Press.

Donnelly, R.A. (1979). On a fundamental difference between energy functionals based on a first- and on a second-order density matrices. *J. Chem. Phys.* **71**: 2874–2879.

Donnelly, R.A. and Parr, R.G. (1978). Elementary properties of an energy functional of the first-order reduced density matrix. *J. Chem. Phys.* **69**: 4431–4439.

Dreizler, R.M. and Providencia, J. da (eds.) (1985). *Density Functional Methods in Physics*. New York: Plenum.

Dunlap, B.I. (1983). $X\alpha$, Cr_2, and the symmetry dilemma. *Phys. Rev. A* **27**: 2217–2219.

Ebner, C. and Saam, W.F. (1975). Renormalized density-functional theory of nonuniform superfluid ^4He at zero temperature. *Phys. Rev. B* **12**: 923–939.

Ebner, C., Saam, W.F., and Stroud, D. (1976). Density functional theory of simple classical fluids. I. Surfaces. *Phys. Rev. A*. **14**: 2264.

Edmiston, C. and Ruedenberg, K. (1963). Localized atomic and molecular orbitals. *Rev. Mod. Phys.* **35**: 457–465.

Engel, E. and Dreizler, R.M. (1988). Solution of the relativistic Thomas–Fermi–Weizsacker model for the case of neutral atoms and positive ions. *Phys. Rev A* **38**: 3909–3917.

Engel, E. and Dreizler, R.M. (1989). Extension of the Thomas–Fermi–Dirac–Weizsächer model: fourth order gradient corrections to the kinetic energy. *J. Phys. B.* **22**: 000–000.

Englert, B.-G. and Schwinger, J. (1982). Thomas–Fermi revisited: The outer regions of the atom. *Phys. Rev. A* **26**: 2322–2329.

Englert, B.-G. and Schwinger, J. (1984a). Statistical atoms: Handling the strongly-bound electrons. *Phys. Rev. A* **29**: 2331–2338.

Englert, B.-G. and Schwinger, J. (1984b). New statistical atom: Some quantum improvements. *Phys. Rev. A* **29**: 2339–2352.

Englert, B.-G. and Schwinger, J. (1984c). New statistical atom: A numerical study. *Phys. Rev. A* **29**: 2353–2363.

Englert, B.-G. and Schwinger, J. (1985a). Semiclassical atom. *Phys. Rev. A* **32**: 26–35.

Englert, B.-G. and Schwinger, J. (1985b). Linear degeneracy in the semiclassical atom. *Phys. Rev. A* **32**: 36–46.

Englert, B.-G. and Schwinger, J. (1985c). Atomic-binding-energy oscillations. *Phys. Rev. A* **32**: 47–63.

Englisch, H. and Englisch, R. (1983). Hohenberg–Kohn theorem and non-v-representable densities. *Physica* **121A**: 253–268.

Englisch, H. and Englisch, R. (1984a). Exact density functionals for ground-state energies. I General results. *Phys. Stat. Sol. (6)* **124**: 711–721.

Englisch, H. and Englisch, R. (1984b). Exact density functionals for ground-state energies. II Details and remarks. *Phys. Stat. Sol. (6)* **124**: 373–379.

Englisch, H., Fieseler, H., and Haufe, A. (1988). Density-functional calculations for excited-state energies. *Phys. Rev. A* **37**: 4570–4576.

Epstein, S.T., Hurley, A.C., Wyatt, R.E., and Parr, R.G. (1967). Integrated and integral Hellmann–Feynman formulas. *J. Chem. Phys.* **47**: 1275–1286.

Erdahl, R. and Smith, V.H. Jr., Eds. (1987). *Density Matrices and Density Functionals*. Dordrecht: Reidel.

Evans, G.C. (1964). *Functionals and Their Applications*. New York: Dover.

Evans, R. (1979). The nature of the liquid-vapour interface and other topics in the statistical mechanics of non-uniform, classical fluids. *Adv. Phys.* **28**: 143–200.

Falicov, L.M. and Somorjai, G.A. (1985). Correlation between catalytic activity and bonding and coordination number of atoms and molecules on transition metal surfaces: Theory and experimental evidence. *Proc. Natl. Acad. Sci. USA* **82**: 2207–2211.

Feibelman, P.J. (1985). First-principles total-energy calculation for a single adatom on a crystal. *Phys. Rev. Lett.* **54**: 2627–2630.

Fermi, E. (1927). Un metodo statistice per la determinazione di alcune proprieta dell'atomo. *Rend. Accad., Lincei* **6**: 602–607.

Fermi, E. (1928a). A statistical method for the determination of some atomic properties and the application of this method to the theory of the periodic system of elements. *Z. Phys.* **48**: 73–79. [A translation into English may be found in March (1975).]

Fermi, E. (1928b). Sulla deduzione statistica di alcune proprieta dell'atomo. Applicazione alla teoria del systema periodico degli elementi. *Rend. Accad. Lincei* **7**: 342–346.

Fermi, E. and Amaldi, E. (1934). Le orbite oos degli elementi. *Accad. Ital. Rome* **6**: 117–149.

Feynman, R.P. (1939). Forces in molecules. *Phys. Rev.* **56**: 340–343.

Feynman, R.P. (1972). *Statistical Mechanics*. Reading, Mass.: Benjamin.

Feynman, R.P. and Hibbs, A.R. (1965). *Quantum Mechanics and Path Integrals*. New York: McGraw-Hill.

Fischer, C.F., Lagowski, J.B. and Vosko, S.H. Ground states of Ca^- and Sc^- from two theoretical points of view. *Phys. Rev. Lett.* **59**: 2263–2266.

Fois, E.S., Selloni, A., Parrinello, M., and Car, R. (1988). Bipolarons in metal–metal halide solutions. *J. Phys. Chem.* **92**: 3268–3273.

Fraga, S. (1964). Non-relativistic self-constant field theory. II. *Theor. Chim. Acta* **2**: 406–410.

Freed, K.F. and Levy, M. (1982). Direct first principles algorithm for the universal electron density functional. *J. Chem. Phys.* **77**: 396–398.

Freeman, D.L. (1975). The interaction of rare gas atoms with graphite surfaces. I. Single adatom energies. *J. Chem. Phys.* **62**: 941–949.

Fritsche, L. (1986). Generalized Kohn–Sham theory for electronic excitations in realistic systems. *Phys. Rev. B* **33**: 3976–3989.

Fu, C.L. and Freeman, A.J. (1988). Multilayer reconstruction and vibrational properties of W(001). *Phys. Rev. B* **37**: 2685–2688.

Fuentealba, P., Stoll, H., and Savin, A. (1988). Atomic correlation energy differences by means of a polarization potential. *Phys. Rev. A* **38**: 483–486.

Fujiwara, Y., Osborn, T.A., and Wilk, S.F.T. (1982). Wigner–Kirkwood expansions. *Phys. Rev. A* **25**: 14–34.

Fukui, K. (1975). *Theory of Orientation and Stereoselection*. Berlin: Springer-Verlag.

Fukui, K. (1987). Role of frontier orbitals in chemical reactions. *Science* **218**: 747–754.

Fukui, K., Yonezawa, T., and Shingu, H. (1952). A molecular orbital theory of reactivity in aromatic hydrocarbons. *J. Chem. Phys.* **20**: 722–725.

Fukui, K., Yonezawa, T., Nagata, C., and Shingu, H. (1954). Molecular orbital theory of orientation in aromatic heteroaromatic, and other conjugated molecules. *J. Chem. Phys.* **22**: 1433–1442.

Gadre, S.R. and Chakravorty, S.J. (1987). Use of a nonlocal density approximation for transformation from electron density to electron momentum density. *J. Chem. Phys.* **86**: 2224–2228.

Garrod, G. and Fusco, M.A. (1976). A density matrix variational calculation for atomic Be. *Int. J. Quantum Chem.* **10**: 495–510.

Garrod, G. and Percus, J.K. (1964). Reduction of the N-particle variational problem. *J. Math. Phys.* **5**: 1756–1776.

Gáspár, R. (1954). Uber dine approximation des Hartree–Fock schen potentials durch eine universelle potential funktion. *Acta Phys. Acad. Sci. Hung.* **3**: 263–286.

Gáspár, R. (1974). Statistical exchange for electron in shell and the $X\alpha$ method. *Acta Phys. Hung.* **35**: 213–218.

Gáspár, R. and Nagý, A. (1982). Comparison of the orbitals of neon. argon and krypton calculated by the Hartree–Fock and the $X\alpha$ methods with several values of α. *Acta Phys. Hung.* **53**: 247–254.

Gáspár, R. and Nagý, A. (1985). The $X\alpha$ method with *ab initio* exchange parameters, diamagnetic susceptibility and nuclear magnetic shielding constants for several atoms. *Acta Phys. Hung.* **58**: 107–111.

Gáspár, R. and Nagý, A. (1986). $X\alpha$ method with theoretically determined parameter α: calculation of shake-up and multi-electron x-ray transition energies. *J. Phys. B* **19**: 2793–2797.

Gáspár, R. and Nagý, A. (1987). $X\alpha$ method with theoretically determined parameter α—exchange parameter α and ionization energies of multiply charged ions. *J. Phys. B* **20**: 3631–3637

Gaydaenko, V.I. and Nikulin, V.K. (1970). Born–Mayer interatomic potential for atoms with $Z = 2$ to $Z = 36$. *Chem. Phys. Lett.* **7**: 360–362.

Gázquez, J.L., Galvan, M., Ortiz, E., and Vek, A. (1987). Atoms and ions in the limit of large nuclear charge. In *Density Matrices and Density Function-*

als. Erdahl, R. and Smith, V.H., Jr., (eds.). Dordrecht: Riedel, pp. 643–662.

Gázquez, J.L. and Ludena, E.V. (1981). The Weizsacker term in density functional theory. *Chem. Phys. Lett.* **83**: 145–148.

Gázquez, J.L. and Ortiz, E. (1984). Electronegativities and hardness of open shell atoms. *J. Chem. Phys.* **81**: 2741–2748.

Gázquez, J.L. and Parr, R.G. (1978). Two-parameter statistical model for atoms. *J. Chem. Phys.* **68**: 2323–2326.

Gázquez, J.L., Ray, N.K., and Parr, R.G. (1978). Simple electrostatic models for vibrating unsymmetrical triatomic molecules and triatomic ions. *Theor. Chim. Acta* **49**: 1–11.

Gázquez, J.L. and Robles, J. (1981). Local electron–electron energy density functionals. *J. Chem. Phys.* **74**: 5927–5928.

Gázquez, J.L. and Robles, J. (1982). On the atomic kinetic energy functionals with full Weizsacker correction. *J. Chem. Phys.* **76**: 1467–1472.

Gázquez, J.L., Vela, A., and Galvan, M. (1986). Behavior of the chemical potential of neutral atoms in the limit of large nuclear charge. *Phys. Rev. Lett.* **56**: 2606–2609.

Gelfand, I.M. and Fomin, S.V. (1963). *Calculus of Variations*. Englewood Cliffs: Prentice-Hall.

Gell-Mann, M. and Brueckner, K.A. (1957). Correlation energy of an electron gas at high density. *Phys. Rev.* **106**: 364–368.

Ghazali, A. and Hugon, P.L. (1978). Density-functional approach to the metal–insulator transition in doped semiconductors. *Phys. Rev. Lett.* **41**: 1569–1572.

Ghosh, S.K. (1987). A unified approach to local force laws in quantum chemistry. *J. Chem. Phys.* **87**: 3513–3517.

Ghosh, S.K. and Balbás, L.C. (1985). Study of the kinetic energy density functional in the locally linear potential approximation. *J. Chem. Phys.* **83**: 5778–5783.

Ghosh, S.K. and Berkowitz, M. (1985). A classical fluid-like approach to the density-functional formalism of many-electron systems. *J. Chem. Phys.* **83**: 2976–2983.

Ghosh, S.K., Berkowitz, M., and Parr, R.G. (1984). Transcription of ground-state density functional theory into a local thermodynamics. *Proc. Natl. Acad. Sci. USA* **81**: 8028–8031.

Ghosh, S.K. and Deb, B.M. (1982a). Densities, density functionals, and electron fluids. *Phys. Rep.* **92**: 1–44.

Ghosh, S.K. and Deb, B.M. (1982b). Quantum fluid dynamics of many-electron systems in three-dimensional space. *Int. J. Quantum Chem.* **22**: 871–888.

Ghosh, S.K. and Deb, B.M. (1982c). Dynamic polarizability of many-electron systems within a time-dependent density-functional theory. *Chem. Phys.* **71**: 295–306.

Ghosh, S.K. and Dhara, A.K. (1988). Density-functional theory of many-electron systems subjected to time-dependent electric and magnetic fields. *Phys. Rev. A* **38**: 1149–1158.

Ghosh, S.K. and Parr, R.G. (1985). Density-determined orthonormal orbital approach to atomic energy functionals. *J. Chem. Phys.* **82:** 3307–3315.
Ghosh, S.K. and Parr, R.G. (1986). Phase-space approach to the exchange energy functional of density-functional theory. *Phys. Rev. A* **34:** 785–791.
Ghosh, S.K. and Parr, R.G. (1987a). Toward a semiempirical density functional theory of chemical binding. *Theor. Chim. Acta* **72:** 379–391.
Ghosh, S.K. and Parr, R.G. (1987b). Improved Thomas-Fermi theory for atoms. In *Density Matrices and Density Functionals*. Erdahl, R. and Smith, V.H. Jr., (eds.) Dordrecht: Riedel, pp. 663–676.
Gianturco, F.A. (1976). Potential energy curves from the electron gas model. II. The ion–rare gas interaction. *J. Chem. Phys.* **64:** 1973–1976.
Gibbs, J.W. (1931). *Collected Works,* Vol. II. New York: Longmans, Green.
Gilbert, T.L. (1975). Hohenberg–Kohn theorem for nonlocal external potentials. *Phys. Rev. B* **12:** 2111–2120.
Golden, S. (1957a). Statistical theory of many-electron systems. General considerations pertaining to the Thomas–Fermi theory. *Phys. Rev.* **105:** 604–615.
Golden, S. (1957b). Statistical theory of many-electron systems. Discrete bases of representation. *Phys. Rev.* **107:** 1283–1290.
Golden, S. (1960). Statistical theory of electronic energies. *Rev. Mod. Phys.* **32:** 322–327.
Goldstein, J.A. and Rieder, G.R. (1987). A rigorous modified Thomas–Fermi theory for atomic systems. *J. Mater. Phys.* **28:** 1198–1202.
Gombas, P. (1949). *Die statistischen Theorie des Atomes und Ihre Anwendungen.* Wein: Springer-Verlag.
Goodgame, M.M. and Goddard, W.A. III (1982). Nature of Mo–Mo and Cr–Cr multiple bonds: A challenge for the local-density approximation. *Phys. Rev. Lett* **48:** 135–138.
Goodisman, J. (1970). Modified Weizsacker corrections in Thomas–Fermi theories. *Phys. Rev. A* **1:** 1574–1575.
Goodisman, J. (1987). Thomas–Fermi–Dirac–jellium model of the metal surface: Change of surface potential with charge. *J. Chem. Phys.* **86:** 882–886.
Gopinathan, M.S. (1977). Improved approximate representation of the Hartree–Fock potential in atoms. *Phys. Rev. A* **15:** 2135–2142.
Gopinathan, M.S., Whitehead, M.A., and Bogdanovic, R. (1976). The Fermi hole and the exchange parameter in $X\alpha$ theory. *Phys. Rev. A* **14:** 1–10.
Gopinathan, M.S. and Whitehead, M.A. (1980). On the dependence of total energy on occupation numbers. *Israel J. Chem.* **19:** 209–214.
Gordon, R.G. and Kim, Y.S. (1972). Theory for the forces between closed-shell atoms and molecules. *J. Chem. Phys.* **56:** 3122–3133.
Gordy, W. (1946). A new method of determining electronegativity from other atomic properties. *Phys. Rev.* **69:** 604–607.
Grammaticos, B. and Voros, A. (1979). Semiclassical approximations for nuclear Hamiltonians. I. Spin-independent potentials. *Ann. Phys.* **123:** 359–380.

Green, S., Garrison, B.J. and Lester, W.A. Jr. (1975). Hartree–Fock and Gordon–Kim interaction potentials for scattering of closed-shell molecules by atoms: (H_2CO, He) and (H_2, Li^+). *J. Chem. Phys.* **63:** 1154–1161.

Grimaldi, F., Grimaldi-Lecourt, A., and Dharma-wardana, M.W.C. (1985). Time-dependent density-functional theory of light absorption in dense plasma: Application to iron plasma. *Phys. Rev. A* **32:** 1063–1071.

Gritsenko, O.V. and Zhidomirov, G.M. (1987). Correlation energy correction as a density functional. A model of the pair distribution function and its application to the first- and second-row atoms and hydrides. *Chem. Phys.* **116:** 21–32.

Groot, R.D. and van der Eerden, J.P. (1988). Molecular model of fused salts near an electrode. *Phys. Rev. A* **38:** 296–300.

Gross, E.K.U. and Dreizler, R.M. (1979). Thomas–Fermi approach to diatomic systems. I. Solution of the Thomas–Fermi and Thomas–Fermi–Dirac–Weizsacker equations. *Phys. Rev. A* **20:** 1798–1807.

Gross, E.K.U. and Dreizler, R.M. (1981). Gradient expansion of the Coulomb exchange energy. *Z. Phys. A* **302:** 103–106.

Gross, E.K.U. and Dreizler, R.M. (1985). Density functional approach to time-dependent and to relativistic systems. In *Density Functional Methods in Physics*. (Dreizler, R.M. and Providencia, J. da, eds.). New York: Plenum, pp. 81–140.

Gross, E.K.U. and Kohn, W. (1985). Local density-functional theory of frequency-dependent linear response. *Phys. Rev. Lett.* **55:** 2850–2852.

Gross, E.K.U., Oliveira, L.N., and Kohn, W. (1988a). Raleigh–Ritz variational principle for ensembles of fractionally occupied states. *Phys. Rev. A* **37:** 2805–2808.

Gross, E.K.U., Oliveira, L.N., and Kohn, W. (1988b). Density functional theory for ensembles of fractionally occupied states II: Basic formalism. *Phys. Rev. A* **37:** 2809–2820.

Gunnarsson, O., Harris, J., and Jones, R.O. (1977). Density functional theory and molecular bonding. I. First-row diatomic molecules. *J. chem. Phys.* **67:** 3970–3979.

Gunnarsson, O. and Jones, R.O. (1980). Density functional calculations for atoms, molecules and clusters. *Phys. Scr.* **21:** 294–401.

Gunnarsson, O. and Jones, R.O. (1981). Self-interaction corrections in the density functional formalism. *Solid State Commun.* **37:** 249–252.

Gunnarsson, O. and Jones, R.O. (1985). Total-energy differences: sources of error in local-density approximations. *Phys. Rev. B* **31:** 7588–7602.

Gunnarsson, O., Jonson, M., and Lundqvist, B.I. (1976). Exchange and correlation in atoms, molecules and solids. *Phys. Lett.* **59A:** 177–179.

Gunnarsson, O., Jonson, M., and Lundqvist, B.I. (1977). Exchange and correlation in inhomogeneous electron systems. *Solid State Commun.* **24:** 765–768.

Gunnarsson, O., Jonson, M., and Lundqvist, B.I. (1979). Descriptions of exchange and correlation effects in inhomogeneous electron systems. *Phys. Rev. B* **20:** 3136–3164.

Gunnarsson, O. and Lundqvist, B.I. (1976). Exchange and correlation in atoms, molecules and solids by the spin-density-functional formalism. *Phys. Rev. B* **13**: 4274–4298.

Gunnarsson, O., Lundqvist, B.I., and Lundqvist, S. (1972). Screening in a spin-polarized electron liquid. *Solid State Commun.* **11**: 149–153.

Gunnarsson, O., Lundqvist, B.I., and Wilkens, J.W. (1974). Contribution to the cohesive energy of simple metals: spin-dependent effect. *Phys. Rev. B* **10**: 1319–1327.

Gunnarsson, O. and Schönhammer, K. (1986). Density-functional treatment of an exactly solvable semiconductor model. *Phys. Rev. Lett.* **56**: 1968–1971.

Gunnarsson, O. and Schönhammer, K. (1988). Gunnarsson and Schönhammer reply. *Phys. Rev. Lett.* **60**: 1583.

Guse, M.P. (1981). An atoms in molecules approach to density functional theory. *J. Chem. Phys.* **75**: 828–833.

Gyftopoulos, E.P. and Hatsopoulos, G.N. (1965). Quantum-thermodynamic definition of electronegativity. *Proc. Natl. Acad. Sci. USA* **60**: 786–793.

Hadjisavvas, N. and Theophilou, A. (1984). Rigorous formulation of the Kohn and Sham theory. *Phys. Rev. A* **32**: 2183–2186.

Hadjisavvas, N. and Theophilou, A. (1985). Rigorous formulation of Slater's transition-state theory for excited states. *Phys. Rev. A* **32**: 720–724.

Handler, G.S. (1965). On a second-order Thomas–Fermi theory. *J. Chem. Phys.* **43**: S253–S255.

Handler, G. (1973). Exponential Hamiltonian: Convergence in an iterated product representation for a linear harmonic oscillator. *J. Chem. Phys.* **58**: 1–4.

Handler, G.S. and March, N. (1975). Linear response theory and molecular vibrations: the Thomas–Fermi model. *J. Chem. Phys.* **63**: 438–442.

Handler, G.S. and Wang, P.S.C. (1972). Convergence of a product representation of the exponential Hamiltonian operator. An example from statistical theory. *J. Chem. Phys.* **56**: 1546–1548.

Handy, N.C., Marron, M.T., and Silverstone, H.J. (1969). Long-range behavior of Hartree–Fock orbitals. *Phys. Rev.* **180**: 45–48.

Hansen, J.P. and McDonald, I.R. (1976). *Theory of Simple Liquids*. London: Academic Press.

Haq, S., Chattaraj, P.K., and Deb, B.M. (1984). A new form for the kinetic energy-density functional for many-electron systems. *Chem. Phys. Lett.* **111**: 79–81.

Harriman, J.E. (1978). Geometry of density matrices. I. Definitions, N matrices and 1 matrices. *Phys. Rev. A* **17**: 1249–1256; II. Reduced density matrices and N representability. *Phys. Rev. A.* **17**: 1257–1268.

Harriman, J.E. (1980). Orthonormal orbitals for the representation of an arbitrary density. *Phys. Rev. A* **24**: 680–682.

Harriman, J.E. (1985). A kinetic energy density functional. *J. Chem. Phys.* **83**: 6283–6287).

Harris, J. (1979). The role of occupation numbers in HKS theory. *Int. J. Quantum Chem.* **13**: 189–193.

Harris, J. (1984). Adiabatic-connection approach to Kohn–Sham theory. *Phys. Rev. A* **29**: 1648–1659.

Harris, J. and Jones. R.O. (1974). The surface energy of a bounded electron gas. *J. Phys. F* **4**: 1170–1186.

Harris, J., Jones, R.O., and Miller, J.E. (1981). Force calculations in the density functional formalism. *J. Chem. Phys.* **75**: 3904–3908.

Harris, R.A. (1975). On a density functional theory of Van der Waals forces. *Chem. Phys. Lett.* **33**: 495–498.

Harris, R.A. (1984). Induction and dispersion forces in the electron gas theory of interacting closed shell systems. *J. Chem. Phys.* **81**: 2403–2405.

Harris, R.A. and Heller, D.F. (1975). Density functional theory of interacting closed shell systems. II. The determination of charge densities and the Hellmann–Feynman theorem. *J. Chem. Phys.* **62**: 3601–3604.

Harris, R.A. and Pratt, L.R. (1985). Discretized propagators, Hartree, and Hartree–Fock equations, and the Hohenberg–Kohn theorem. *J. Chem. Phys.* **82**: 856–859.

Harrison, J.G. (1983). An improved self-interaction-corrected local spin density functional for atoms. *J. Chem. Phys.* **78**: 4562–4566.

Harrison, J.G. (1987). Electron affinities in the self-interaction-corrected local spin density approximation. *J. Chem. Phys.* **86**: 2849–2853.

Haymet, A.D.J. (1987). Theory of the equilibrium liquid-vapor transition. *Annu. Rev. Phys. Chem.* **38**: 89–108.

Heaton, R.A., Harrison, R.A., and Lin, C.C. (1983). Self-interaction correction for density-functional theory of electronic energy bands of solids. *Phys. Rev. B* **28**: 5992–6607.

Hedin, L. and Lundqvist, B.I. (1971). Explicit local exchange correlation potentials. *J. Phys. C: Solid State Phys.* **4**: 2064–2083.

Heller, D.F., Harris, R.A., and Gelbart, W.M. (1975). Density functional formulation of collisional polarizabilities. *J. Chem. Phys.* **62**: 1947–1953.

Henderson, G. (1981). Variational theorems for the single-particle probability density and density matrix in momentum space. *Phys. Rev. A* **23**: 19–20.

Herman, F., Van Dyke, J.P. and Ortenburger, I.B. (1969). Improved statistical exchange approximation for inhomogeneous many-electron systems. *Phys. Rev. Lett.* **32**: 807–811.

Herring, C. (1986). Explicit estimation of the ground-state kinetic energies from electron densities. *Phys. Rev. A* **34**: 2614–2631.

Herring, C. and Chopra, M. (1988). Some tests of an approximate density functional for the ground-state kinetic energy of a fermion system. *Phys. Rev. A* **37**: 31–42.

Hillery, M., O'Connell, R.F., Scully, M.O. and Wigner, E.P. (1984). Distribution functions in physics: fundamentals. *Phys. Rep.* **106**: 121–167.

Hinze, J. and Jaffe, H.H. (1962). Electronegativity. I. Orbital electronegativity of neutral atoms. *J. Am. Chem. Soc.* **84**: 540–546.

Hodges, C.H. (1973). Quantum corrections to the Thomas–Fermi approximation—the Kirzhnits method. *Can. J. Phys.* **51**: 1428–1437.

Hoffman, G.G., Pratt, L.R., and Harris, R.A. (1988). Monte Carlo integration of density-functional theory: fermions in a harmonic well. *Chem. Phys. Lett.* **148**: 313–316.

Hohenberg, P., and Kohn, W. (1964). Inhomogeneous electron gas. *Phys. Rev.* **136**: B864–B871.
Hu, C.D. and Langreth, D.C. (1985). A spin dependent version of the Langreth–Mehl exchange-correlation functional. *Phys. Scr.* **32**: 391–396.
Hu, C.D. and Langreth, D.C. (1986). Beyond the random-phase approximation in nonlocal-density-functional theory. *Phys. Rev. B.* **33**: 943–959.
Hubbard, J. (1963). Electron correlations in narrow energy bands. *Proc. Roy. Soc. (London)* **A276**: 238–257.
Huheey, J.E. (1965). The electronegativity of groups. *J. Phys. Chem.* **69**: 3284–3291.
Huheey, J.E. (1971). Electronegativity, acids, and bases. IV. Concerning the inductive effect of alkyl groups. *J. Org. Chem.* **36**: 204–205.
Huheey, J.E. (1978). *Inorganic Chemistry: Principles of Structure and Reactivity*, 2d edn. New York: Harper and Row.
Hunter, G. (1987). The exact Schröndinger equation for the electron density. In *Density Matrices and Density Functionals*. Erdahl, R. and Smith, V.H. Jr., (eds.). Dordrecht: Riedel, pp. 583–596.
Hurley, A.C. (1976). *Electron Correlation in Small Molecules*. London: Academic Press.
Iczkowski, R.P. and Margrave, J.L. (1961). Electronegativity. *J. Am. Chem. Soc.* **83**: 3547–3551.
Janak, J.F. (1978). Proof that $\partial E/\partial n_i = \varepsilon_i$ in density functional theory. *Phys. Rev. B* **18**: 7165–7168.
Jennings, B.K. and Bhaduri, R.K. (1975). Smoothing methods in the independent particle model. *Nucl. Phys. A* **237**: 149–156.
Jennings, B.K., Bhaduri, R.K., and Brack, M. (1975). Semiclassical approximation in a realistic one-body potential. *Nucl. Phys. A* **253**: 29–44.
Johnson, K.H. (1973). Scattered-wave theory of the chemical bond. *Adv. Quantum Chem.* **7**: 143–185.
Jones, R.O. (1987). Molecular calculations with the density functional formalism. In *Ab Initio Methods in Quantum Chemistry*. (Lawley, K.P., ed.). New York: Wiley, pp. 413–437.
Jones, R.O. and Gunnarsson, O. (1985). Density-functional formalism: sources of error in local-density approximations. *Phys. Rev. Lett.* **55**: 107–110.
Kalos, M.H. (1984). *Monte Carlo Methods in Quantum Problems*. Dordrecht: Reidel.
Kanhere, D.G., Panat, P.V., Rajagopal, A.K., and Callaway, J. (1986). Exchange-correlation potentials for spin-polarized systems at finite temperatures. *Phys. Rev. A* **33**: 490–497.
Katriel, J. (1980). An alternative interpretation of Theophilou's extension of the Hohenberg–Kohn theorem to excited states. *J. Phys. Chem.* **13**: L375–L376.
Keller, J. and Gázquez, J.L. (1979). Self-consistent-field electron-gas local-spin-density model including correlation for atoms. *Phys. Rev. A* **20**: 1289–1291.
Kemister, G. (1986). Comment on the exchange energy formula of Ghosh and Parr. *Phys. Rev. A* **34**: 4480–4481.
Kemister, G. and Nordholm, S. (1985). Local spin density functional analysis of

exchange and correlation effects on some first row diatomic molecules. *J. Chem. Phys.* **83**: 5163–5173.

Kim, Y.S. (1978). Study of the Ar–N_2 interaction. I. Electron gas model (Gordon–Kim model) potential calculation. *J. Chem. Phys.* **68**: 5001–5005.

Kim, Y.S. and Gordon, R.G. (1974a). Study of the electron gas approximation. *J. Chem. Phys.* **60**: 1842–1850.

Kim, Y.S. and Gordon, R.G. (1974b). Ion–rare gas interactions on the repulsive part of the potential curves. *J. Chem. Phys.* **60**: 4323–4331.

Kim, Y.S. and Gordon, R.G. (1974c). Unified theory for the intermolecular forces between closed shell atoms and ions. *J. Chem. Phys.* **61**: 1–16.

Kirzhnits, D.A. (1957). Quantum corrections to the Thomas–Fermi equation. *Sov. Phys.—JETP* **5**: 64–71 [(1957) *Zh. Eksp. Teor. Fiz.* **32**; 115–123.]

Kirzhnits, D.A. (1967). *Field Theoretical Methods in Many-Body Systems.* Oxford: Pergamon.

Kitaura, K., Satoko, C., and Morokuma, K. (1979). Total energies of molecules with the local density functional approximation and Gaussian basis sets. *Chem. Phys. Lett.* **65**: 206–211.

Kleinman, L. (1984). Exchange density-functional gradient expansion. *Phys. Rev. B* **30**: 2223–2225.

Klopman, G. (1965). Electronegativity. *J. Chem. Phys.* **43**: S124–S129.

Klopman, G. (1968). Chemical reactivity and the concept of charge and frontier-controlled reactions. *J. Am. Chem. Soc.* **90**, 223–234. [Reprinted in Pearson (1973).]

Kohl, H. and Driezler, R.M. (1986). Time-dependent density-functional theory: conceptual and practical aspects. *Phys. Rev. Lett.* **56**: 1993–1995.

Kohn, W. (1983). v-Representability and density functional theory. *Phys. Rev. Lett.* **51**: 1596–1598.

Kohn, W. (1986). Density-functional theory for excited states in a quasi-local-density approximation. *Phys. Rev. A* **34**: 737–741.

Kohn, W. and Sham, L.J. (1965). Self-consistent equations including exchange and correlation effects. *Phys. Rev.* **140**: A1133–A1138.

Kohn, W. and Vashishta, P. (1983). General density functional theory. In *Theory of the Inhomogeneous Electron Gas.* (Lundqvist, S. and March, N.H., eds.). New York: Plenum, pp. 79–147.

Kolos, W. and Radzio, E. (1978). Application of the statistical method in the theory of intermolecular interactions. *Int. J. Quantum Chem.* **13**: 627–634.

Komorowski, L. (1987). Empirical evaluation of chemical hardness. *Chem. Phys. Lett.* **134**: 536–540.

Koopmans, T. (1934). Uber die zuordnung von wellen funktionen und eigenwerten zu den einzelnen elektronen eines atom. *Physica* **1**: 104–113.

Kress, J.D. and DePristo, A.E. (1987). Corrected effective medium method. I. One-body formulation with applications to atomic chemisorption and diatomic molecular potentials. *J. Chem. Phys.* **87**: 4700–4715.

Kress, J.D. and DePristo, A.E. (1988). Corrected effective medium method. II. N-body formulation. *J. Chem. Phys.* **88**: 2596–2608.

Kryachko, E.S., Petkov, I.Z., and Stoitsov, M.V. (1987a). Method of local scaling transformations and density functional theory in quantum chemistry. II. The procedure for reproducing a many-electron wave function from X-ray diffraction data on one-electron density. *Int. J. Quantum Chem.* **32**: 467–472.

Kryachko, E.S., Petkov, I.Z., and Stoitsov, M.V. (1987b). Method of local scaling transformations and density functional theory in quantum chemistry. III. The energy density functional: spin-restricted approach. *Int. J. Quantum Chem.* **33**: 473–489.

Kutzler, F.W. and Painter, G.S. (1987). Energies of atoms with nonspherical charge densities calculated with non-local density-functional theory. *Phys. Rev. Lett.* **59**: 1285–1287.

Lackner, K.S. and Zweig, G. (1983). Introduction to the chemistry of fractionally charged atoms: Electronegativity. *Phys. Rev. D* **28**: 1671–1691.

Landau, L.D. and Lifshitz, E.M. (1958). *Quantum Mechanics*. London: Pergamon.

Lang, N.D. and Kohn, W. (1970). Theory of metal surfaces: charge density and surface energy. *Phys. Rev. B* **1**: 4555–4568.

Langer, R.E. (1937). On the connection formulas and the solution of the wave equation. *Phys. Rev.* **51**: 669–676.

Langreth, D.C. and Mehl, M.J. (1981). Easily implementable nonlocal exchange-correlation energy functional. *Phys. Rev. Lett.* **47**: 446–450.

Langreth, D.C. and Mehl, M.J. (1983). Beyond the local-density approximation in calculations of ground-state electronic properties. *Phys. Rev. B* **28**: 1809–1834. Erratum (1984): *Phys. Rev. B* **29**: 2310.

Langreth, D.C. and Perdew, J.P. (1977). Exchange-correlation energy of a metallic surface: Wave-vector analysis. *Phys. Rev. B* **15**: 2884–2901.

Langreth, D.C. and Perdew, J.P. (1980). Theory of nonuniform electronic systems. I. Analysis of the gradient approximation and a generalization that works. *Phys. Rev. B* **21**: 5469–5493.

Langreth, D.C. and Vosko, S.H. (1987). Exact electron-gas response functions at high density. *Phys. Rev. Lett.* **59**: 497–500.

Langreth, D.C. and Vosko, S.H. (1988). Langreth and Vosko reply. *Phys. Rev. Lett.* **60**: 1984.

Lannoo, M., Schlüter, M., and Sham. L.J. (1985). Calculation of the Kohn–Sham potential and its discontinuity for a model-semiconductor. *Phys. Rev. B* **32**: 3890–3899.

Laufer, P.M. and Krieger, J.B. (1986). Test of density-functional approximations in an exactly soluble model. *Phys. Rev. A* **33**: 1480–1491.

Lawley, K.P. (ed.) (1987). *Ab Initio Methods in Quantum Chemistry, Parts I and II* (*Advances in Chemical Physics* **67, 68**). Chichester: Wiley-Interscience.

Lebowitz, J.L. and Percus, J.K. (1963). Statistical mechanics of nonuniform fluids. *J. Math. Phys.* **4**: 116–123.

Lee, C. and Ghosh, S.K. (1986). Density gradient expansion of the kinetic energy functional for molecules. *Phys. Rev. A* **33**: 3506–3507.

Lee, C. and Parr, R.G. (1987). Gaussian and other approximations to the first-order density matrix of electronic systems, and the derivation of various local-density theories. *Phys. Rev. A* **35**: 2377–2383.

Lee, C., Yang, W., and Parr, R.G. (1987). Local softness and chemical reactivity in the molecules CO, SCN^- and H_2CO. *Theochem* **163**: 305–313.

Lee, C., Yang, W., and Parr, R.G. (1988). Development of the Colle–Salvetti correlation energy formula into a functional of the electron density. *Phys. Rev. B* **37**: 785–789.

Lee, S., Bylander, D.M., and Kleinman, L. (1988). Dissociation energy of Mo_2 and Mo_2^+. *Phys. Rev. B* **37**: 10035–10038.

Lee, S. and Kim, Y.S. (1979). Study of the $Ar-N_2$ interaction. II. Modification of the electron gas model potential at intermediate and large distances. *J. Chem. Phys.* **70**: 4856–4863.

Leite, J.R., Ferreira, L.G., and Pereira, J.R. (1972). A correction to the Liberman approximation for the exchange energy. *Phys. Lett.* **40A**: 315–316.

LeSar, R. and Gordon, R.G. (1982). Electron gas model for molecular crystals. Application to the alkali and alkaline-earth hydroxides. *Phys. Rev. B* **25**: 7221–7237.

Levine, I.N. (1983). *Quantum Chemistry*, 3d edn. Boston: Allyn and Bacon.

Levine, Z.H. and Soven, P. (1984). Time-dependent local-density theory of dielectric effects in small molecules. *Phys. Rev. A* **29**: 625–635.

Levy, M. (1979). Universal variational functionals of electron densities, first-order density matrices, and natural spin-orbitals and solution of the v-representability problem. *Proc. Natl. Acad. Sci. USA* **76**: 6062–6065.

Levy, M. (1982). Electron densities in search of Hamiltonians. *Phys. Rev. A* **26**: 1200–1208.

Levy, M. (1987a). Correlation energy functionals of one-matrices and Hartree–Fock densities. In *Density Matrices and Density Functionals*. (Erdahl, R. and Smith, V.H. Jr., eds.). Dordrecht: Reidel, pp. 479–498.

Levy, M. (1987b). The coordinate scaling requirements in density functional theory. In *Single-Particle Density in Physics and Chemistry*. (March, N.H. and Deb, B.M., eds.). London: Academic Press, pp. 45–57.

Levy, M. and Goldstein, J.A. (1987). Electron density-functional theory and X-ray structure factors. *Phys. Rev. B* **35**: 7887–7890.

Levy, M., Nee, T.-S., and Parr, R.G. (1975). Method for direct determination of localized orbitals. *J. Chem. Phys.* **63**: 316–318.

Levy, M. and Ou-Yang, H. (1988). Exact properties of the Pauli potential for the square-root of the electron density and the kinetic energy functional. *Phys. Rev. A* **38**: 625–629.

Levy, M. and Perdew, J.P. (1985a). Hellmann–Feynman, virial and scaling requisites for the exact universal density functional. Shape of the correlation potential and diamagnetic susceptibility for atoms. *Phys. Rev. A* **32**: 2010–2021.

Levy, M. and Perdew, J.P. (1985b). The constrained search formulation of density functional theory. In *Density Functional Methods in Physics*. (Dreizler, R.M. and Providencia, J. da, eds.). New York: Plenum, pp. 11–30.

Levy, M., Perdew, J.P. and Sahni, V. (1984). Exact differential equation for the density and ionization energy of a many-particle system. *Phys. Rev. A* **30**: 2745–2748.

Levy, M., Yang, W. and Parr, R.G. (1985). A new functional with homogeneous coordinate scaling in density functional theory: $F[\rho, \lambda]$. *J. Chem. Phys.* **83**: 2334–2336.

Li, L. and Parr, R.G. (1986). The atom in a molecule: A density matrix approach. *J. Chem. Phys.* **84**: 1704–1711.

Li, T. and Tong, P. (1985). Hohenberg–Kohn theorem for time-dependent ensembles. *Phys. Rev. A* **31**: 1950–1951.

Lieb, E. H. (1981a). Variational principle for many-fermion systems. *Phys. Rev. Lett.* **46**: 457–459.

Lieb, E.H. (1981b). Thomas–Fermi and related theories of atoms and molecules. *Rev. Mod. Phys.* **53**: 603–641.

Lieb, E.H. (1982). Density functionals for Coulomb systems. In *Physics as Natural Philosophy, Essays in Honor of Laszlo Tisza on his 75th Birthday*. (Feshbach, M. and Shimony, A., eds.). Cambridge: MIT Press, pp. 111–149. [For a revised version, see (1983). *Int. J. Quantum Chem.* **24**: 243–277.]

Lieb, E.H. (1985). Density functionals for Coulomb systems. In *Density Functional Methods in Physics*. (Dreizler, R.M. and Providencia, J. da, eds.). New York, Plenum, pp. 31–80.

Lieb, E.H. and Simon, B. (1973). Thomas–Fermi theory revisited. *Phys. Rev. Lett.* **31**: 681–683.

Lieb, E.H. and Simon, B. (1977). The Thomas–Fermi theory of atoms, molecules and solids. *Adv. in Math.* **23**: 22–116.

Linderberg, J. (1977). Differential ionization energy: connection between quantum chemistry and chemical thermodynamics. *Int. J. Quantum Chem.* **12**: Suppl. 1, 267–276.

Lindgren, I. (1971). A statistical exchange approximation for localized electrons. *Int. J. Quant. Chem.* **5**: 411–420.

Lindholm, E. and Asbrink, L. (1985). *Molecular Orbitals and Their Energies, Studied by the Semiempirical HAM Method*. Berlin: Springer-Verlag.

Lindholm, E. and Lundqvist, S. (1985). Semiempirical MO methods, deduced from density functional theory. *Phys. Scr.* **32**: 220–224.

Lloyd, J. and Pugh, D. (1977). Some applications of a self exchange corrected electron gas model for intermolecular forces. *J. Chem. Soc., Faraday II* **73**: 234–244.

Löwdin, P.-O. (1955a). Quantum theory of many-particle systems. I. Physical interpretations by means of density matrices, natural spin-orbitals, and convergence problems in the method of configuration interaction. *Phys. Rev.* **97**: 1474–1489.

Löwdin, P.-O. (1955b). Quantum theory of many-particle systems. II. Study of the ordinary Hartree–Fock Approximation. *Phys. Rev.* **97**: 1490–1508.

Löwdin, P.-O. (1959). Correlation problem in many-electron quantum mechanics. I. Review of different approaches and discussion of some current ideas. In *Advances in Chemical Physics*, Vol. II. (Prigogine, I., ed.). New York: Interscience, pp. 207–322.

Lohr, L.L. (1984). Protonic counterpart of electronegativity as an organizing principle for acidity and basicity. *J. Phys. Chem.* **88:** 3607–3611.
Longuet-Higgins, H.C. and Salem, L. (1960). The alternation of bond lengths in large conjugated molecules. III. The cyclic polyenes $C_{18}H_{18}$, $C_{24}H_{24}$ and $C_{30}H_{30}$. *Proc. Roy. Soc. (London)* **A257:** 445–456.
Ludena, E.V. (1983). An approximate universal energy functional in density functional theory. *J. Chem. Phys.* **79:** 6174–6181.
Ludena, E.V. and Sierraalta, A. (1985). Necessary conditions for the mapping of γ into ρ. *Phys. Rev. A* **32:** 19–25.
Lundqvist, S. and March, N.H. Eds. (1983). *Theory of the Inhomogeneous Electron Gas.* New York: Plenum.
MacDonald, J.K.L. (1933). Successive approximations by the Raleigh–Ritz variation method. *Phys. Rev.* **43:** 830–833.
MacDonald, J.K.L. (1934). On the modified Ritz method. *Phys. Rev.* **46:** 828.
Macke, W. (1955a). Zur wellermechanischen Behandlung von Vielkorperproblemen *Ann. Phys.* **17:** 1–9.
Macke, W. (1955b). Wave-mechanical treatment of the Fermi gas. *Phys. Rev.* **100:** 992–993.
Malone, J.G. (1933). The electric moment as a measure of the ionic nature of covalent bonds. *J. Chem. Phys.* **1:** 197–199.
Manoli, S. and Whitehead, M.A. (1984). Electronegativities of the elements from density-functional theory. I. The Hartree–Fock–Slater theory. *J. Chem. Phys.* **81:** 841–846.
Manoli, S. and Whitehead, M.A. (1988). Generalized-exchange local-spin-density-functional theory: self-interaction correction. *Phys. Rev. A* **38:** 630–635.
Marc, G. and McMillan, W.G. (1985). The virial theorem. *Adv. Chem. Phys.* **58:** 209–361.
March, N.H. (1957). The Thomas–Fermi approximation in quantum mechanics. *Adv. in Phys.* **6:** 1–98.
March, N.H. (1975). *Self-Consistent Fields in Atoms.* Oxford: Pergamon.
March, N.H. (1986a). Differential equation for the electron density in large molecules. *Int. J. Quantum Chem: Quantum Biology Symp.* **13:** 003–008.
March, N.H. (1986b). The local potential determining the square root of the ground-state electron density of atoms and molecules from the Schrödinger equation. *Phys. Lett.* **113A:** 476–478.
March, N.H. and Deb, B.M. (eds.) (1987). *Single-Particle Density in Physics and Chemistry.* London: Academic Press.
March, N.H. and Nalewajski, R.F. (1987). Exact density-potential relation for the ground state of the Be atom. *Phys. Rev. A* **35:** 525–528.
March, N.H. and Parr, R.G. (1980). Chemical potential, Teller's theorem, and the scaling of atomic and molecular energies. *Proc. Natl. Acad. Sci. USA,* **77:** 6285–6288.
Martin, R.M. (1968). A simple bond charge for vibrations in covalent crystals. *Chem. Phys. Lett.* **2:** 268–270.
Massa, L., Goldberg, M., Frishberg, C., Boehme, R.F., and LaPlaca, S.J. (1985). Wave functions derived from quantum modeling of the electron

density from coherent X-ray diffraction: beryllium metal. *Phys. Rev. Lett.* **55**: 622–625.
Massey, H.S.W. and Sida, D.W. (1955). Collision processes in meteor trails. *Phil. Mag.* **46**: 190–198.
McClendon, M. (1988). Real-space diffusion theory of multiparticle quantum systems. *Phys. Rev. A* **38**: 5851–5855.
McQuarrie, D.A. (1976). *Statistical Mechanics.* New York: Harper & Row.
McWeeny, R. (1960). Some recent advances in density matrix theory. *Rev. Mod. Phys.* **32**: 335–369.
McWeeny, R. (1976). Present state of the correlation problem. In *Proceedings of the Second International Congress of Quantum Chemistry.* (Pullman, B. and Parr, R.G., eds.). Dordrecht: Reidel, pp. 3–31.
McWeeny, R. and Sutcliffe, B.T. (1969). *Methods of Molecular Quantum Mechanics.* London: Academic Press.
Mearns, D. and Kohn, W. (1987). Frequency-dependent v-representability in density-functional theory. *Phys. Rev. A* **35**: 4796–4799.
Mermin, N.D. (1965). Thermal properties of the inhomogeneous electron gas. *Phys. Rev.* **137**: A1441–A1443.
Merzbacher, E. (1970). *Quantum Mechanics,* 2nd edn. New York: Wiley.
Messiah, A. (1961). *Quantum Mechanics,* Vol. I. Amsterdam: North Holland.
Meyer, J., Bartel, J., Brack, M., Quentin, P., and Aicher, S. (1986). A simple Gaussian approximation for the one-body density matrix. *Phys. Lett. B* **172**: 122–128.
Miller, T.M. and Bederson, B. (1978). Atomic and molecular polarizabilities—A review of recent advances. *Adv. Atom. Mol. Phys.* **13**: 1–55.
Moffitt, W. (1951). Atoms in molecules and crystals. *Proc. Roy. Soc. (London)* **A210**: 245–268.
Mohammed, A.-R.E., and Sahni, V. (1985). Density-functional-theory of correlation-energy effects at metallic surfaces. *Phys. Rev. B* **31**: 4879–4889.
Morrell, M.M., Parr, R.G., and Levy, M. (1975). Calculation of ionization potentials from density matrices and natural functions, and the long-range behavior of natural orbitals. *J. Chem. Phys.* **62**: 549–554.
Mortier, W.J. (1987). Electronegativity equalization and its applications. In *Electronegativity.* (Sen, K. D. and Jorgensen, C.K., ed.). Berlin: Springer-Verlag, pp. 125–143.
Mortier, W.J., van Genechten, K., and Gasteiger, J. (1985). Electronegativity equalization: application and parameterization. *J. Am. Chem. Soc.* **107**: 829–835.
Mortier, W.J., Ghosh, S.K., and Shankar, S. (1986). Electronegativity equalization method for the calculation of atomic charges in molecules. *J. Am. Chem. Soc.* **108**: 4315–4320.
Moruzzi, V.L., Janak, J.F., and Williams, A.R. (1978). *Calculated Electronic Properties of Metals.* New York: Pergamon, p. 26.
Mullay, J. (1987). Estimation of atomic and group electronegativities. In *Electronegativity.* (Sen, K.D. and Jorgensen, C.K., eds.). Berlin: Springer-Verlag, pp. 1–25.
Mulliken, R.S. (1934). A new electroaffinity scale: together with data on valence

states and an ionization potential and electron affinities. *J. Chem. Phys.* **2:** 782–793.

Mulliken, R.S. (1952). Molecular compounds and their spectra. II. *J. Am. Chem. Soc.* **64:** 811–824.

Murphy, D.R. (1981). The sixth-order term of the gradient expansion of the kinetic energy density functional. *Phys. Rev. A* **24:** 1682–1688.

Murphy, D.R. and Parr, R.G. (1979). Gradient expansion of the kinetic energy density functional: local behavior of the kinetic energy density. *Chem. Phys. Lett.* **60:** 377–379.

Murphy, D.R. and Wang, W.-P. (1980). Comparative study of the gradient expansion of the atomic kinetic energy functional-neutral atoms. *J. Chem. Phys.* **72:** 429–433.

Nakatsuji, H. (1976). Equation for the direct determination of the density matrix. *Phys. Rev. A* **14:** 41–50.

Nalewajski, R.F. (1984a). On theoretical explanation of the hard and soft acids and bases principle. *J. Am. Chem. Soc.* **106:** 944–945.

Nalewajski, R.F. (1984b). Atoms-in-a-molecule model of the chemical bond. *J. Phys. Chem.* **88:** 6234–6240.

Nalewajski, R.F. (1984c). Internal density functional theory of molecular systems. *J. Chem. Phys.* **81:** 2088–2102.

Nalewajski, R.F. (1984d). Electrostatic effects in interactions between hard (soft) acids and bases. *J. Am. Chem. Soc.* **106:** 944–945.

Nalewajski, R. (1985). A study of electronegativity equalization. *J. Phys. Chem.* **89:** 2831–2837.

Nalewajski, R.F. (1988). General relations between molecular sensitivities and their physical content. *Z. Naturforsch.* **43a:** 65–72.

Nalewajski, R.F. and Capitani, J.F. (1982). Density functional theory: non Born-Oppenheimer Legendre transformations and Maxwell relations, equilibrium and stability conditions. *J. Chem. Phys.* **77:** 2514–2526.

Nalewajski, R.F. and Koniski (1987). Density polarization and chemical reactivity. *z. Naturforsch.* **42a:** 451–462.

Nalewajski, R.F. and Kozlowski, P.M. (1988). Density-potential relation and the Pauli potential for model systems. *Acta Phys. Pol. A* **74:** 287–294.

Nalewajski, R.F. and Parr, R.G. (1982). Legendre transforms and Maxwell relations in density functional theory. *J. Chem. Phys.* **77:** 399–407.

Nikulin, V.K. and Tsarev, Y.N. (1975). On the calculation of repulsive interatomic interaction potentials by the statistical theory. Pairs of positive alkali metal ions and noble gas atoms. *Chem. Phys.* **10:** 433–437.

Nordholm, S. (1987). Analysis of covalent bonding by nonergodic Thomas–Fermi theory. *J. Chem. Phys.* **86:** 363–369.

Norskov, J.K. and Lang, N.D. (1980). Effective-medium theory of chemical binding: Application to chemisorption. *Phys. Rev. B* **21:** 2131–2136.

Nozieres, P. and Pines, D. (1966). *The Theory of Quantum Liquids*. I. New York: Benjamin.

Nyden, M.R. (1983). An orthogonality constrained generalization of the Weizacker density functional method. *J. Chem. Phys.* **78:** 4048–4051.

Nyden, M.R. and Acharya, P.K. (1981). Coreless Thomas–Fermi models of atomic structure. *J. Chem. Phys.* **75:** 4567–4571.

Nyden, M.R. and Parr, R.G. (1983). Restatement of conventional electronic wavefunction determination as a density functional procedure. *J. Chem. Phys.* **78:** 4044–4047.
Oliveira, L.N., Gross, E.K.W., and Kohn, W. (1988a). Density-functional theory for ensembles of fractionally occupied states II: Application to the He atom. *Phys. Rev. A* **37:** 2821–2833.
Oliveira, L.N., Gross, E.K.W., and Kohn, W. (1988b). Density-functional theory for superconductors. *Phys. Rev. Lett.* **60:** 2430–2433.
Oliver, G.L. and Perdew, J.P. (1979). Spin-density gradient expansion for the kinetic energy. *Phys. Rev. A* **20:** 397–403.
Orsky, A.R. and Whitehead, M.A. (1987). Electronegativities in density functional theory: diatomic bond energies and hardness parameters. *Can. J. Chem.* **65:** 1970–1979.
Ossicini, S. and Bertoni, C.M. (1985). Density-functional calculation of atomic structure with nonlocal exchange and correlation. *Phys. Rev. A* **31:** 3550–3556.
Palke, W.E. (1980). The electronic chemical potential and the H atom in the H_2 molecule. *J. Chem. Phys.* **72:** 2511–2514.
Pant, M.M. and Rajagopal, A.K. (1972). Theory of inhomogeneous magnetic electron gas. *Solid State Commun.* **10:** 1157–1160.
Pariser, R. (1953). An improvement in the π-electron approximation in LCAO MO theory. *J. Chem. Phys.* **21:** 568–569.
Parker, G.W. (1988). Numerical solution of the Thomas–Fermi equation for molecules. *Phys. Rev. A* **38:** 2205–2210.
Parr, R.G. (1963). *The Quantum Theory of Molecular Electronic Structure.* New York: Benjamin.
Parr, R.G. (1975). The description of molecular structure. *Proc. Natl. Acad. Sci. USA* **72:** 763–771.
Parr, R.G. (1984). Remarks on the concept of an atom in a molecule and on charge transfer between atoms on molecule formation. *Int. J. Quantum Chem.* **26:** 687–692.
Parr, R.G. (1988). Derivation of a local formula for electron–electron repulsion energy. *J. Phys. Chem.* **92:** 3060–3061.
Parr, R.G. and Bartolotti, L.J. (1982). On the geometric mean principle of electronegativity equalization. *J. Am. Chem. Soc.* **104:** 3801–3803.
Parr, R.G. and Bartolotti, L.J. (1983). Some remarks on the density functional theory of few-electron systems. *J. Phys. Chem.* **87:** 2810–2815.
Parr, R.G. and Berk, A. (1981). The bare-nuclear potential as harbinger for the electron density in a molecule. In *Chemical Applications of Atomic and Molecular Electrostatic Potentials.* (Polizer, P. and Truhlar, D.G., eds.). New York: Plenum, pp. 51–62.
Parr, R.G. and Brown, J.E. (1968). Toward understanding vibrations of polyatomic molecules. *J. Chem. Phys.* **49:** 4849–4852.
Parr, R.G. and Chen, M.-B. (1981). Circulant orbitals for atoms and molecules. *Proc. Natl. Acad. Sci. USA* **78:** 1323–1326.
Parr, R.G., Craig, D.P., and Ross, I.G. (1951). Molecular orbital calculations of the lower excited levels of benzene, configuration interaction included. *J. Chem. Phys.* **18:** 1561–1563.

Parr, R.G., Donnelly, R.A., Levy, M., and Palke, W.E. (1978). Electronegativity: the density functional viewpoint. *J. Chem. Phys.* **68:** 3801–3807.

Parr, R.G. and Gadre, S.R. (1980). On the basic homogeneity characteristic of atomic and molecular electronic energies. *J. Chem. Phys.* **72:** 3669–3673.

Parr, R.G., Gadre, S.R., and Bartolotti, L.J. (1979). Local density functional theory of atoms and molecules. *Proc. Natl. Acad. Sci. USA* **76:** 2522–2526.

Parr, R.G. and Ghosh, S.K. (1986). Thomas–Fermi theory for atomic systems. *Proc. Natl. Acad. Sci. USA* **83:** 3577–3579.

Parr, R.G. and Pearson, R.G. (1983). Absolute hardness: companion parameter to absolute electronegativity. *J. Am. Chem. Soc.* **105:** 7512–7516.

Parr, R.G., Rupnik, K., and Ghosh, S.K. (1986). Phase-space approach to the density-functional calculation of Compton profiles of atoms and molecules. *Phys. Rev. Lett.* **56:** 1555–1558.

Parr, R.G. and Yang, W. (1984). Density functional approach to the frontier-electron theory of chemical reactivity. *J. Am. Chem. Soc.* **106:** 4049–4050.

Pasternak, A. (1977). Electronegativity based on the simple bond charge model. *Chem. Phys.* **26:** 101–112.

Pasternak, A. (1980). Electronegativity: The bond charge and the chemical potential approaches. *J. Chem. Phys.* **73:** 593–594.

Pasternak, A., Anderson, A., and Leech, J.W. (1977). A bond charge model for the lattice dynamics of iodine. *J. Phys. C* **10:** 3261–3271.

Pasternak, A., Anderson, A., and Leech, J.W. (1978). Lattice dynamics of bromine and chlorine using the simple bond charge model. *J. Phys. C* **11:** 1563.

Pathak, R. (1984). Bound excited states within the density-functional formalism: The Levy functional. *Phys. Rev. A* **29:** 978–979.

Pathak, R.K., Sharma, B.S., and Thakkar, A.J. (1986). Approximate relationships between density power integrals, moments of the momentum density, and interelectronic repulsions in diatomic molecules. *J. Chem. Phys.* **85:** 958–962.

Pauling, L. (1960). *The Nature of the Chemical Bond*, 3d edn. Ithaca: Cornell.

Payne, M.C., Joannopoulos, J.D., Alan, D.C., Teter, M.P., and Vanderbilt, D.H. (1986). Molecular dynamics and ab initio total energy calculations. *Phys. Rev. Lett.* **56:** 2656.

Payne, P.W. and Fujiwara, F.G. (1979). Numerical tests of a local exchange potential for atomic systems. *Chem. Phys. Lett.* **67:** 408–411.

Pearson, E.W. and Gordon, R.G. (1985). Local asymptotic gradient corrections to the energy functional of an electron gas. *J. Chem. Phys.* **82:** 881–889.

Pearson, R.G. (1963). Hard and soft acids and bases. *J. Am. Chem. Soc.* **85:** 3533–3539. [Reprinted in Pearson (1973).]

Pearson, R.G. (1966). Acids and Bases. *Science* **151:** 172–177. [Reprinted in Pearson (1973).]

Pearson, R.G. (1967). Hard and soft acids and bases. *Chem. Brit.* **3:** 103–107.

Pearson, R.G., Ed. (1973). *Hard and Soft Acids and Bases*. Stroudsburg: Dowden, Hutchinson, and Ross.

Pearson, R.G. (1988). Absolute electronegativity and hardness: application to inorganic chemistry. *Inorg. Chem.* **27:** 734–740.

Pederson, M.R., Heaton, R.A., and Lin, C.C. (1984). Local-density Hartree–Fock theory of electronic states of molecules with self-interaction correction. *J. Chem. Phys.* **80:** 1972–1975.

Pederson, M.R., Heaton, R.A., and Lin, C.C. (1985). Density-functional theory with self-interaction correction: Application to the lithium molecule. *J. Chem. Phys.* **82:** 2688–2699.

Pedroza, A.C. (1986). Nonlocal density functionals: Comparison with exact results for finite systems. *Phys. Rev. A* **33:** 804–813.

Percus, J.K. (1978). The role of model systems in the few-body reduction of the N-Fermion problem. *Int. J. Quantum Chem.* **13:** 89–124.

Perdew, J.P. (1984). Self-interaction correction. Pages 173–205 in Dahl and Avery (1984).

Perdew, J.P. (1985a). Accurate density functional for the energy: Real-space cutoff of the gradient expansion for the exchange hole. *Phys. Rev. Lett.* **55:** 1665–1668.

Perdew, J.P. (1985b). What do the Kohn–Sham orbital energies mean? How do atoms dissociate? In *Density Functional Methods in Physics.* (Dreizler, R.M. and Providencia, J. da, eds.). New York: Plenum, pp. 265–308.

Perdew, J.P. (1986). Density-functional approximation for the correlation energy of the inhomogeneous electron gas. *Phys. Rev. B* **33:** 8822–8824.

Perdew, J.P. and Levy, M. (1983). Physical content of the exact Kohn–Sham orbital energies: Band gaps and derivative discontinuities. *Phys. Rev. Lett.* **51:** 1884–1887.

Perdew, J.P. and Levy, M. (1984). Density functional theory for open systems, In *Many-Body Phenomena at Surfaces.* (Langreth, D. and Suhl, H., eds.). Orlando: Academic Press, pp. 71–89.

Perdew, J.P. and Levy, M. (1985). Extrema of the density functional for the energy: Excited states from the ground-state theory. *Phys. Rev. B* **31:** 6264–6272.

Perdew, J.P. and Norman, M.R. (1982). Electron removal energies in Kohn–Sham density functional theory. *Phys. Rev. B* **26:** 5445–5450.

Perdew, J.P., Parr, R.G., Levy, M., and Balduz, J.L. Jr. (1982). Density functional theory for fractional particle number: derivative discontinuities of the energy. *Phys. Rev. Lett.* **49:** 1691–1694.

Perdew, J.P. and Yue, W. (1986). Accurate and simple density functional for the electronic exchange energy: Generalized gradient approximation. *Phys. Rev. B* **33:** 8800–8802.

Perdew, J.P. and Zunger, A. (1981). Self-interaction correction to density-functional approximations for many-electron systems. *Phys. Rev. B* **23:** 5048–5079.

Perrot, F. and Dharma-Wardana, M.W.C. (1984). Exchange-correlation potentials for electron-ion systems at finite temperatures. *Phys. Rev. A* **30:** 2619–2626.

Petkov, I.Z., Stoitsov, M.V., and Kryachko, E.S. (1986). Method of local-scaling transformations and density-functional theory in quantum chemistry. *Int. J. Quantum Chem.* **29:** 149–161.

Peukert, V. (1978). A new approximation method for electron systems. *J. Phys. C* **11:** 4945–4956.

Phillips, P. and Davidson, E.R. (1984). The chemical potential for interacting fermions in a harmonic potential. Pages 43–52 in Dahl and Avery (1984).
Pines, D. (1963). *Elementary Excitation in Solids*. New York: Benjamin.
Plindov, G.I. and Pogrebnya, S.K. (1988). A simple approximation to the kinetic energy functional. *Chem. Phys. Lett.* **143**: 535–537.
Plumer, M.L. and Stott, M.J. (1985). Approximate kinetic energy functionals for atoms in extended systems. *J. Phys. C* **18**: 4143–4163.
Politzer, P. (1969a). Bond orders of homonuclear diatomic molecules. *J. Chem. Phys.* **50**: 2780–2781.
Politzer, P. (1969b). Bond orders of heteronuclear diatomic molecules. *J. Chem. Phys.* **51**: 459–460.
Politzer, P. (1970). Constant term in the energy function of a point-charge model of diatomic molecules. *J. Chem. Phys.* **52**: 2157–2158.
Politzer, P. (1987). A relationship between the charge capacity and the hardness of neutral atoms and groups. *J. Chem. Phys.* **86**: 1072–1073.
Politzer, P. and Parr, R.G. (1974). Some new energy formulas for atoms and molecules. *J. Chem. Phys.* **61**: 4258–4262.
Politzer, P., Parr, R.G., and Murphy, D.R. (1983). Relationships between atomic chemical potentials, electrostatic potentials, and covalent radii. *J. Chem. Phys.* **79**: 3859–3861.
Politzer, P., Parr, R.G., and Murphy, D.R. (1985). Approximate determination of Wigner–Seitz radii from free-atom wave functions. *Phys. Rev. B* **31**: 2806–2809.
Politzer, P. and Truhlar, D.G., Eds. (1981). *Chemical Applications of Atomic and Molecular Electrostatic Potentials*. New York: Plenum.
Politzer, P. and Weinstein, H. (1979). Some relations between electronic distribution and electronegativity. *J. Chem. Phys.* **71**: 4218–4220.
Politzer, P. and Zilles, B.A. (1984). Some observations concerning electronic densities, electrostatic potentials and chemical potentials. *Croatica Chem. Acta* **57**: 1055–1064.
Pratt, L.R., Hoffman, G.G., and Harris, R.A. (1988). Statistical theory of electron densities. *J. Chem. Phys.* **88**: 1818–1823.
Pritchard, H.O. and Sumner, F.H. (1956). The application of electronic digital computers to molecular orbital problems. II. A new approximation for hetero-atom systems. *Proc. Roy. Soc. (London)* **A235**: 136–143.
Puska, M.J. and Nieminen, R.M. (1984). Theory of hydrogen and helium impurities in metals. *Phys. Rev. B* **29**: 5382–5397.
Puska, M.J., Nieminen, R.M., and Manninen, M. (1981). Atoms embedded in an electron gas: Immersion energies. *Phys. Rev. B* **24**: 3037–3047.
Radzio-Andzelm, E. (1981). Investigation of intermolecular interactions based on the atomic statistical model. *Int. J. quantum Chem.* **20**: 601–612.
Rae, A.I.M. (1973). A theory for the interactions between closed shell systems. *Chem. Phys. Lett.* **18**: 574–577.
Rae, A.I.M. (1975). A calculation of the interaction between pairs of rare-gas atoms. *Mol. Phys.* **29**: 467–483.
Rajagopal, A.K. (1980). Theory of inhomogeneous electron systems: spin-density-functional formalism. *Adv. Chem. Phys.* **41**: 59–193.

Rajagopal, A.K. and Callaway, J. (1973). Inhomogeneous electron gas. *Phys. Rev. B* **7**: 1912–1919.

Ramana, M.V. and Rajagopal, A.K. (1983). Inhomogeneous relativistic electron systems: a density functional formalism. *Adv. Chem. Phys.* **54**: 231–302.

Rasolt, M. and Geldart, D.J.W. (1988). Comment on "Exact electron-gas response functions at high density". *Phys. Rev. Lett.* **60**: 1983.

Rasolt, M., Malmstrom, G. and Geldart, D.J.W. (1979). Wave-vector analysis of metallic surface energy. *Phys. Rev. B* **20**: 3012–3019.

Ray, N.K. and Parr, R.G. (1980). Diamagnetic shieldings of atoms in molecules and their relation to electronegativity. *J. Chem. Phys.* **73**: 1334–1339.

Ray, N.K., Samuels, L. and Parr, R.G. (1979). Studies of electronegativity equalization. *J. Chem. Phys.* **70**: 3680–3684.

Ree, F.H. and Winter, N.W. (1980). Ab initio and Gordon–Kim intermolecular potentials for two nitrogen molecules. *J. Chem. Phys.* **73**: 322–336.

Reed, M. and Simon, G. (1972). *Functional Analysis,* Volume I. New York: Academic Press.

Robles, J. (1986). A virial expansion of the local pressure in a quantum system within the local thermodynamic transcription of density functional theory. *J. Chem. Phys.* **85**: 7245–7250.

Rooney, R., Tseng, T.J., and Whitehead, M.A. (1980). Dependence of the one-electron eigenvalues, ε_i and the total energy, E, on the α parameter in the Hartree–Fock–Slater scheme for atoms. *Phys. Rev. A* **22**: 1375–1382.

Roothaan, C.C.J. (1951). New developments in molecular orbital theory. *Rev. Mod. Phys.* **23**: 69–89.

Rowlinson, J.S. and Widom, B. (1982). *Molecular Theory of Capillarity*. Oxford: Clarendon Press.

Ruedenberg, K. (1977). An approximation relation between orbital SCF energies and total SCF energy in molecules. *J. Chem. Phys.* **66**: 375–376.

Runge, E. and Gross, E.K.U. (1984). Density-functional theory for time-dependent systems. *Phys. Rev. Lett.* **52**: 997–1000.

Rychlewski, J. and Parr, R.G. (1986). The atom in a molecule: a wave function approach. *J. Chem. Phys.* **84**: 1696–1703.

Sahni, V., Bohnen, K.-P., and Harbola, M.K. (1988). Analysis of the local-density approximation of density-functional theory. *Phys. Rev. A* **37**: 1895–1907.

Sahni, V., Gruenebaum, J. and Perdew, J.P. (1982). Study of the density-gradient expansion for the exchange energy. *Phys. Rev. B* **26**: 4371–4377.

Sahni, V. and Levy, M. (1986). Exchange and correlation energies in density functional theory: bounds from available data. *Phys. Rev. B* **33**: 3869–3872.

Sanderson, R.T. (1951). An interpretation of bond lengths and a classification of bonds. *Science* **114**: 670–672.

Sanderson, R.T. (1976). *Chemical Bonds and Bond Energy*, 2d edn. New York: Academic Press.

San-Fabián, E. and Moscardó, F. (1984). Electron correlation in the Coulomb hole model. Comparison of methods. *J. Chem. Phys.* **81**: 4008–4013.

Savin, A., Stoll, H., and Preuss, H. (1986). An application of correlation energy density functionals to atoms and molecules. *Theor. Chim. Acta* **70**: 407–419.

Savin, A., Wedig, V., Preuss, H., and Stoll, H. (1984). Molecular correlation energies obtained with a nonlocal density functional. *Phys. Rev. Lett.* **53:** 2087–2089.

Schaefer, H.F. III. (1972). *The Electronic Structure of Atoms and Molecules.* Reading, Mass.: Addison-Wesley.

Schaefer, H.F. Ed. (1977). *Methods of Electronic Structure Theory and Applications of Electronic Structure Theory.* New York: Plenum.

Schönhammer, K. and Gunnarsson, O. (1987). Discontinuity of the exchange-correlation potential in density-functional theory. *J. Phys. C.* **20:** 3675–3689.

Schwarz, K. (1972). Optimization of the statistical exchange parameter α for the free atoms H through Nb. *Phys. Rev. B* **5:** 2466–2468.

Sears, S.B., Parr, R.G., and Dinur, U. (1980). On the quantum-mechanical kinetic energy as a measure of the information in a distribution. *Israel J. Chem.* **19:** 165–173.

Segal, G.A., Ed. (1977). *Semiempirical Methods of Electronic Structure Calculation, Part A: Techniques; Part B: Applications.* New York: Plenum.

Senatore, G. and Subbaswamy, K.R. (1987). Nonlinear response of closed-shell atoms in the density-functional formalism. *Phys. Rev. A* **35:** 2440–2447.

Sham, L.J. (1971). Approximations of the exchange and correlation potentials. In *Computational Methods in Bond Theory* (Marcus, P.J., Janak, J.F., and Williams, A.R., eds.). New York: Plenum, pp. 458–468.

Sham, L.J. (1985). Exchange and correlation in density-functional theory. *Phys. Rev. B* **32:** 3876–3882.

Sham, L.J. (1986). Density functionals beyond the local density approximation. *Int. J. Quantum Chem.: Quantum Chem. Symp.* **19:** 491–495.

Sham, L.J. and Schlüter, M. (1983). Density-functional theory of the energy gap. *Phys. Rev. Lett.* **51:** 1888–1891.

Sham, L.J. and Schlüter, M. (1985). Density-functional theory of the band gap. *Phys. Rev. B* **32:** 3883–3889.

Sham, L.J. and Schlüter, M. (1988). Comment on density-functional treatment of an exactly solvable semiconductor model. *Phys. Rev. Lett.* **60:** 1582.

Shankar, S. and Parr, R.G. (1985). Electronegativity and hardness as coordinates in structure stability diagrams. *Proc. Natl. Acad. Sci. USA* **82:** 264–266.

Sharma, B.S. and Thakkar, A.J. (1985). Correlation energy generating potentials for molecular hydrogen. *J. Chem. Phys.* **83:** 3577–3583.

Shen, Y.-T., Bylander, D.M., and Kleinman, L. (1988). Nonlocal Kohn–Sham exchange corrections to Si band gaps and binding energy. *Phys. Rev. B* **36:** 3465–3468.

Shih, C.C. (1976). Hartree–Fock densities in Thomas–Fermi–Dirac formulas including the inhomogeneity term. *Phys. Rev. A* **14:** 919–921.

Shih, C.C., Murphy, D.R., and Wang, W.-P. (1980). Gradient expansion of the exchange energy density functional: a complementary expansion of the atomic energy functional. *J. Chem. Phys.* **73:** 1340–1343.

Simons, G. (1972). Point charge-point dipole model for vibrating triatomic molecules. *J. Chem. Phys.* **56:** 4310–4313.

Simons, G. and Finlan, J.M. (1974). Advantages of a nonharmonic-oscillator analysis of molecular vibrations. *Phys. Rev. Lett.* **33:** 131–133.

Simons, G. and Parr, R.G. (1971). Development of the bond-charge model for vibrating diatomic molecules. *J. Chem. Phys.* **55**: 4197–4202.

Simons, G., Parr, R.G., and Finlan, J.M. (1973). New alternative to the Dunham potential for diatomic molecules. *J. Chem. Phys.* **59**: 3229–3234.

Sinanoglu, A. and Brueckner, K.A. (1970). *Three Approaches to Electron Correlation in Atoms*. New Haven, Conn.: Yale University Press.

Slater, J.C. (1951). A simplification of the Hartree–Fock method. *Phys. Rev.* **81**: 385–390.

Slater, J.C. (1972). Statistical exchange-correlation in the self-consistent field. *Adv. Quantum Chem.* **6**: 1–92.

Slater, J.C. (1974). *The Self-Consistent Field for Molecules and Solids: Quantum Theory of Molecules and Solids*, Vol. 4. New York: McGraw-Hill.

Slater, J.C. (1975). Comparison of TFD and Xα methods for molecules and solids. *Int. J. Quantum. Chem. Symp.* **9**: 7–21.

Smith, V.H. Jr. (1984). Density functional theory and local potential approximations from momentum space considerations. Pages 1–19 in Dahl and Avery (1984).

Smith, V.H. Jr. and Absar, I. (1977). Basic concepts of quantum chemistry for electron density studies. *Israel J. Chem.* **16**: 87–102.

Spackman, M.A. (1986). Atom–atom potentials via electron gas theory. *J. Chem. Phys.* **85**: 6579–6586.

Stanton, R.E. (1962). Hellmann–Feynman theorem and correlation energies. *J. Chem. Phys.* **36**: 1298–1300.

Stich, W., Gross, E.K.U., Malzacher, P., and Dreizler, R.M. (1982). Accurate solution of the Thomas–Fermi–Dirac–Weizsacker variational equations for the case of neutral atoms and positive ions. *Z. Phys. A* **309**: 5–11.

Stillinger, F.H. Jr. and Buff, F.P. (1962). Equilibrium statistical mechanics of inhomogenous fluids. *J. Chem. Phys.* **37**: 1–2.

Stoddart, J.C. and March, N.H. (1971). Density-functional theory of magnetic instabilities in metals. *Ann. Phys. (NY)* **64**: 174–210.

Stoll, H., Golka, E., and Preuss, H. (1980). Correlation energies in the spin-density formalism. II. Applications and empirical corrections. *Theor. Chim. Acta.* **55**: 29–41.

Stoll, H., Pavlidou, C.M.E., and Preuss, H. (1978). On the calculation of correlation energies in the spin-density functional formalism. *Theor. Chim. Acta.* **49**: 143–149.

Stott, M.J. and Zaremba, E. (1980a). Linear-response theory within the density-functional formalism: Application to atomic polarizabilities. *Phys. Rev. A* **21**: 12–23. Erratum (1980). *Phys. Rev. A* **22**: 2293.

Stott, M.J. and Zaremba, E. (1980b). Quasiatoms: An approach to atoms in nonuniform electronic systems. *Phys. Rev. B* **22**: 1564–1583.

Stott, M.J. and Zaremba, E. (1982). Energies of light atoms in uniform electronic host systems. *Can. J. Phys.* **60**: 1145–1151.

Szabo, A. and Ostlund, N.S. (1982). *Modern Quantum Chemistry*. New York: Macmillan.

Szasz, L., Berrios-Pagan, I., and McGinn, G. (1975). Density-functional formalism. *Z. Naturforsch.* **30a**: 1516–1534.

Tachibana, A. (1988a). Density functional theory for hidden high-T_c superconductivity. In *High-Temperature Superconducting Materials*. (Hatfield, W.E. and Miller, J.H. Jr., eds.). New York: Dekker, pp. 99–106.

Tachibana, A. (1988b). Shape wave in density functional theory. *Int. J. Quantum Chem.* **34**: 309–323.

Tachibana, A. (1989). Application of the Mermin entropy principle to the "apparatus" density functional theory. *Int. J. Quantum Chem.* **35**: 000–000.

Tachibana, A. and Parr, R.G. (1989). Regional density functional theory of electron transferability. Unpublished.

Tal, Y. and Bader, R.F.W. (1978). Studies of the energy density functional approach. I. Kinetic energy. *Int. J. Quantum Chem. Symp.* **12**: 153–168.

Talman, J.D. and Shadwick, W.F. (1976). Optimized effective atomic central potential. *Phys. Rev. A* **14**: 36–40.

Teller, E. (1962). On the stability of molecules in the Thomas–Fermi theory. *Rev. Mod. Phys.* **34**: 627–631.

Theophilou, A. (1979). The energy density functional formalism for excited states. *J. Phys. C* **12**: 5419–5430.

Thomas, L.H. (1927). The calculation of atomic fields. *Proc. Camb. Phil. Soc.* **23**: 542–548. [Reprinted in March (1975).]

Tomishima, Y. and Yonei, K. (1966). Solution of the Thomas–Fermi–Dirac equation with a modified Weizsacker correction. *J. Phys. Soc. Jpn.* **21**: 142–153.

Tong, B.Y. (1972). Kohn–Sham self-consistent calculation of the structure of metallic sodium. *Phys. Rev. B* **6**: 1189–1198.

Tong, B.Y. and Sham, L.J. (1966). Application to a self-consistent scheme including exchange and correlation effects to atoms. *Phys. Rev.* **144**: 1–4.

Tseng, T.J., Hong, S.H., and Whitehead, M.A. (1980). The SCF-E_a-SW method: the one-electron eigenvalues and total energies for even-Z atoms, Ne to Ar. *J. Phys. B* **13**: 4101–4109.

Tseng, T.J. and Whitehead, M.A. (1981a). Self-consistent-field-Ξa method: parameterization of a in the modified Hartree–Fock–Slater method for atoms. *Phys. Rev. A* **24**: 16–20.

Tseng, T.J. and Whitehead, M.A. (1981b). Self-consistent-field-Ξa method: the atomic properties of several atoms using theoretical a parameters derived from the Fermi hole. *Phys. Rev. A* **24**: 21–28.

Vaidehi, N. and Gopinathan, M.S. (1984). Ab initio local-density potential for atoms: the modified Ξ method. *Phys. Rev. A* **29**: 1679–1690.

Valone, S.M. (1980a). Consequences of extending 1 matrix energy functionals from pure-state representable to all ensemble representable 1 matrices. *J. Chem. Phys.* **73**: 1344–1349.

Valone, S.M. (1980b). A one-to-one mapping between one-particle densities and some n-particle ensembles. *J. Chem. Phys.* **73**: 4653–4655.

Valone, S.M. and Capitani, J.F. (1981). Bound excited states in density functional theory. *Phys. Rev. A* **23**: 2127–2133.

van Genechten, K.A., Mortier, W.J., and Geerlings, P. (1987). Intrinsic framework electronegativity: A novel concept in solid state chemistry. *J. Chem. Phys.* **86**: 5063–5071.

Vignale, G. and Rasolt, M. (1987). Density-functional theory in strong magnetic fields. *Phys. Rev. Lett.* **59:** 2360–2363.
Vignale, G. and Rasolt, M. (1988). Current and spin-density-functional theory for inhomogeneous electronic systems in strong magnetic fields. *Phys. Rev. B* **37:** 10685–10696.
Vinayagam, S.C. and Sen, K.D. (1988). Absolute hardness parameter: finite difference versus density functional theoretical definition. *Chem. Phys. Lett.* **144:** 178–179.
Volterra, V. (1959). *Theory of Functionals*. New York: Dover.
von Barth, U. (1979). Local-density theory of multiplet structure. *Phys. Rev. A* **20:** 1693–1703.
von Barth, U. and Hedin, L. (1972). A local exchange-correlation potential for the spin polarized case: I. *J. Phys. C* **5:** 1629–1642.
von Barth, U. and Pedroza, A.C. (1985). The cohesive energy and charge-density form factors of beryllium as a test on the Langreth–Perdew–Mehl approximation. *Phys. Scr.* **32:** 353–358.
Vosko, S.H. and Wilk, L. (1983). A comparison of self-interaction-corrected local correlation energy formulas. *J. Phys. B* **16:** 3687–3702.
Vosko, S.J., Wilk, L., and Nusair, M. (1980). Accurate spin-dependent electron liquid correlation energies for local spin density calculations: a critical analysis. *Can. J. Phys.* **58:** 1200–1211.
Waldman, M. and Gordon, R.G. (1979). Scaled electron gas approximation for intermolecular forces. *J. Chem. Phys.* **71:** 1325–1339.
Wang, J.S.Y. and Rasolt, M. (1976). Study of the density gradient expansion in the surface energy calculation for metals. *Phys. Rev. B* **13:** 5330–5337.
Wang, W.-P. (1980). Comparative study of the gradient expansion of the atomic kinetic energy functional-isoelectronic series. *J. Chem. Phys.* **73:** 416–418.
Wang, W.-P. (1982). Fixed-shell statistical atomic models with piecewise decaying electron densities. *Phys. Rev. A* **25:** 2901–2912.
Wang, W.-P. and Parr, R.G. (1977). Statistical atomic numbers with piecewise exponentially decaying electron densities. *Phys. Rev. A* **16:** 891–902.
Wang, W.-P., Parr, R.G., Murphy, D.R., and Henderson, G. (1976). Gradient expansion of the atomic kinetic energy functional. *Chem. Phys. Lett.* **43:** 409–412.
Wang, Y.R. and Overhauser, A.W. (1986). Comparison of exact exchange energies with local-density approximations. *Phys. Rev. B* **34:** 6839–6842.
Weinert, W., Freeman, A.J., and Ohnishi, S. (1986). H and the W(001) surface reconstructions: Local bonding to surface states. *Phys. Rev. Lett.* **56:** 2295–2298.
Weinstein, H., Politzer, P., and Srebrenik, G. (1975). A misconception concerning the electronic density distribution of an atom. *Theor. Chim. Acta* **38:** 159–163.
Weissbluth, M. (1978). *Atoms and Molecules*. New York: Academic Press.
Weizsacker, C.F. von (1935). Zur theorie dier kernmassen. *Z. Physik* **96:** 431–458.
Wigner, E. (1932). On the quantum correction for thermodynamic equilibrium. *Phys. Rev.* **40:** 749–759.

Williams, A.R. and von Barth, U. (1983). Applications of density functional theory to atoms, molecules, and solids. In *Theory of the Inhomogeneous Electron Gas.* (Lundqvist, S. and March, N.H., eds.). New York: Plenum, pp. 189–308.

Wilson, E.B. Jr. (1962). Four-dimensional electron density function. *J. Chem. Phys.* **36**: 2232–2233.

Wilson, S. (1984). *Electron Correlation in Molecules.* Oxford: Clarendon Press.

Wimmer, E., Krakauer, H., Weinert, M., and Freeman, A.J. (1981). Full-potential self-consistent linearized-augmented-plane-wave method for calculating the electronic structure of molecules and surfaces: O_2 molecule. *Phys. Rev. B* **24**: 864–875.

Wood, C.P. and Pyper, N.C. (1981). Electron gas predictions of interatomic potentials tested by *ab initio* calculations. *Mol. Phys.* **43**: 1371–1383.

Xu, B.-X. and Rajagopal, A.K. (1985). Current-density-functional theory for time-dependent systems. *Phys. Rev. A* **31**: 2682–2684.

Yang, W. (1986). Gradient correction in Thomas–Fermi theory. *Phys. Rev. A* **34**: 4575–4585.

Yang, W. (1987). Ab initio approach for many-electron systems without invoking orbitals: An integral formulation of the density functional theory. *Phys. Rev. Lett.* **59**: 1569–1572.

Yang, W. (1988a). *Ab initio* approach for many-electron systems without invoking orbitals: An integral formulation of density functional theory. *Phys. Rev. A* **38**: 5494–5503.

Yang, W. (1988b). Thermal properties of many-electron systems: An integral formulation of density functional theory. *Phys. Rev. A* **38**: 5504–5511.

Yang, W. (1988c). Dynamic linear response of many-electron systems: An integral formulation of density functional theory. *Phys. Rev. A* **38**: 5512–5519.

Yang, W. and Harriman, J.E. (1986). Analysis of the kinetic energy functional in density functional theory. *J. Chem. Phys.* **84**: 3320–3323.

Yang, W., Lee, C., and Ghosh, S.K. (1985). Molecular softness as the average of atomic softnesses: companion principle to the geometric mean principle for electronegativity equalization. *J. Phys. Chem.* **89**: 5413–5415.

Yang, W. and Mortier, W. (1986). The use of global and local molecular parameters for the analysis of the gas-phase basicity of amines. *J. Am. Chem. Soc.* **108**: 5708–5711.

Yang, W. and Parr, R.G. (1985). Hardness, softness, and the fukui function in the electronic theory of metals and catalysis. *Proc. Natl. Acad. Sci. USA* **82**: 6723–6726.

Yang, W., Parr, R.G., and Lee, C. (1986). Various functionals for the kinetic energy density of an atom or a molecule. *Phys. Rev. a* **34**: 4586–4590.

Yang, W., Parr, R.G., and Pucci, R. (1984). Electron density, Kohn–Sham frontier orbitals, and fukui functions. *J. Chem. Phys.* **81**: 2862–2863.

Ying, S.C. (1974). Hydrodynamic response of inhomogeneous metallic systems. *Nuovo Cimento* **B23**: 270–281.

Yonei, K. (1971). An extended Thomas–Fermi–Dirac theory for diatomic molecules. *J. Phys. Soc. Jpn.* **31**: 882–894.

Yonei, K. and Tomishima, Y. (1965). On the Weizsacker correction to the Thomas–Fermi theory of atoms. *J. Phys. Soc. Jpn.* **20:** 1051–1057.
Zangwill, A. and Soven, P. (1980). Density-functional approach to local-field effects in finite systems: Photoabsorption in the rare gases. *Phys. Rev. A* **21:** 1561–1572.
Zhou, Z., Parr, R.G., and Garst, J.F. (1988). Absolute hardness as a measure of aromaticity. *Tetrahedron Lett.* **29:** 4843–4846.
Ziegler, T., Tschinke, V., and Becke, A. (1987a). A theoretical study on the strength of multiple metal–metal bonds in binuclear complexes and transition-metal dimers by a non-local density functional method. *Polyhedron* **6:** 685–693.
Ziegler, T., Tschinke, V., and Becke, A. (1987b). A theoretical study on the relative strengths of the metal–hydrogen and metal–methyl bonds in complexes of middle to late transition metals. *J. Am. Chem. Soc.* **109:** 1351–1358.
Zorita, M.L., Alonso, J.A., and Balbás, L.C. (1985). Local behavior of the kinetic energy in density functional theory. *Int. J. Quantum Chem.* **27:** 393–406.
Zumbach, G. and Maschke, K. (1983). New approach to the calculation of density functionals. *Phys. Rev. A* **28:** 544–554; Erratum, *Phys. Rev. A* **29:** 1585–1587.
Zumbach, G. and Maschke, K. (1985). Density-matrix functional theory for the N-particle ground state. *J. Chem. Phys.* **82:** 5604–5607.

AUTHOR INDEX

Absar, I. 14, 313
Acharya, P. K. 106, 138, 139, 140, 285, 287, 307
Adams, W. H. 224, 285
Aicher, S. 118, 305
Akai, H. 244, 290
Alan, D. C. 244, 308
Alder, B. J. 275, 288
Allan, N. L. 136, 285
Almbladh, C.-O. 152, 153, 285
Alonso, J. A. 138, 191, 233, 285, 286, 317
Alvarellos, J. E. 191, 285
Amaldi, E. 181, 292
Amaral, O. A. V. 199, 285
Anderson, A. 230, 308
Arfken, G. 161, 246, 285
Aryasetiawan, F. 152, 285
Asbrink, L. 234, 303
Ashby, N. 117, 285
Ashcroft, N. W. 83, 217, 285, 290
Averill, F. W. 239, 285
Avery, J. 281, 290

Bader, R. F. W. 14, 138, 140, 200, 222, 286, 288, 314
Baird, N. C. 93, 286
Balazs, N. L. 114, 115, 265, 286
Balbás, L. C. 138, 233, 286, 294, 317
Balduz, J. L., Jr. 70, 75, 224, 309
Baltin, R. 138, 140, 244, 286
Bamzai, A. S. 281
Baroni, S. 186, 212, 243, 286
Bartel, J. 118, 305
Bartolotti, L. J. 84, 85, 93, 124, 126, 127, 138, 139, 140, 149, 166, 167, 208, 212, 240, 285, 286, 287, 307, 308
Baskes, M. I. 243, 287
Becke, A. D. 179, 198, 200, 287, 317
Bederson, B. 210, 305
Berk, A. 115, 135, 287, 308
Berkowitz, M. 102, 104, 119, 125, 227, 239, 241, 243, 287, 294
Bernholc, J. 179, 287
Berrios-Pagan, I. 239, 314
Bertoni, C. M. 200, 307

Bhaduri, R. K. 128, 299
Blaizot, J.-P. 35, 71, 77, 203, 259, 263, 287
Blügel, S. 244, 290
Boehme, R. F. 243, 305
Bogdanovic, R. 183, 295
Bohnen, K.-P. 191, 311
Borkman, R. F. 229, 231, 236, 287
Boyd, R. J. 95, 287
Brack, M. 118, 128, 204, 287, 299, 305
Brown, J. E. 229, 230, 236, 288, 308
Brual, G., Jr. 221, 288
Brueckner, K. A. 12, 274, 294, 313
Buff, F. P. 67, 313
Bylander, D. M. 86, 179, 302, 312

Calder, G. V. 229, 288
Callaway, J. 170, 204, 281, 288, 299, 311
Capitani, J. F. 205, 217, 244, 288, 306, 315
Car, R. 244, 288, 292
Carr, W. J., Jr. 274, 288
Carroll, M. T. 200, 288
Cedillo, A. 141, 288
Ceperley, D. M. 275, 288
Chacón, E. 191, 285
Chakrovorty, S. J. 244, 293
Chandler, D. 217, 288, 291
Chattaraj, P. K. 138, 140, 288, 290, 297
Chayes, J. T. 54, 69, 288
Chayes, L. 54, 69, 288
Chen, M.-B. 11, 307
Chen, Z. 116, 288
Chopra, M. 138, 298
Choquet-Bruhat, Y. 252, 288
Cioslowski, J. 56, 60, 116, 163, 289
Clayton, M. M. 224, 285
Clementi, E. 126, 134, 137, 289
Clugston, M. J. 220, 289
Cohen, L. 32, 199, 243, 266, 270, 289
Cohen, J. S. 220, 289
Coleman, J. 31, 32, 41, 43, 289
Colle, R. 199, 289
Connolly, J. W. D. 156, 178, 281, 289
Cook, K. 156, 289
Cooper, D. L. 136, 285
Cortona, P. 183, 289

Coulson, C. A. 97, 289
Courant, R. 246, 252, 254, 290
Craig, D. P. 13, 308
Csavinsky, P. 116, 290
Cummins, P. L. 140, 290
Curtin, W. A. 217, 290

Dahl, J. P. 281, 290
Dalgarno, A. 210, 290
Das, G. P. 244, 290
Datta, D. 98, 290
Davidson, E. R. 14, 28, 35, 72, 281, 290, 310
Deb, B. M. 17, 138, 139, 140, 208, 212, 243, 244, 281, 282, 288, 290, 294, 297, 304
Dederichs, P. H. 244, 290
Delley, G. 179, 290
DePristo, A. E. 138, 141, 198, 290, 300
Dewitt-Morette, C. 252, 288
Dhar, S. 179, 208, 291
Dhara, A. K. 209, 243, 291, 294
Dharma-wardana, M. W. C. 204, 212, 291, 296, 310
Dillard-Bleick, M. 252, 288
Ding, K. 217, 291
Dinur, U. 138, 312
Dirac, P. A. M. 20, 105, 108, 291
Donnelly, R. A. 74, 213, 214, 222, 291, 308
Dreizler, R. M. 133, 135, 136, 140, 141, 208, 210, 243, 282, 291, 296, 300, 313
Dunlap, B. I. 179, 291

Ebner, C. 66, 291
Edmiston, C. 11, 291
Ellis, D. E. 179, 290
Engel, E. 136, 243, 291
Englert, B.-G. 117, 140, 243, 291, 292
Englisch, H. 71, 187, 205, 208, 217, 292
Englisch, R. 71, 187, 205, 217, 292
Erdahl, R. 31, 282, 292
Evans, G. C. 246, 292
Evans, R. 67, 69, 292

Falicov, L. M. 102, 292
Feibelman, P. J. 243, 292
Fermi, E. 47, 105, 181, 292
Ferreira, L. G. 200, 302
Feynman, R. P. 17, 44, 45, 161, 259, 260, 292
Fieseler, H. 208, 292
Finlan, J. M. 230, 313
Fischer, C. F. 179, 292
Fois, E. S. 244, 292
Fomin, S. V. 246, 248, 253, 294
Fraga, S. 114, 116, 292

Freed, K. F. 187, 292
Freeman, A. J. 179, 221, 290, 293, 315, 316
Freeman, D. L. 243, 293
Frishberg, C. 32, 199, 243, 290, 305
Fritsche, L. 208, 293
Fu, C. L. 243, 293
Fuentealba, P. 200, 293
Fujiwara, Y. 160, 200, 293, 308
Fukui, K. 101, 293
Fusco, M. A. 41, 293

Gadre, S. R. 116, 124, 126, 127, 166, 167, 244, 287, 293, 308
Galvan, M. 94, 95, 293, 294
Garrison, B. J. 221, 296
Garrod, G. 41, 293
Garst, J. F. 98, 317
Gáspár, R. 155, 183, 293
Gasteiger, J. 235, 305
Gaydaenko, V. I. 220, 293
Gázquez, J. L. 94, 95, 116, 126, 138, 141, 200, 230, 236, 283, 288, 293, 294, 299
Geerlings, P. 235, 315
Gelbart, W. M. 220, 221, 298
Geldart, D. J. W. 195, 197, 311
Gelfand, I. M. 246, 248, 253, 294
Gell-Mann, M. 274, 294
Ghazali, A. 204, 294
Ghosh, S. K. 55, 60, 92, 98, 104, 113, 117, 119, 120, 136, 138, 139, 141, 208, 209, 212, 235, 236, 239, 240, 241, 242, 243, 244, 282, 287, 290, 291, 294, 295, 301, 305, 308, 316
Giannozzi, P. 212, 286
Gianturco, R. A. 220, 295
Gibbs, J. W. 44, 46, 257, 295
Gilbert, T. L. 55, 214, 295
Girifalco, L. A. 138, 191, 285
Goddard, W. A., III 179, 295
Goldberg, M. 243, 305
Golden, S. 158, 295
Goldstein, J. A. 117, 243, 295, 302
Golka, E. 313
Gombás, P. 105, 110, 282, 285
Goodgame, M. M. 179, 295
Goodisman, J. 138, 243, 295
Gopinathan, M. S. 164, 183, 295, 314
Gordon, R. G. 117, 138, 157, 219, 220, 221, 289, 295, 300, 302, 309, 315
Gordy, W. 93, 230, 295
Grammaticos, B. 128, 132, 295
Green, S. 221, 296
Grimaldi, F. 212, 296
Grimaldi-Lecourt, A. 212, 296
Gritsenko, O. V. 199, 200, 296
Groot, R. D. 243, 296

Gross, E. K. U. 133, 135, 136, 140, 141, 207, 208, 209, 211, 243, 296, 307, 311, 313
Grout, P. J. 136, 285
Gruenebaum, J. 198, 311
Gunnarsson, O. 60, 86, 157, 173, 178, 179, 182, 186, 190, 191, 192, 193, 197, 205, 275, 296, 297, 299, 312
Guse, M. P. 224, 297
Gyftopoulos, E. P. 75, 297

Hadjisavvas, N. 149, 207, 297
Handler, G. S. 104, 158, 297
Handy, N. C. 11, 297
Hansen, J. P. 257, 297
Haq, S. 138, 297
Harbola, M. K. 191, 311
Harriman, J. E. 55, 149, 215, 297, 316
Harris, J. 112, 157, 176, 186, 296, 297, 298
Harris, R. A. 158, 163, 220, 221, 298
Harrison, J. G. 183, 298
Harrison, R. A. 183, 298, 310
Hatsopoulos, G. N. 75, 297
Haufe, A. 208, 292
Haymet, A. D. J. 217, 291, 298
Heaton, R. A. 183, 298, 309
Hedin, L. 170, 177, 178, 274, 298, 315
Heller, D. F. 220, 221, 298
Henderson, G. 136, 244, 298, 315
Herman, F. 140, 298
Herring, C. 138, 298
Hibbs, A. R. 161, 292
Hilbert, D. 246, 252, 254, 290
Hillery, M. 265, 267, 270, 298
Hinze, J. 93, 298
Hodges, C. H. 128, 132, 298
Hoffman, G. G. 158, 163, 298, 310
Hohenberg, P. 50, 51, 52, 54, 299
Holzman, M. A. 117, 285
Holzwarth, N. A. W. 179, 287
Hong, S. H. 183, 314
Hu, C. D. 197, 299
Hubbard, J. 97, 299
Hugon, P. L. 204, 294
Huheey, J. E. 98, 225, 235, 299
Hunter G. 243, 299
Hurley, A. C. 14, 16, 292, 299

Iczkowski, R. P. 93, 299

Jaffe, H. H. 93, 298
Janak, J. F. 164, 272, 275, 299, 306
Jennings, B. K. 128, 265, 286, 299
Joannopoulos, J. D. 244, 308
Johnson, K. H. 156, 278, 299

Jones, R. O. 112, 157, 176, 182, 186, 193, 197, 282, 296, 298, 299
Jonson, M. 180, 191, 192, 193, 296
Jorgensen, C. K. 284, 306

Kalos, M. H. 163, 299
Kanhere, D. G. 204, 299
Karplus, M. 156, 289
Katriel, J. 207, 299
Keller, J. 200, 283, 299
Kemister, G. 119, 185, 242, 299
Kestner, N. R. 179, 291
Kim, Y. S. 117, 157, 219, 220, 295, 300, 302
Kirzhnits, D. A. 128, 300
Kitaura, K. 180, 300
Kleinman, L. 86, 179, 300, 302, 312
Klopman, G. 93, 98, 300
Kohl, H. 210, 300
Kohn, W. 50, 51, 52, 54, 141, 142, 144, 154, 165, 173, 183, 201, 207, 208, 211, 243, 296, 299, 300, 301, 305, 307
Kolos, W. 221, 300
Komorowski, L. 233, 300
Koninski, M. 226, 228, 233, 306
Koopmans, T. 10, 300
Kozlowski, P. M. 244, 306
Krakauer, H. 243, 316
Kress, J. D. 138, 141, 198, 243, 290, 300
Krieger, J. B. 157, 301
Kryachko, E. S. 243, 244, 284, 301, 310
Kutzler, F. W. 205, 301

Lackner, K. S. 275, 301
Lagowski, J. B. 179, 292
Landau, L. D. 111, 301
Lang, N. D. 243, 301, 307
Langer, R. E. 140, 301
Langreth, D. C. 195, 196, 197, 198, 199, 283, 299, 301
Lannoo, M. 199, 301
LaPlaca, S. J. 243, 305
Laufer, P. M. 157, 301
Lawley, K. P. 13, 301
Lebowitz, J. L. 67, 301
Lee, C. 98, 101, 118, 120, 121, 122, 123, 136, 138, 199, 243, 289, 301, 302, 316, 317
Lee, S. 179, 220, 302
Leech, J. W. 230, 308
Leite, J. R. 200, 294
LeSar, R. 221, 302
Lester, W. A., Jr. 221, 296
Levine, I. N. 3, 302
Levine, Z. H. 212, 302
Levy, M. 12, 15, 30, 54, 56, 58, 70, 74, 75, 77, 86, 133, 140, 148, 149, 151, 152, 164,

Levy, M. (*cont.*)
  176, 187, 189, 205, 213, 214, 216, 222, 224, 237, 238, 239, 243, 244, 292, 302, 303, 305, 308, 309, 312
Li, L. 224, 303
Li, T. 210, 303
Lieb, E. H. 39, 54, 55, 58, 69, 82, 105, 114, 116, 133, 149, 150, 163, 206, 213, 288, 303
Lifshitz, E. M. 111, 301
Lin, C. C. 183, 298, 309
Linderberg, J. 75, 303
Lindgren, I. 181, 303
Lindholm, E. 234, 303
Lloyd, J. 221, 303
Löwdin, P.-O. 18, 28, 29, 32, 35, 36, 303, 304
Lohr, L. L. 217, 304
Longuet-Higgins, H. C. 97, 304
Ludena, E. V. 138, 140, 214, 284, 294, 304
Lundqvist, B. I. 173, 178, 179, 186, 190, 191, 192, 193, 205, 274, 275, 283, 296, 297, 298
Lundqvist, S. 234, 297, 303, 304

MacDonald, J. K. L. 205, 207, 304
Macke, W. 55, 304
Mallion, R. B. 97, 289
Malmstrom, G. 195, 311
Malone, J. G. 91, 304
Malzacher, P. 133, 135, 136, 313
Manninen, M. 243, 310
Manoli, S. 166, 182, 304
Maradudin, A. A. 274, 288
Marc, G. 18, 304
March, N. H. 47, 95, 104, 105, 110, 136, 140, 170, 243, 281, 283, 284, 285, 288, 297, 304, 305, 313
Margrave, J. L. 93, 299
Markus, G. E. 95, 287
Marron, M. T. 11, 297
Martin, R. M. 230, 305
Mascardó, R. 200
Maschke, K. 56, 60, 214, 317
Massa, L. J. 243, 289, 305
Massey, H. S. W. 220, 305
McClendon, M. 244, 305
McCoy, J. D. 217, 288
McDonald, I. R. 257, 297
McGinn, G. 239, 314
McMillan, W. G. 18, 304
McQuarrie, D. A. 44, 49, 203, 305
McWeeny, R. 3, 28, 32, 35, 199, 285, 305
Mearns, D. 211, 305
Mehl, M. J. 196, 198, 199, 301
Mermin, N. D. 46, 66, 83, 207, 285, 305
Merzbacher, E. 3, 305

Messiah, A. 20, 305
Meyer, J. 118, 305
Miller, J. E. 112, 298
Miller, T. M. 210, 305
Moffitt, W. 97, 222, 305
Mohammed, A.-R. E. 197, 305
Morokuma, K. 180, 300
Morrell, M. M. 15, 29, 305
Mortier, W. J. 92, 225, 236, 305, 315, 316
Moruzzi, V. L. 272, 275, 306
Moscardo, F. 200, 312
Mullay, J. 276, 306
Mulliken, R. S. 74, 93, 98, 306
Murphy, D. R. 95, 117, 128, 132, 136, 138, 140, 233, 306, 310, 313, 315

Nagata, C. 101, 293
Nagy, A. 183, 293
Nakatsuji, H. 32, 306
Nalewajski, R. F. 90, 104, 217, 225, 226, 228, 233, 243, 244, 288, 304, 306
Nee, T.-S. 12, 302
Nguyen-Dong, T. T. 222, 286
Nieminen, R. M. 243, 310
Nikulin, V. K. 220, 293, 306
Nordholm, S. 140, 185, 244, 290, 299, 307
Norman, M. R. 243, 309
Norskov, J. K, 243, 307
Nozières, P. 274, 307
Nusair, M. 154, 178, 199, 275, 315
Nyden, M. R. 11, 60, 140, 307

O'Connell, R. F. 265, 267, 270, 298
O'Leary, B. 97, 289
Ohnishi, S. 243, 315
Oliveira, L. N. 207, 208, 243, 296, 307
Oliver, G. L. 176, 307
Orsky, A. R. 276, 307
Ortenburger, I. B. 140, 298
Ortiz, E. 95, 233, 293, 294
Osborn, T. A. 160, 293
Ossicini, S. 200, 307
Ostlund, N. S. 3, 10, 12, 13, 14, 20, 314
Ou-Yang, H. 244, 302
Overhauser, A. W. 197, 315

Pack, R. T. 220, 289
Painter, G. S. 205, 239, 285, 301
Palke, W. E. 74, 222, 224, 307, 308
Panat, P. V. 204, 299
Pant, M. M. 170, 307
Pariser, R. 97, 307
Parker, G. W. 113, 307
Parrinello, M. 244, 288, 292

AUTHOR INDEX

Pasternak, A. 230, 235, 308
Pathak, R. K. 126, 207, 308
Pauling, L. 91, 92, 308
Pavlidou, C. M. E. 185, 199, 313
Payne, M. C. 244, 308
Payne, P. W. 200, 308
Pearson E. W. 138, 309
Pearson, R. G. 81, 95, 98, 225, 227, 276, 277, 278, 279, 280, 308, 309
Pederson, M. R. 183, 309
Pedroza, A. C. 152, 153, 194, 200, 285, 309, 315
Percus, J. K. 31, 41, 58, 67, 149, 293, 301, 309
Perdew, J. P. 56, 70, 75, 76, 77, 86, 133, 140, 141, 148, 149, 151, 152, 164, 166, 172, 176, 181, 182, 183, 186, 189, 195, 196, 197, 198, 199, 203, 205, 216, 224, 237, 238, 243, 301, 302, 303, 307, 309, 310, 311
Pereira, J. R. 200, 302
Perrot, F. 204, 310
Petkov, I. Z. 243, 244, 300, 301, 310
Peukert, V. 208, 310
Phillips, P. 72, 310
Pines, D. 157, 274, 307, 310
Plindov, G. I. 138, 310
Plumer, M. L. 140, 310
Pogrebnya, S. K. 138, 310
Politzer, P. 14, 17, 95, 98, 115, 230, 233, 234, 310, 316
Pratt, L. R. 158, 163, 298, 310
Preuss, H. 185, 197, 312, 313
Pritchard, H. O. 93, 310
Providencia, J. da, 282, 291
Pucci, R. 168, 317
Pugh, D. 221, 303
Puska, M. J. 243, 310
Pyper, N. C. 221, 316

Quentin, P. 118, 305

Radzio (Radzio-Andzelm), E. 221, 300, 311
Rae, A. I. M. 220, 221, 311
Rajagopal, A. K. 164, 168, 204, 208, 209, 243, 284, 299, 307, 311, 316
Ramana, M. V. 243, 311
Rasolt, M. 136, 195, 197, 243, 311, 315
Ray, N. K. 93, 95, 230, 231, 232, 236, 294, 311
Ree, F. H. 220, 311
Reed, M. 254, 311
Rieder, G. R. 117, 295
Ripka, G. 25, 71, 203, 259, 263, 287
Robles, J. 126, 139, 141, 241, 288, 294, 311
Roetti, E. 126, 134, 137, 289

Rooney, R. 183, 311
Roothaan, C. C. J. 7, 8, 10, 311
Ross, I. G. 13, 308
Rothstein, S. M. 221, 288
Rowlinson, J. S. 67, 217, 311
Ruedenberg, K. 11, 116, 229, 288, 291, 311
Runge, E. 209, 217, 290, 311
Rupnik, K. 241, 242, 308
Ruskai, M. B. 54, 288
Rychlewski, J. 224, 311

Saam, W. F. 66, 291
Sahni, V. 133, 140, 191, 197, 198, 216, 239, 243, 244, 290, 303, 305, 311, 312
Salahub, D. R. 284
Salem, L. 97, 304
Salvetti, O. 199, 289
Samuels, L. 93, 230, 231, 232, 311
Sanderson, R. T. 90, 93, 235, 312
San-Fabián, E. 200, 312
Santhanam, P. 199, 289
Satoko, C. 180, 300
Savin, A. 197, 199, 200, 293, 312
Schaefer, H. F., III 12, 13, 312
Schlüter, M. 86, 198, 199, 284, 301, 312
Schönhammer, K. 60, 86, 297, 312
Schwarz, K. 155, 312
Schwinger, J. 117, 140, 243, 291, 292
Scully, M. O. 265, 267, 270, 298
Sears, S. B. 138, 139, 285, 312
Segal, G. A. 13, 97, 312
Selloni, A. 244, 292
Sen, K. D. 277, 284, 306, 315
Senatore, G. 212, 312
Settle, F. A., Jr. 236, 287
Shadwick,W. F. 185, 198, 314
Sham, L. J. 86, 140, 141, 142, 144, 154, 156, 164, 178, 180, 183, 197, 198, 199, 201, 284, 300, 301, 312, 314
Shankar, S. 92, 235, 236, 305, 312
Sharma, B. S. 126, 200, 308, 312
Shen, Y.-T. 86, 312
Shih, C. C. 117, 136, 140, 313
Shingu, H. 101, 293
Sichel, J. M. 93, 286
Sida, D. W. 220, 305
Sierraalta, A. 214, 304
Silverstone, H. J. 11, 197
Simon, B. 114, 116, 254, 303, 311
Simons, G. 230, 231, 287, 313
Sinanoğlu, O. 14, 313
Singer, S. J. 217, 288
Slater, J. C. 34, 154, 155, 156, 164, 166, 178, 284, 313
Smith, V. H., Jr. 14, 31, 244, 282, 292, 313
Smithline, S. J. 217, 291

Somorjai, G. A. 102, 292
Soven, P. 210, 211, 212, 302, 317
Spackman, M. A. 221, 313
Spruch, L. 116, 288
Srebrenik, G. 14, 316
Stanton, R. E. 17, 313
Stich, W. 133, 135, 136, 313
Stillinger, F. H., Jr. 67, 313
Stoddart, J. C. 170, 313
Stoitsov, M. V. 243, 244, 300, 301, 310
Stoll, H. 185, 197, 199, 200, 293, 312, 313
Stott, M. J. 140, 152, 210, 243, 285, 310, 313, 314
Stroud, D. 66, 291
Subbaswamy, K. R. 212, 312
Suhl, H. 283
Sumner, F. H. 93, 310
Sutcliffe, B. T. 3, 35, 305
Szabo, A. 3, 10, 12, 13, 14, 20, 314
Szasz, L. 239, 314

Tachibana, A. 86, 228, 243, 244, 314
Tal, Y. 138, 140, 314
Talman, J. D. 185, 198, 314
Tarazona, P. 191, 285
Teller, E. 50, 314
Testa, A. 212, 286
Teter, M. P. 244, 308
Thakkar, A. J. 126, 200, 308, 312
Theophilou, A. 149, 152, 207, 297, 314
Thomas, L. H. 47, 105, 314
Tomishima, Y. 133, 134, 135, 314, 317
Tong, B. Y. 156, 157, 178, 180, 197, 210, 303, 314
Truhlar, D. G. 310
Tsarev, Y. N. 220, 306
Tschinke, V. 179, 317
Tseng, T. J. 183, 311, 314
Turcel, E. 186, 243, 286

Vaidehi, N. 183, 314
Valone, S. M. 82, 205, 213, 214, 215, 314, 315
van der Eerden, J. R. 243, 296
Van Dyke, J. P. 140, 298
van Genechten, K. 235, 305, 315
Vanderbilt, D. H. 244, 308
Vashishta, P. 173, 201, 300
Vega, L. A. 233, 286
Vek, A. 95, 293
Vela, A. 94, 95, 294
Vignale, G. 243, 315
Vinayagan, S. C. 277, 315
Volterra, V. 246, 249, 315

von Barth, U. 170, 177, 178, 200, 205, 243, 274, 275, 315, 316
Voros, A. 128, 132, 295
Vosko, S. H. 154, 178, 179, 180, 197, 199, 200, 275, 288, 292, 301, 315

Waldman, M. 220, 315
Wang, J. S. Y. 136, 315
Wang, P. S. C. 158, 297
Wang, W.-P. 14, 117, 136, 140, 306, 313, 315
Wang, Y. R. 197, 315
Wedig, V. 197, 312
Weinert, M. 243, 315, 316
Weinstein, H. 14, 243, 310, 316
Weissbluth, M. 20, 316
Weizsäcker, C. F. v. 172, 316
West, C. G. 136, 285
Whitehead, M. A. 93, 164, 166, 182, 183, 277, 286, 295, 304, 307, 311, 314
Widom, B. 67, 217, 311
Wigner, E. P. 265, 267, 270, 298, 316
Wilk, L. 154, 178, 180, 186, 199, 275, 315
Wilk, S. F. T. 160, 293
Wilkens, J. W. 178, 297
Williams, A. R. 243, 272, 275, 306, 316
Wilson, E. B., Jr. 17, 316
Wilson, S. 14, 316
Wimmer, E. 243, 316
Winter, N. W. 220, 311
Wood, C. P. 221, 316
Wyatt, R. E. 16, 292

Xu, B.-X. 208, 209, 316

Ying, S. C. 208, 317
Yonei, K. 133, 134, 135, 314, 317
Yonezawa, T. 101, 293
Yue, W. 197, 198, 309

Zangwill, A. 210, 211, 212, 317
Zaparovanny, Y. I. 266, 270, 289
Zaremba, E. 210, 243, 313, 314
Zeller, R. 244, 290
Zhidomirov, G. M. 199, 200, 296
Zhou, Z. 98, 302, 317
Ziegler, T. 179, 317
Zilles, B. A. 115, 310
Zorita, M. L. 138, 317
Zumbach, G. 56, 60, 214, 317
Zunger, A. 149, 164, 166, 172, 181, 182, 183, 186, 310
Zweig, G. 275, 301

SUBJECT INDEX

Absolute hardness, *see* hardness
Acids and bases 98
  see HSAB principle
Adiabatic connection 186–189, 194, 203
Alpha parameter in Xα theory 154–155, 183
  recommended values for atoms 155
Annihilation and creation operators, 41, 259–264
Antisymmetry principle 4, 23, 29
Aromaticity 98
Atom in a molecule 90, 221–224
  Bader definition 222, 224
  density-functional definition 222–223
  density-matrix model 224
  in HSAB principle 225
  need for the construct 221–222
  promotion energy of 223–224
  valence state of 222, 223–224
  wave-function viewpoint 222, 224
Atomic charges 92, 235–236
Atomic radius 95
Atoms, electronegativities and hardnesses of 277
Atoms and molecules, long-range interactions among 210
  see Gordon–Kim method

Band gap 86, 96
Basis functions or set 12
  continuous 22
Binding energy, from charge transfer 91–92, 93, 225, 226
Bloch equation 269
Boltzmann density matrix 269
Bond distance, relation to electronegativity and softness 232
Bond electronegativity 233–234
Bond hardness 233
Bond-charge model 229–234
  bond-charge values 230–231
  for diatomic force constants 229
  delocalization length in 230
  as density-functional model 235
  electronegativities in 230–234
  hardness and softness in 232–234

Born–Oppenheimer approximation 3, 217
Boson systems 205, 216
Bra, ket notation 20–24

Canonical ensemble 27, 44, 269
  Helmholtz free energy in 45, 61
  partition function in 45
  zero-temperature limit 81–84
Catalysis, local softness and 102
Cations, electronegativity and hardness of 278
Charge-current conservation 209
Charge transfer
  driven by chemical potential difference 91, 224, 225–226
  energy associated with 91
Chemical binding, 218, 222, 229
  in Kohn–Sham theory 218
  see bond-charge model, HSAB principle
Chemical potential 70–86
  analogy with classical quantity 91, 214, 239
  of atoms in a molecule 223
  in bond-charge model 232
  canonical ensemble 83–84
  classical systems 67
  in density-functional theory 52, 142
  in density-matrix-functional theory 213–214
  differential expression for 88
  as electronegativity 74–75, 88, 92, 276
  grand canonical ensemble 45
  in grand-canonical ensemble dft 64, 66, 70–74
  role in HSAB principle 225–228
  from numerical interpolation 84, 89
  pure state 81–83
  in TFDW model 113
  in Thomas–Fermi–Dirac theory 109, 110
  in Thomas–Fermi theory 48, 110
  see electronegativity, electronegativity equalization
Chemical potential derivatives 87–104
  see hardness, Fukui function
Chemisorption 243

325

326 SUBJECT INDEX

Classical approximation for Wigner distribution 130, 269–270
Classical systems 66, 217
Closed-shell atoms, interaction energy 157
Closure relation 22, 23
Compton profiles 241–242
Configuration interaction 14
Constrained search 57, 59, 62, 65, 68, 70, 82, 150, 164, 171, 172, 187, 205, 207, 208, 213, 214, 215, 216, 237
Convex (concave) functions and functionals 62, 255–258
  Jensen's inequality for 256, 257
Convex set 32, 43
  extreme element of 43
Correlation (energy) 11, 13–14, 40, 106, 176, 177–178, 183, 199
Correlation-energy functional 177, 185, 186, 190
  Colle–Salvetti formula 199
  diatomic molecules 185
  in Gordon–Kim method 219
  LSD calculation of, for atoms 183, 185
  local spin-density approximation for 178, 185
  need for improved approximations for 244
  random phase approximation 275
  scaling behavior 239
  uniform electron gas, 177, 274–275
Correlation hole 190
Correlation potential 154, 184–185
Covalent radius 93
Crystal structure prediction by dft 235
Current vector 209
Cusp condition, see electron density, density matrix, first and second order

Density, see electron density
Density-density response 210
Density-functional theory (dft) 47–69, 281–284
  for boson systems 205
  canonical ensemble 60–64
  classical systems 66–69, 217
  connection with stochastic mechanics 244
  for constrained systems 244
  formulation using scaling transformations 244
  fundamental differential expression 87
  grand canonical ensemble 64–66
  ground state 51–60
  and hydrodynamics 244
  inequalities 244
  and molecular dynamics 244
  momentum-space 244

  spin-polarized 157
  stability conditions in 244
  transcription into local thermodynamics 239–243
  see spin-density-functional theory
Density matrix 25, 281
  for single determinant 36
  pth-order 28, 30
  reduced 27–32
  second-quantized representation 263–264
  spinless 32–35
  and Wigner distribution function 265
Density matrix, first order 28, 30
  cusp conditions on 35
  in terms of field operators 42
  Gaussian approximation 120, 122
  in terms of local potential 130, 159
  normalization condition 28, 118
  path-integral representation 161
  for single determinant 35–36
  spinless 32
  for uniform electron gas 107
  see density-matrix-functional theory
Density matrix, second-order 28, 194
  cusp conditions on 35
  for single determinant 36, 39
  normalization condition 28
  spinless 33
  in terms of field operators 42
Density-matrix-functional theory 213–215
  for atoms in molecules 224
  derivation by constrained search 213
  in natural orbital representation 214
Density operator 24–27
  commutator with Hamiltonian 27
  determined by density for ground state 63, 65
  ensemble 25, 26, 27, 44, 202
  Fock-space 45, 64
  idempotent, for pure state 26
  time-dependent 27
  and Weyl classical function 266–267
  and Wigner distribution function 265
Density operator, first order 29, 36
  eigenfunctions and eigenvalues 29
  idempotency condition 36–37
  positive semidefinite 29
Density operator, second-order 29
  eigenfunctions and eigenvalues 29
  positive semidefinite 29
Density of states, 48, 108
  as related to softness 102
Dipole moment, 91, 93
Dispersion forces 219, 220

Electrochemistry 243

Electron affinity 74, 166, 224, 243, 276
Electron density 14–16, 33, 281
  atomic, calculated by various methods 135
  atomic, piecewise exponential approximation for 116
  for atoms in molecules 223
  cusp at nucleus in 14
  differential equation for 243
  in Gordon–Kim method 219
  example of non-v-representable 54
  in LDA method 156
  in terms of Kohn–Sham orbitals 145, 203
  approximation in terms of local potential 131
  long-range behavior of 15
  in terms of natural orbitals 143
  N-representability of 55–56
  operator 68, 170
  path-integral representation 162
  scaled 109, 237
  second-quantized formula 264
  in TFDW model 133
  Thomas–Fermi theory 110–112
  Thomas–Fermi–Dirac model 110
  as variable in semiempirical dft 234
  effect of Weizsacker correction 127
Electron repulsion energy functional
  determination by constrained search 214
  Hartree–Fock 106
  universal 60, 63
Electron magnetization density 170
Electron spin density 169
Electron spin susceptibility 173
Electronegativity
  alternative definitions 93
  in bond-charge model 230–234
  as chemical potential 74–75, 92, 276
  difference, as coordinate in structure-stability diagrams 235
  as driving force for charge transfer 92
  equalization schemes 92
  finite-difference definition 83, 95–96
  history of concept 92
  intrinsic framework 235
  Mulliken's formula 74, 93, 166
  periodic dependence on atomic number 94–95
  transition-state calculation 166–167
  values for 95, 276–279
  valence-state dependence 93
Electronegativity equalization 90–95
  in bond-charge model 232–234
  in HSAB analysis 226–228
  in polyatomic molecules 235
Electron-electron repulsion energy
  classical 34, 238, 248
  in Hartree–Fock theory 8, 11
  in terms of exchange-correlation hole 35
  local model 123–127
  operator for 4, 263
  in terms of second-order density matrix 124, 177
Electronic energy (functional) 3
  atoms, calculated by various methods 134, 156
  canonical ensemble 82
  change in charge transfer process 91, 226–227
  convexity of 72, 78, 80, 81, 84, 95
  in density-functional theory 52
  density-matrix formula 31, 33, 37, 39, 60
  density-matrix-functional theory 213
  differentiability of 52
  differential expression for 87
  excited state theory 207
  Gordon–Kim method 219
  grand canonical ensemble dft 71
  Hartree–Fock theory 7, 8, 9
  Hartree–Fock–Kohn–Sham theory 184
  Kohn–Sham theory 145, 147, 165
  multicomponent systems 216
  purely local model 126
  spin-density-functional theory 171
  spin-polarized Kohn–Sham theory 172
  Thomas–Fermi–Dirac model 109–110
  Thomas–Fermi theory 49
  TFDW model 133–135
  uniform electron gas 271–275
Electrophilic reagent 101
Electrostatic potential 50
  used to define electronegativity 93
Electrostatic theorem 17
Embedded atoms (effective medium theory) 243
Energy functional, *see* electronic energy (functional)
Entropy functional 63
  for non-interacting reference system 201
Entropy maximization, 44, 45, 239–241
Entropy of a mixed state 44
Euler–Lagrange (Euler) equation
  density-functional theory 52, 71, 74, 142, 204
  finite-temperature dft 68
  in finite-temperate Kohn–Sham theory 203
  for extremal of a functional 252–253
  Hartree–Fock theory 38
  Hartree–Fock–Kohn–Sham method 184
  Kohn–Sham method 142, 144, 146, 158, 165
  modified Thomas–Fermi model 117
  natural boundary conditions for 254

Euler–Lagrange (Euler) equation (*cont.*)
  in purely local theory 127
  spin-polarized Kohn–Sham method 172
  TFDW model 133, 135
  Thomas–Fermi–Dirac model 109
  Thomas–Fermi theory 50
  wave function theory 6
Exchange-correlation free-energy functional 202
Exchange-correlation functional (energy)
  average density approximation 191, 194
  in density-functional theory 52, 144
  formula in terms of exchange-correlation hole 186–194
  Langreth–Mehl approximation 196–197
  local density approximation for 153, 190, 196
  modified weighted density approximation 193–194
  need for improved approximations for 244
  in spin-density functional theory 173
  for uniform electron gas 275
  wave-vector analysis of 194–197
  weighted density approximation 191–193, 194
Exchange-correlation hole (charge) 34
  in Hartree–Fock theory 40
  local density approximation for 190
  spherical average 189
  sum rule 34, 189
  weighted-density approximation for 191–194
  see exchange-correlation energy functional
Exchange-correlation potential 145
  local density approximation for 153
  self-interaction in 182
Exchange energies of atoms 120, 123, 182, 198
Exchange energy (functional)
  Dirac formula 108, 113, 154, 272
  in terms of exchange hole 190
  Gaussian formula 120
  in Gordon–Kim method 219
  gradient correction 140, 197
  in Hartree–Fock theory 106
  phase-space approach to 141, 242–243
  rational fraction approximation 141
  scaling behavior 238
  in spin-density-functional theory 176–177, 273
  for uniform electron gas 272–274
Exchange hole 40, 189, 193
  calculated for neon atom 191, 192, 193
  spherical average 193
  sum rule 190

Exchange potential 154, 185
Excited states
  conversion to ground state by apparatus 86
  density-functional theory for 204–208
  lowest state of given symmetry 205
  orthogonality difficulties with 6, 204
External potential 3
Extra-ionic resonance energy 92, 93

$F[\rho]$, Hohenberg–Kohn functional
  bounds 60
  constrained search definition 57, 62, 65, 66, 68, 187
  definition 52, 54
  differential of 89
  Fock space 70
  multicomponent systems 215, 216
  scaling behavior 237–238
  spin-polarized generalization 171
  subtleties in 60
  universality of 53, 60, 63, 66, 68
Fermi energy 48, 211
Fermi function 202
Fermi level 102
Fermi–Amaldi correction 181, 182
Fermi–Dirac distribution 48, 202
Ferromagnetic electron gas 177
Feynman path integral 159, 204
Finite-temperature dft 60–69
  Kohn–Sham version 201–204
Fock space 26, 46, 202, 259–264
Force concept 281
  see Hellman–Feynman theorem
Fraga relations 114, 116
Frontier orbital, *see* HOMO and LUMO
Frontier-electron theory of chemical reactivity 101
Fukui function 88, 99–101, 168, 236
  role in HSAB principle 225–228
Functional 5, 7, 246–254
  chain rule for differentiation 251
  convex (concave) 257–258
  derivative 15, 88, 246–248
  differential 246
  homogeneous 251–252
  inverse of derivative 250–251
  local 109, 248
  necessary condition for extremal in 252–253
  necessary and sufficient conditions for minimum 253, 257
  second-order derivative 249
  Taylor (Volterra) expansion 249
  universal 53

SUBJECT INDEX

Functional derivative 15, 88, 246–250
  chain rule for 250
  inverse of 250–251

Gaussian model 118–123
Gaussian resummation 120, 125
Gibbs inequality 62, 65, 68, 257
Global property 88
Gordon–Kim method 219–221
  alternative ansatz 221
  applied to rare-gas atoms 220
  extensions 220–221
  for ion-rare gas interactions 220
  insensitivity to error 221
  self-interaction correction in 221
Gradient correction or expansion
  Hodges correction 132
  in Thomas–Fermi theory 127–132
  Weizsacker correction 127, 175
Grand canonical ensemble 45–46
  grand partition function in 46, 67
  grand potential in 45
  zero-temperature limit 70–74, 75–81
Grand potential (functional) 66
  differential of 89
  variational principle for 65, 67, 201
Green's function (propagator) 129, 133, 159, 211
  physical meaning of 161

$\hbar$ expansion, see phase-space distribution function
HAM method, as semiempirical Kohn–Sham method 234
Hardness (or absolute hardness) 81, 88, 95–98
  in bond-charge model 232
  of bonds 233
  as energy charge for disproportionation reaction 96–97
  in HSAB principle 225–282
  kernel 103–104, 225, 228
  as measure of aromaticity 98
  state and environmental dependence 98
  sum, as coordinate in structure-stability diagrams 235
  values for, 276–279
Hardness, local 101–104
  in HSAB principle 225, 228
Hartree method, orthogonalized 12, 147, 183
Hartree–Fock method or theory (HF) 7–13
  canonical form 10
  Coulomb-exchange operator in 8
  compared with Kohn–Sham method 147, 148

  density-matrix formulation 35–40, 213, 248
  differential equation 8
  energies for neutral atoms 113
  exchange operator as nonlocal operator 23, 38
  Fock operator 8
  invariance to unitary transformation 9
  noninteracting reference system in 149
  operator formulation 38
  restricted (RHF) 9, 148
  self-interaction in 180
  self consistency in 9
  unrestricted open-shell (UHF) 12
Hartree–Fock orbitals 7, 9
  canonical 10
  circulant 11
  localized 11
  unitary transformation of 10
Hartree–Fock–Kohn–Sham method (HFKS) 183–186, 245
Hartree–Fock Slater method, see $X\alpha$ method
Heaviside step function 128
Hellmann–Feynman theorems 16, 112, 115, 188, 221
Helmholtz free energy functional 62–63
  for non-interacting reference system 201
  variational principle for 63
Hilbert space 20, 46, 259
Hohenberg–Kohn theorems 51–53
  first theorem, proof by contradiction 51–52
  first theorem, restatement as a mapping 53
  first theorem, proof by constrained search 57
  second theroem 52, 54
HOMO and LUMO 99–101, 166, 167
Homogeneity
  of functionals 251–252
  of kinetic energy operator 18
  of potential energy operator 18
  of TFD energy components 109
Homogeneous electron gas, see uniform electron gas
HSAB principle 98, 224–228
  covalent effects in 227
  hard likes hard 225
  polarization and electrostatic effects in 227
  soft likes soft 225
Hubbard model 97
Hydrodynamical formulation of dft 244, 282

Idempotency 22, 26, 36–37, 42

Information theory 138
  see phase-space distribution function
Inner product 20
Insulator, hardness of 96
Integral
  Coulomb 7
  electron-electron repulsion 13
  electron-nucleus attraction 13
  exchange 7
  overlap 13
Integral formulation of dft 157–163
Interatomic forces 219–221
  see Gordon–Kim method
Intrinsic framework electronegativity 235
Ionization energy or potential 10, 72, 166, 178, 224, 243, 276

Janak theorem 149, 165, 173
Jellium 271
Jensen's inequality 62, 256, 257, 258

Kinetic energy, in terms of natural orbitals 143
Kinetic energy density, local 119, 136, 138, 239
Kinetic energy functional
  in Gordon–Kim method 219
  gradient corrections 132, 136–140, 175
  Kohn–Sham 143, 144, 149–152, 162, 164
  spin-density-functional theory 172
  TF or TFD theory 49, 108, 132, 175, 247, 252, 272
  universal 58, 63
  Weizsacker term in 127, 247, 252
Kohn–Sham effective potential 145
  exact 199
  exchange-only 198
  finite temperature 203
  in HFKS method 184
  spin-dependent 173
  time-dependent 209
Kohn–Sham equations 142–145
  canonical form 146, 151
  compared with Hartree–Fock equations 147, 149
  derivation 145–149, 151, 165
  finite-temperature extension 203–204
  local density approximation (LDA) 153, 203
  self-consistent nature of 145, 147
  semiempirical version of 234
  time-dependent extension 209
  for uniform electron gas 271
Kohn–Sham noninteracting reference system 141, 148
  Hamiltonian for 143, 144, 202
  kinetic energy of 143, 144, 149–152, 162, 172, 174–175, 238
  v-representability problem with 144, 148, 187
  statistical mechanics of 202
Kohn–Sham orbitals 143–144
  polar form for time-dependent systems 209
  for uniform electron gas 271
Kohn–Sham spin-polarized (spin-density) theory 148, 172–174, 283
  local spin-density approximation for self-interaction correction to 182
Kohn–Sham theory (method) 142–168
  and chemical binding 218, 234
  exactness in principle 147
  extension to nonintegral occupation numbers 163–168
  finite-temperature extension 201–204
  inclusion of spin 148, 169–174
  intergal formulation 132, 157–163, 204
  orbital energies, physical meaning of 149
  time-dependent extension 208–210
  total energy formula in 147
Koopmans theorem 10

Lagrange undetermined multipliers 6, 254–255
  in Hartree–Fock method 8, 37
  in Kohn–Sham method 146, 151
  in phase-space maximum-entropy approach 240
  in statistical mechanics 44, 45
  in Thomas–Fermi theory 50
  in time-dependent dft 209
Levy constrained-search formulation of dft 56–60
  see constrained search
Ligand-field splitting 220
Linear response function 16, 88, 90, 104
  for non-interacting ground state 211–212
  spectral representation of 210
Linear response, theory of dynamic 210–212
  applications 210, 212
  finite-temperature and non-linear extensions 212
Local-density approximation (LDA) 49, 109, 127, 152–157, 174, 281, 283
  early calculations using 156–157
  for exchange-correlation energy 153, 176, 196
  in finite-temperature KS theory 203
  for local temperature 121
  self-interaction in 180
  see $X\alpha$ method

Long-range behavior
 electron density 11
 Hartree–Fock orbitals 15
 natural orbitals 30
 TFDW density 133
 Thomas–Fermi density 112, 117
 Thomas–Fermi–Weizsacker density 127
Local spin-density approximation (LSD) 174, 175, 176, 178, 181, 275
 assessment 180, 245
 illustrative calculations 178–179

Macroscopic systems 66–69, 217, 239
Magnetic field, Hamiltonian in presence of 169
Maxwell–Boltzmann distribution law 240
Maxwell relations 88, 90
Mean-path approximation 132, 160
Metal-insulator-transition 204
Metal-metal bonds 179
Methods
 ab initio 13
 HF 7
 HFKS 183
 Hückel 97
 LDA 153
 LSD 174
 PPP, CNDO, INDO, MINDO 97
 RHF 9
 semiempirical 13, 234
 SIC 181
 UHF 12
 $X\alpha$ 154
 Ξa 183
Molecular solids 221
Molecules, electronegativities and hardnesses of 280
Momentum space 241
 density 241, 266
 density-functional theory in 244
Multicomponent systems, dft for 215–217
 finite temperature extension 217
 problem of universality with 216
 non-Born–Oppenheimer molecules 217

Natural geminals 29
Natural spin orbitals 29
 in density-matrix-functional theory 214
 differential equation for 214
 occupation numbers of 29, 43, 143, 163–164
Noninteracting system, *see* Kohn–Sham noninteracting reference system compared with Hartree and Hartree–Fock concepts 147

N-representability 40–44
 of electron density 54–56
 first-order density matrix 32, 43
 Pauli necessary conditions for 41
 second-order density matrix 31
Nucleophilic reagent 101

Occupation numbers, nonintegral 163–168
Operator
 adjoint 22
 annihilation and creation 41, 259
 electron-electron repulsion 263
 electron-nucleus attraction 4
 field 41, 261
 Hamiltonian 3, 169
 Hermitian or self-adjoint 5, 22, 26
 idempotent 22
 identity 22
 kinetic energy 4, 263
 local 23
 magnetization density 170
 matrix representation 23
 nonlocal 23
 number 42, 264
 occupation number 42, 264
 one-electron 27, 30, 33, 261–262, 263
 positive semidefinite 26
 projection 22
 representation in phase space of 266–269
 Sturm–Liouville 11
 in second-quantized Hamiltonian 27, 202
 trace 25
 two-electron 27, 31, 33, 262–263, 264
Orbital energies
 Hartree–Fock 8, 116
 Kohn–Sham 145, 146, 147, 149, 165
Orbital, spatial and spin 7

Pair correlation function 34, 191
 sum rule for 34
 see exchange-correlation hole
Paramagnetic electron gas 177
Perturbation theory, first-order 15, 87
Phase-space distribution function 239–243, 265–270
 Compton profiles from 241–242
 determination by constrained entropy maximation (information theory) 239–241
 exchange energy from 242
Photoabsorption cross section 212
Plasma 204
Poisson equation 111
Polarizability, as measure of softness 98, 225

Polarizability, static and dynamic 209, 212
Positive ions, in TF and TFD models 113
Pressure, local 240–241
Promotion energy 222–224
Protonic counterpart of electronegativity 217

Radical reagent 101
Radicals, electronegatives and hardnesses of 279
Random phase approximation 196, 275
Reactivity
 frontier-electron theory of 101, 168
 measured by chemical potential response 101
Relativistic effects 173, 243
r_s definition 272

Sanderson Principle of Electronegativity Equalization 90
 geometric mean equalization priciple 92
Scaling
 derivation of TFD model using 109–110
 of electron density 109, 237
 of electronic coordinates only 18, 237
 of electronic and nuclear coordinates 18
 in Gordon–Kim method 220
 of Hohenberg–Kohn $F[\rho]$ 238
 of Hohenberg–Kohn $T_s[\rho]$ 238
 of Hohenberg–Kohn $E_x[\rho]$ and $E_c[\rho]$ 238–239
 inequalities 238–239
 of nuclear charges 17
 relations in density-functional theory 237–239
 time-dependent 27
Scaling transformations 244
Schrödinger equation 3
Second quantization for fermions 27, 202, 259–264
Self-interaction correction (SIC) 180–183
 exchange only version 182
 in Hartree–Fock–Kohn–Sham method 185
 in modified Kohn–Sham equations 182
Semiconductor, hardness of 96
 metal-insulator transition 204
Semiempirical density-functional theory 234–236
 different levels of modelling 235–236
Sensitivity coefficient 253
Shape factor 85
Slater determinant or single determinant 7
Softness (global) 95
 in HSAB principle 225–228
 measure of polarizability 98

Softness, kernel 103–104, 225
Softness, local 101–104
 as fluctuation in grand ensemble 102
 in HSAB principle 225, 228
 as local density of states at Fermi level 102
Spectroscopic constants 179
Spin density 35, 169
Spin-density (spin-polarized) functional theory 157, 169–174
 advantages of 173–174
Spin-orbit coupling 173
Spin polarization 35
Stability conditions 244
State, pure or mixed 25
 necessary and sufficient conditions for pure state 26
 stationary 27
Stationary principle, see variational principle
Statistical mechanics, basic principles 44–46
Stochastic mechanics 244
Stress tensor 241
Structure-stability diagram 235
Superconductivity 228, 243
Sum rule 34, 40, 189, 191, 192, 194
Surfaces 243, 283
 atoms adsorbed on 221

Teller nonbinding theorem
 nullification by Weizsacker correction 127, 133
 in Thomas–Fermi theory 50, 114–115
Temperature
 in canonical ensemble 45, 46
 in grand canonical ensemble 45, 46
Temperature, local 119, 125, 239, 243
Thermodynamic transcription of density-functional theory 239–243
 local pressure 240–241
 see phase-space distribution function
Thomas–Fermi model (theory) (TF) 47, 282, 283
 assumptions of Thomas 47
 correction for inner shells 117, 138
 derivations 47–51, 106–109
 at finite temperature 210
 high Z limit for atoms 115
 kinetic energy in 49, 175, 272
 modified, to improve cusp behavior 113, 117
 nonergodic 244
 self-interaction in 180
 solution for neutral atoms 110–113
 total energy for neutral atoms 112, 113
 see Teller nonbonding theorem

SUBJECT INDEX

Thomas–Fermi–Dirac–Weizsacker model (TFDW) 132–136
Thomas–Fermi–Dirac theory (model) (TFD) 105, 282
  alternative Gaussian model 118–123
  derivations 106–109, 109–110, 118–123, 239–243
  purely local model 123–127
Time-dependent systems 208–212
  ensembles 210
Transition state 166–168
Transition-metal dimers 179

Uniform electron gas 49, 106, 121, 154, 174, 177, 190, 196, 211, 219, 271–275
Units, atomic 3

Valence state, *see* atom in a molecule
Variational principle
  average of first m states 206–208
  canonical ensemble 44–45, 61–63
  canonical-ensemble dft 63, 82
  density-functional theory 52, 54, 58, 82, 105, 142, 221
  density-matrix-functional theory 213
  excited state 205
  finite-temperature Kohn–Sham theory 201
  grand canonical ensemble 46, 65, 67
  grand-canonical-ensemble dft 67, 71
  ground-state wave function 5–7
  Hartree–Fock theory 39
  multicomponent systems 215, 216

Thomas–Fermi theory 50, 105
Vibrational force constants 229
  see bond charge model
Virial theorem 17, 114, 221, 222, 230
v-representability
  electron density 53–54
  elimination of problem by constrained search 57–58
  ensemble 56
  problem in Kohn–Sham theory 114, 148, 166
  pure-state 56
  in theory of linear response 211

Wave-vector analysis, *see* exchange-correlation functional
Weyl classical function 266
Weizsacker correction, *see* kinetic energy functional, gradient correction
Wigner distribution function 129, 239–242, 265–270
  see phase-space distribution function

$X\alpha$ method (approximation) 154–156, 281, 284
  contrast with TFD method 156
  effective potential in 154
  Hartree–Fock–Slater equation 154
  self-interaction corrected 183
  spin-polarized 178
  see local density approximation

Xray structure factors 243